Two-dimensional Product-Cubic Systems, Vol. VII

Albert C. J. Luo

Two-dimensional Product-Cubic Systems, Vol. VII

Self-Quadratic Vector Fields

Springer

Albert C. J. Luo
Department of Mechanical & Mechatronics Engineering
Southern Illinois University Edwardsville
Edwardsville, IL, USA

ISBN 978-3-031-48482-7 ISBN 978-3-031-48483-4 (eBook)
https://doi.org/10.1007/978-3-031-48483-4

This Springer imprint is published by the registered company Springer Nature Switzerland AG
The registered company address is: Gewerbestrasse 11, 6330 Cham, Switzerland

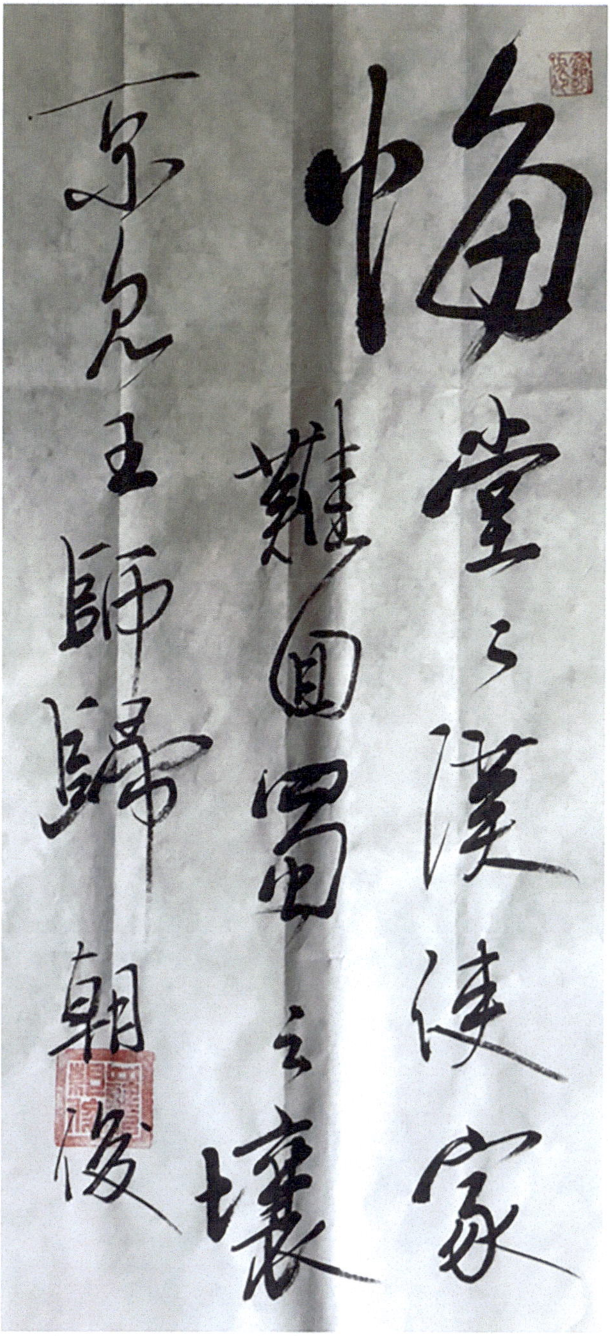

悔，堂堂汉使家难回，蜀之壤，不见王师归。

——朝俊【十六字令「悔」】己丑秋

Preface

In this book, nonlinear dynamics and singularity of cubic dynamical systems possessing a product cubic vector field and a self-univariate quadratic vector field are discussed. The equilibrium singularity and bifurcation dynamics are discussed through the appearing and switching bifurcations. The saddle-source (sink) is the appearing bifurcations for saddle and source (sink). The double-saddle equilibriums are the appearing bifurcations of the saddle-source and saddle-sink, and such double-saddle equilibriums are also the appearing bifurcations of the network of saddles, sink, and source. The infinite-equilibriums for the switching bifurcations exist in such cubic nonlinear systems. Inflection upper-saddle and lower-saddle infinite-equilibriums are the switching bifurcations of saddle-source (sink) with connected inflection-source (sink) flows. The hyperbolic (hyperbolic-secant)-sink and source are for the switching bifurcation of saddle, source, and sink with connected inflection-source (sink) flows. Up-down and down-up lower-saddle and upper-saddle infinite-equilibriums are the switching bifurcations of saddle-source and saddle-sink with hyperbolic and hyperbolic-secant flows. The inflection-source (sink) infinite-equilibriums are the switching bifurcations of saddle, sink, and source with hyperbolic and hyperbolic-secant flows.

In Chap. 1, a theory for nonlinear dynamics and singularity of cubic dynamical systems possessing self-quadratic and product-cubic vector fields is developed. The self-linear and crossing-quadratic product vector fields in such cubic dynamical systems are discussed first, and then the crossing-linear and self-quadratic product vector fields are also discussed. The corresponding appearing and switching bifurcations in such product-cubic systems with a self-quadratic vector field are developed through two theorems. The appearing bifurcations include saddle-source (sink), double-saddles, hyperbolic-to-hyperbolic-secant flows and inflection-source (sink) flows. The switching bifurcations on the infinite-equilibrium include inflection upper-saddle (lower-saddle), hyperbolic (hyperbolic-secant) source and sink, up-down (down-up) lower-saddle and upper-saddle infinite-equilibriums, inflection-source (sink).

In Chap. 2, the appearing and switching bifurcations for self-linear and crossing-quadratic product system with a self-quadratic vector field are discussed, and the corresponding network of equilibriums are presented. The inflection-source (sink) flows and saddle-source (sink) for the appearing bifurcations are presented first. The inflection upper-saddle and lower-saddle on the infinite equilibriums are for inflection-source (sink) flows with connected saddle-source (sink). The inflection-source (sink) flows with connected saddles, source, and sink are presented. The hyperbolic and hyperbolic-secant sources (sinks) are the switching bifurcations of the inflection-source (sink) flows with connected saddle, source, and sink. The up-down-to-down-up upper-saddle (lower-saddle) are the switching bifurcations of hyperbolic and hyperbolic-secant flows with connected saddle-source (sink). The series of connected hyperbolic and hyperbolic-secant flows with saddle, sink and source are presented.

In Chap. 3, the appearing and switching bifurcations in a crossing-linear and self-quadratic product system with a self-quadratic vector field are discussed, and the corresponding network of equilibriums are presented. The double-saddles with hyperbolic-to-hyperbolic flows are presented first for the appearing bifurcations. The up-down and down-up saddles are the switching bifurcations for the hyperbolic-to-hyperbolic-secant flow with connected double-saddles. The series of hyperbolic-to-hyperbolic-secant flows with connected saddle-source and saddle-sink are presented. The inflection-source (sink) on the infinite-equilibrium is for the switching bifurcations of a hyperbolic-to-hyperbolic-secant flow with connected saddle-source (sink). The network of the paralleled hyperbolic and hyperbolic-secant flows with paralleled saddle-source and saddle-sink. The up-down-to-down-up and down-up-to-up-down upper-saddle (lower-saddle) are presented for the switching bifurcation of a hyperbolic (hyperbolic-secant) flow with saddle-source (sink). The networks of paralleled hyperbolic and hyperbolic-secant flows with connected saddle, sink, and source are presented. The connected inflection-sources (sinks) are presented for the paralleled hyperbolic and hyperbolic-secant flows with saddle and source (sink).

Finally, the author hopes the materials presented herein can provide a better understanding of cubic nonlinear systems in science and engineering.

Edwardsville, IL, USA Albert C. J. Luo

Contents

Chapter 1
Self-Quadratic and Product-Cubic Systems

In this chapter, nonlinear dynamics and singularity of self-quadratic and product-cubic systems are discussed. The equilibrium singularity and bifurcations are discussed through the appearing and switching bifurcations. The saddle-sources (sink) equilibriums are the appearing bifurcations for saddle and source (sink) equilibriums. The double-saddle equilibriums are the appearing bifurcations of the saddle-source and saddle-sinks, and such double-saddle equilibriums are also the appearing bifurcations of the network of saddles, sink, and source. The infinite-equilibriums for the switching bifurcations exist in such cubic nonlinear systems. Inflection upper-saddle and lower-saddle infinite-equilibriums are the switching bifurcation of saddle-sink and saddle-source with inflection-sink and source flows. Hyperbolic and hyperbolic-secant, sink and source infinite-equilibriums are for the switching bifurcations of saddle, source, and sink with inflection-sink and source flows. Up-down and down-up lower-saddle and upper-saddle infinite-equilibriums are the switching bifurcations of saddle-source and saddle-sink with hyperbolic and hyperbolic-secant flows.

1.1 Self-Linear and Crossing-Quadratic Product Vector Fields

In this section, cubic dynamical systems with a self-quadratic vector field and a self-linear and crossing-quadratic product vector field will be discussed. The corresponding dynamical behaviors will be presented through the following theorem.

Theorem 1.1 *Consider a two-dimensional, cubic dynamical system with a product-cubic vector field and a self-quadratic vector field as*

A. C. J. Luo, *Two-dimensional Product-Cubic Systems, Vol. VII*,
https://doi.org/10.1007/978-3-031-48483-4_1

$$\begin{aligned}\dot{x}_{j_2} &= a_{j_2j_20}(x_{j_2}^2 + B_{j_2j_2}x_{j_2} + C_{j_2j_2}),\\ \dot{x}_{j_1} &= a_{j_1j_10}(x_{j_1} - a_{j_1j_11})(x_{j_2}^2 + B_{j_1j_2}x_{j_2} + C_{j_1j_2})\end{aligned} \tag{1.1}$$

with

$$\Delta_{j_1j_2} = B_{j_1j_2}^2 - 4C_{j_1j_2}, \Delta_{j_2j_2} = B_{j_2j_2}^2 - 4C_{j_2j_2}. \tag{1.2}$$

(i) For $\Delta_{j_1j_2} < 0$ and $\Delta_{j_2j_2} < 0$, the standard form is

$$\begin{aligned}\dot{x}_{j_2} &= a_{j_2j_20}\left[(x_{j_2} - a_{j_2j_2})^2 + b_{j_2j_2}\right],\\ \dot{x}_{j_1} &= a_{j_1j_10}(x_{j_1} - a_{j_1j_11})\left[(x_{j_2} - a_{j_1j_2})^2 + b_{j_1j_2}\right]\end{aligned} \tag{1.3}$$

where

$$\begin{aligned}a_{j_1j_2} &= -\frac{1}{2}B_{j_1j_2}, b_{j_1j_2} = \frac{1}{4}(-\Delta_{j_1j_2}),\\ a_{j_2j_2} &= -\frac{1}{2}B_{j_2j_2}, b_{j_2j_2} = \frac{1}{4}(-\Delta_{j_2j_2}).\end{aligned} \tag{1.4}$$

The first integral manifold is

$$\begin{aligned}\ln\frac{|x_{j_1} - a_{j_1j_11}|}{|x_{j_10} - a_{j_1j_11}|} = \frac{a_{j_1j_10}}{a_{j_2j_20}}\Big\{ &\left[(x_{j_2} - a_{j_2j_2}) - (x_{j_20} - a_{j_2j_2})\right]\\ &+(a_{j_2j_2} - a_{j_1j_2})\ln\frac{|(x_{j_2} - a_{j_2j_2})^2 + b_{j_2j_2}|}{|(x_{j_20} - a_{j_2j_2})^2 + b_{j_2j_2}|}\\ &+\left[(a_{j_2j_2} - a_{j_1j_2})^2 + b_{j_1j_2} - b_{j_2j_2}\right]\\ &\times\frac{1}{\sqrt{b_{j_2j_2}}}\left[\arctan\frac{x_{j_2} - a_{j_2j_2}}{\sqrt{b_{j_2j_2}}} - \arctan\frac{x_{j_20} - a_{j_2j_2}}{\sqrt{b_{j_2j_2}}}\right]\Big\}.\end{aligned} \tag{1.5}$$

The equilibrium of $x_{j_1}^* = a_{j_1j_11}$ has the following properties.

- *For $a_{j_1j_10} > 0$ and $a_{j_2j_20} > 0$,*

$$(a_{j_1j_11}, \dot{x}_{j_2}) = \underbrace{(\mathrm{SO}, \mathrm{pF})}_{\text{source flow }(+)}. \tag{1.6}$$

The equilibrium of $x_{j_1}^* = a_{j_1j_11}$ is an (SO,pF)*-positive source flow.*

- *For* $a_{j_1j_10}<0$ *and* $a_{j_2j_20}>0$,

$$(a_{j_1j_11},\dot{x}_{j_2})=\underbrace{(\mathrm{SI},\mathrm{pF})}_{\text{sink flow }(+)}. \tag{1.7}$$

The equilibrium of $x^*_{j_1}=a_{j_1j_11}$ *is an* (SI,pF)*-positive sink flow.*

- *For* $a_{j_1j_10}>0$ *and* $a_{j_2j_20}<0$,

$$(a_{j_1j_11},\dot{x}_{j_2})=\underbrace{(\mathrm{SO},\mathrm{nF})}_{\text{source flow }(-)}. \tag{1.8}$$

The equilibrium of $x^*_{j_1}=a_{j_1j_11}$ *is an* (SO,nF)*-negative source flow.*

- *For* $a_{j_1j_10}<0$ *and* $a_{j_2j_20}<0$,

$$(a_{j_1j_11},\dot{x}_{j_2})=\underbrace{(\mathrm{SI},\mathrm{nF})}_{\text{sink flow }(-)}. \tag{1.9}$$

The equilibrium of $x^*_{j_1}=a_{j_1j_11}$ *is an* (SI,nF)*-negative sink flow.*

(ii) For $\Delta_{j_1j_2}<0$ *and* $\Delta_{j_2j_2}=0$, *the standard form is*

$$\begin{aligned}\dot{x}_{j_2}&=a_{j_2j_20}(x_{j_2}-a_{j_2j_21})^2,\\ \dot{x}_{j_1}&=a_{j_1j_10}(x_{j_1}-a_{j_1j_11})\big[(x_{j_2}-a_{j_1j_2})^2+b_{j_1j_2}\big]\end{aligned} \tag{1.10}$$

where

$$\begin{aligned}a_{j_1j_2}&=-\frac{1}{2}B_{j_1j_2},\, b_{j_1j_2}=\frac{1}{4}(-\Delta_{j_1j_2}),\\ a_{j_2j_21}&=-\frac{1}{2}B_{j_2j_2}.\end{aligned} \tag{1.11}$$

The first integral manifold is

$$\begin{aligned}\ln\frac{|x_{j_1}-a_{j_1j_11}|}{|x_{j_10}-a_{j_1j_11}|}=&\frac{a_{j_1j_10}}{a_{j_2j_20}}\Big\{\big[(x_{j_2}-a_{j_2j_21})-(x_{j_20}-a_{j_2j_21})\big]\\ &+2(a_{j_2j_21}-a_{j_1j_2})\ln\frac{|x_{j_2}-a_{j_2j_21}|}{|x_{j_20}-a_{j_2j_21}|}\\ &-\big[(a_{j_2j_21}-a_{j_1j_2})^2+b_{j_1j_2}\big]\big(\frac{1}{x_{j_2}-a_{j_2j_21}}-\frac{1}{x_{j_20}-a_{j_2j_21}}\big)\Big\}.\end{aligned} \tag{1.12}$$

The equilibrium of $(x^*_{j_1},x^*_{j_2})=(a_{j_1j_11},a_{j_2j_21})$ *has the following properties.*

- *For $a_{j_1j_10} > 0$ and $a_{j_2j_20} > 0$,*

$$(a_{j_1j_11}, a_{j_2j_21}) = \underbrace{(\mathrm{SO}, \mathrm{US})}_{\text{upper-saddle source}}. \tag{1.13}$$

The equilibrium of $(x^*_{j_1}, x^*_{j_2}) = (a_{j_1j_11}, a_{j_2j_21})$ is an (SO,US)-upper-saddle source.

- *For $a_{j_1j_10} < 0$ and $a_{j_2j_20} > 0$,*

$$(a_{j_1j_11}, a_{j_2j_21}) = \underbrace{(\mathrm{SI}, \mathrm{US})}_{\text{upper-saddle-sink}}. \tag{1.14}$$

The equilibrium of $(x^*_{j_1}, x^*_{j_2}) = (a_{j_1j_11}, a_{j_2j_21})$ is an (SI,US)-upper-saddle sink.

- *For $a_{j_1j_10} > 0$ and $a_{j_2j_20} < 0$,*

$$(a_{j_1j_11}, a_{j_2j_21}) = \underbrace{(\mathrm{SO}, \mathrm{LS})}_{\text{lower-saddle source}}. \tag{1.15}$$

The equilibrium of $(x^*_{j_1}, x^*_{j_2}) = (a_{j_1j_11}, a_{j_2j_21})$ is an (SO,LS)-lower-saddle source.

- *For $a_{j_1j_10} < 0$ and $a_{j_2j_20} < 0$,*

$$(a_{j_1j_11}, a_{j_2j_21}) = \underbrace{(\mathrm{SI}, \mathrm{LS})}_{\text{lower-saddle sink}}. \tag{1.16}$$

The equilibrium of $(x^*_{j_1}, x^*_{j_2}) = (a_{j_1j_11}, a_{j_2j_21})$ is an (SI,LS)-lower-saddle sink.

– *The saddle-source and saddle-sink are the appearing bifurcations of saddle-source and saddle-sink.*

(iii) For $\Delta_{j_1j_2} < 0$ and $\Delta_{j_2j_2} > 0$, the standard form is

$$\begin{aligned} \dot{x}_{j_2} &= a_{j_2j_20}(x_{j_2} - a_{j_2j_21})(x_{j_2} - a_{j_2j_22}), \\ \dot{x}_{j_1} &= a_{j_1j_10}(x_{j_1} - a_{j_1j_11})\left[(x_{j_2} - a_{j_1j_2})^2 + b_{j_1j_2}\right] \end{aligned} \tag{1.17}$$

where

$$\begin{aligned} & a_{j_1j_2} = -\frac{1}{2}B_{j_1j_2}, b_{j_1j_2} = \frac{1}{4}(-\Delta_{j_1j_2}), \\ & b_{j_2j_21}, b_{j_2j_22} = -\frac{1}{2}(B_{j_2j_2} \pm \sqrt{\Delta_{j_2j_2}}), \\ & \{a_{j_2j_21}, a_{j_2j_22}\} = \mathrm{sort}\{b_{j_2j_21}, b_{j_2j_22}\}, a_{j_2j_21} < a_{j_2j_22}. \end{aligned} \tag{1.18}$$

The first integral manifold is

$$
\begin{aligned}
\ln \frac{|x_{j_1} - a_{j_1j_1 1}|}{|x_{j_1 0} - a_{j_1j_1 1}|} = \frac{a_{j_1j_1 0}}{a_{j_2j_2 0}} & \Big\{ \big[(x_{j_2} - a_{j_2j_2 1}) - (x_{j_2 0} - a_{j_2j_2 1})\big] \\
& + \textstyle\sum_{s_1=1}^{2} (a_{j_2j_2 s_1} - a_{j_1j_2}) \ln \frac{|x_{j_2} - a_{j_2j_2 s_1}|}{|x_{j_2 0} - a_{j_2j_2 s_1}|} \\
& + \Big[\textstyle\prod_{s_1=1}^{2} (a_{j_2j_2 s_1} - a_{j_1j_2}) + b_{j_1j_2}\Big] \\
& \times (\textstyle\sum_{s_2=1}^{2} \frac{1}{(a_{j_2j_2 s_2} - a_{j_1j_2})} \ln \frac{|x_{j_2} - a_{j_2j_2 s_2}|}{|x_{j_2 0} - a_{j_2j_2 s_2}|}) \Big\}.
\end{aligned}
\tag{1.19}
$$

The equilibrium of $(x_{j_1}^*, x_{j_2}^*) = (a_{j_1j_1 1}, a_{j_2j_2 s_1})$ $(s_1, s_2 \in \{1,2\}, s_1 \neq s_2)$ *has the following properties.*

- *For* $a_{j_1j_1 0} > 0$ *and* $a_{j_2j_2 0}(a_{j_2j_2 s_1} - a_{j_2j_2 s_2}) > 0$,

$$(a_{j_1j_1 1}, a_{j_2j_2 s_1}) = \underbrace{(\mathrm{SO}, \mathrm{SO})}_{\text{source}}. \tag{1.20}$$

  The equilibrium of $(x_{j_1}^*, x_{j_2}^*) = (a_{j_1j_1 1}, a_{j_2j_2 s_1})$ *is an* (SO,SO)*-source.*
- *For* $a_{j_1j_1 0} < 0$ *and* $a_{j_2j_2 0}(a_{j_2j_2 s_1} - a_{j_2j_2 s_2}) > 0$,

$$(a_{j_1j_1 1}, a_{j_2j_2 s_1}) = \underbrace{(\mathrm{SI}, \mathrm{SO})}_{\text{saddle}}. \tag{1.21}$$

  The equilibrium of $(x_{j_1}^*, x_{j_2}^*) = (a_{j_1j_1 1}, a_{j_2j_2 s_1})$ *is an* (SI,SO)*-saddle.*
- *For* $a_{j_1j_1 0} > 0$ *and* $a_{j_2j_2 0}(a_{j_2j_2 s_1} - a_{j_2j_2 s_2}) < 0$,

$$(a_{j_1j_1 1}, a_{j_2j_2 s_1}) = \underbrace{(\mathrm{SO}, \mathrm{SI})}_{\text{saddle}}. \tag{1.22}$$

  The equilibrium of $(x_{j_1}^*, x_{j_2}^*) = (a_{j_1j_1 1}, a_{j_2j_2 s_1})$ *is an* (SO,SI)*-saddle.*
- *For* $a_{j_1j_1 0} < 0$ *and* $a_{j_2j_2 0}(a_{j_2j_2 s_1} - a_{j_2j_2 s_2}) < 0$,

$$(a_{j_1j_1 1}, a_{j_2j_2 s_1}) = \underbrace{(\mathrm{SI}, \mathrm{SI})}_{\text{sink}}. \tag{1.23}$$

  The equilibrium of $(x_{j_1}^*, x_{j_2}^*) = (a_{j_1j_1 1}, a_{j_2j_2 s_1})$ *is an* (SI,SI)*-sink.*

(iv) For $\Delta_{j_1j_2} = 0$ *and* $\Delta_{j_2j_2} < 0$, *the standard form is*

$$
\begin{aligned}
\dot{x}_{j_2} &= a_{j_2j_2 0}\big[(x_{j_2} - a_{j_2j_2})^2 + b_{j_2j_2}\big], \\
\dot{x}_{j_1} &= a_{j_1j_1 0}(x_{j_1} - a_{j_1j_1 1})(x_{j_2} - a_{j_1j_2 1})^2
\end{aligned}
\tag{1.24}
$$

where

$$
\begin{aligned}
a_{j_1j_2 1} &= -\frac{1}{2}B_{j_1j_2},\\
a_{j_2j_2} &= -\frac{1}{2}B_{j_2j_2}, b_{j_2j_2}=\frac{1}{4}(-\Delta_{j_2j_2}).
\end{aligned}
\tag{1.25}
$$

The first integral manifold is

$$
\begin{aligned}
\ln\frac{|x_{j_1}-a_{j_1j_1 1}|}{|x_{j_1 0}-a_{j_1j_1 1}|} &= \frac{a_{j_1j_1 0}}{a_{j_2j_2 0}}\Big\{\big[(x_{j_2}-a_{j_2j_2})-(x_{j_2 0}-a_{j_2j_2})\big]\\
&+(a_{j_2j_2}-a_{j_1j_2 1})\ln\frac{|(x_{j_2}-a_{j_2j_2})^2+b_{j_2j_2}|}{|(x_{j_2 0}-a_{j_2j_2})^2+b_{j_2j_2}|}\\
&+\big[(a_{j_2j_2}-a_{j_1j_2 1})^2-b_{j_2j_2}\big]\frac{1}{\sqrt{b_{j_2j_2}}}\\
&\times\Big(\arctan\frac{x_{j_2}-a_{j_2j_2}}{\sqrt{b_{j_2j_2}}}-\arctan\frac{x_{j_2 0}-a_{j_2j_2}}{\sqrt{b_{j_2j_2}}}\Big)\Big\}.
\end{aligned}
\tag{1.26}
$$

The equilibrium of $(x_{j_1}^*, x_{j_2}^*) = (a_{j_1j_1 1}, a_{j_1j_2 1})$ *has the following properties.*

- *For* $a_{j_1j_1 0} > 0$ *and* $a_{j_2j_2 0} > 0$,

$$
(a_{j_1j_1 1}, a_{j_1j_2 1}) = \underbrace{(\mathrm{DI{:}II}\mathrm{SO}, \mathrm{pF})}_{\text{inflection source flow }(+)}\ . \tag{1.27}
$$

  The equilibrium of $(x_{j_1}^*, x_{j_2}^*) = (a_{j_1j_1 1}, a_{j_1j_2 1})$ *is a* ($_{\mathrm{DI:II}}$SO, pF)*-positive inflection source flow.*

- *For* $a_{j_1j_1 0} < 0$ *and* $a_{j_2j_2 0} > 0$,

$$
(a_{j_1j_1 1}, a_{j_1j_2 1}) = \underbrace{(\mathrm{II{:}DI}\mathrm{SI}, \mathrm{pF})}_{\text{inflection sink flow }(+)}\ . \tag{1.28}
$$

  The equilibrium of $(x_{j_1}^*, x_{j_2}^*) = (a_{j_1j_1 1}, a_{j_1j_2 1})$ *is an* ($_{\mathrm{II:DI}}$SI, pF)*-positive inflection sink flow.*

- *For* $a_{j_1j_1 0} > 0$ *and* $a_{j_2j_2 0} < 0$,

$$
(a_{j_1j_1 1}, a_{j_1j_2 1}) = \underbrace{(\mathrm{II{:}DI}\mathrm{SO}, \mathrm{nF})}_{\text{inflection source flow }(-)}\ . \tag{1.29}
$$

  The equilibrium of $(x_{j_1}^*, x_{j_2}^*) = (a_{j_1j_1 1}, a_{j_1j_2 1})$ *is an* ($_{\mathrm{II:DI}}$SO, nF)*-negative inflection source flow.*

- *For* $a_{j_1j_10}<0$ *and* $a_{j_2j_20}<0$,

$$(a_{j_1j_11},a_{j_1j_21})=\underbrace{({}_{\mathrm{DI:II}}\mathrm{SI},\mathrm{nF})}_{\text{inflection sink flow }(-)}. \tag{1.30}$$

The equilibrium of $(x_{j_1}^*,x_{j_2}^*)=(a_{j_1j_11},a_{j_1j_21})$ *is a* $({}_{\mathrm{DI:II}}\mathrm{SI},\mathrm{nF})$*-negative inflection sink flow.*

(v) For $\Delta_{j_1j_2}=0$ *and* $\Delta_{j_2j_2}=0$, *the standard form is*

$$\begin{aligned}\dot{x}_{j_2}&=a_{j_2j_20}(x_{j_2}-a_{j_2j_21})^2,\\ \dot{x}_{j_1}&=a_{j_1j_10}(x_{j_1}-a_{j_1j_11})(x_{j_2}-a_{j_1j_21})^2\end{aligned} \tag{1.31}$$

where

$$a_{j_1j_21}=-\frac{1}{2}B_{j_1j_2},a_{j_2j_21}=-\frac{1}{2}B_{j_2j_2}. \tag{1.32}$$

(v_1*) The first integral manifold for* $a_{j_2j_21}\neq a_{j_1j_21}$ *is*

$$\begin{aligned}\ln\frac{|x_{j_1}-a_{j_1j_11}|}{|x_{j_10}-a_{j_1j_11}|}&=\frac{a_{j_1j_10}}{a_{j_2j_20}}\Big\{\big[(x_{j_2}-a_{j_2j_21})-(x_{j_20}-a_{j_2j_21})\big]\\ &\quad+2(a_{j_2j_21}-a_{j_1j_21})\ln\frac{|x_{j_2}-a_{j_2j_21}|}{|x_{j_20}-a_{j_2j_21}|}\\ &\quad-(a_{j_2j_21}-a_{j_1j_21})^2(\frac{1}{x_{j_2}-a_{j_2j_21}}-\frac{1}{x_{j_20}-a_{j_2j_21}})\Big\}.\end{aligned} \tag{1.33}$$

(v_{1a}*) The equilibrium of* $(x_{j_1}^*,x_{j_2}^*)=(a_{j_1j_11},a_{j_2j_21})$ *with* $a_{j_2j_21}\neq a_{j_1j_21}$ *has the following properties.*

- *For* $a_{j_1j_10}(a_{j_2j_21}-a_{j_1j_21})^2>0$ *and* $a_{j_2j_20}>0$,

$$(a_{j_1j_11},a_{j_2j_21})=\underbrace{(\mathrm{SO},\mathrm{US})}_{\text{upper-saddle source}}. \tag{1.34}$$

The equilibrium of $(x_{j_1}^*,x_{j_2}^*)=(a_{j_1j_11},a_{j_2j_21})$ *is an* (SO,US)*-upper-saddle source.*

- *For* $a_{j_1j_10}(a_{j_2j_21}-a_{j_1j_21})^2<0$ *and* $a_{j_2j_20}>0$,

$$(a_{j_1j_11},a_{j_2j_21})=\underbrace{(\mathrm{SI},\mathrm{US})}_{\text{upper-saddle-sink}}. \tag{1.35}$$

The equilibrium of $(x_{j_1}^*,x_{j_2}^*)=(a_{j_1j_11},a_{j_2j_21})$ *is an* (SI,US)*-upper-saddle sink.*

- *For* $a_{j_1j_10}(a_{j_2j_21}-a_{j_1j_21})^2>0$ *and* $a_{j_2j_20}<0$,

$$(a_{j_1j_11},a_{j_2j_21})=\underbrace{(\mathrm{SO},\mathrm{LS})}_{\text{lower-saddle source}}. \tag{1.36}$$

The equilibrium of $(x^*_{j_1},x^*_{j_2})=(a_{j_1j_11},a_{j_2j_21})$ *is an* (SO,LS)*-lower-saddle source.*

- *For* $a_{j_1j_10}(a_{j_2j_21}-a_{j_1j_21})^2<0$ *and* $a_{j_2j_20}<0$,

$$(a_{j_1j_11},a_{j_2j_21})=\underbrace{(\mathrm{SI},\mathrm{LS})}_{\text{lower-saddle sink}}. \tag{1.37}$$

The equilibrium of $(x^*_{j_1},x^*_{j_2})=(a_{j_1j_11},a_{j_2j_21})$ *is an* (SI,LS)*-lower-saddle sink.*

– *The saddle-source and saddle-sink are the appearing bifurcations of saddle-source and saddle-sink.*

(v_{Ib}) *The equilibrium of* $(x^*_{j_1},x^*_{j_2})=(a_{j_1j_11},a_{j_1j_21})$ *with* $a_{j_2j_21}\neq a_{j_1j_21}$ *has the following properties.*

- *For* $a_{j_1j_10}>0$ *and* $a_{j_2j_20}(a_{j_1j_21}-a_{j_2j_21})^2>0$,

$$(a_{j_1j_11},a_{j_1j_21})=\underbrace{({}_{\mathrm{DI:II}}\mathrm{SO},\,\mathrm{pF})}_{\text{inflection source flow }(+)}. \tag{1.38}$$

The equilibrium of $(x^*_{j_1},x^*_{j_2})=(a_{j_1j_11},a_{j_1j_21})$ *is a* (${}_{\mathrm{DI:II}}$SO, pF)*-positive inflection source flow.*

- *For* $a_{j_1j_10}<0$ *and* $a_{j_2j_20}(a_{j_1j_21}-a_{j_2j_21})^2>0$,

$$(a_{j_1j_11},a_{j_1j_21})=\underbrace{({}_{\mathrm{II:DI}}\mathrm{SI},\,\mathrm{pF})}_{\text{inflection sink flow }(+)}. \tag{1.39}$$

The equilibrium of $(x^*_{j_1},x^*_{j_2})=(a_{j_1j_11},a_{j_1j_21})$ *is an* (${}_{\mathrm{II:DI}}$SI, pF)*-positive inflection sink flow.*

- *For* $a_{j_1j_10}>0$ *and* $a_{j_2j_20}(a_{j_1j_21}-a_{j_2j_21})^2<0$,

$$(a_{j_1j_11},a_{j_1j_21})=\underbrace{({}_{\mathrm{II:DI}}\mathrm{SO},\,\mathrm{nF})}_{\text{inflection source flow }(-)}. \tag{1.40}$$

The equilibrium of $(x^*_{j_1},x^*_{j_2})=(a_{j_1j_11},a_{j_1j_21})$ *is an* (${}_{\mathrm{II:DI}}$SO, nF)*-negative inflection sink flow.*

- *For* $a_{j_1j_10}<0$ *and* $a_{j_2j_20}(a_{j_1j_21}-a_{j_2j_21})^2<0$,

$$(a_{j_1j_11},a_{j_1j_21})=\underbrace{({}_{\mathrm{DI:II}}\mathrm{SI},\,\mathrm{nF})}_{\text{inflection sink flow }(-)}. \tag{1.41}$$

The equilibrium of $(x^*_{j_1}, x^*_{j_2}) = (a_{j_1j_1 1}, a_{j_1j_2 1})$ *is a* (${}_{\mathrm{DI:II}}$SI, nF)*-negative inflection sink flow.*

(v_2) The first integral manifold for $a_{j_2j_2 1} = a_{j_1j_2 1}$ *is*

$$\ln \frac{|x_{j_1} - a_{j_1j_1 1}|}{|x_{j_1 0} - a_{j_1j_1 1}|} = \frac{a_{j_1j_1 0}}{a_{j_2j_2 0}} \left[(x_{j_2} - a_{j_2j_2 1}) - (x_{j_2 0} - a_{j_2j_2 1})\right]. \tag{1.42}$$

(v_{2a}) The infinite-equilibrium of $x^*_{j_2} = a_{j_1j_2 1} = a_{j_2j_2 1}$ *with* $\overline{x}_{j_1} \neq a_{j_1j_1 1}$ *has the following properties.*

- *For* $a_{j_1j_1 0}(\overline{x}_{j_1} - a_{j_1j_1 1}) > 0$ *and* $a_{j_2j_2 0} > 0$,

$$(\overline{x}_{j_1}, a_{j_2j_2 1}) = \underbrace{(\mathrm{II, US})}_{\text{increasing-inflection upper-saddle}}. \tag{1.43}$$

The infinite-equilibrium of $x^*_{j_2} = a_{j_1j_2 1} = a_{j_2j_2 1}$ *is an* (II,US)*-increasing-inflection upper-saddle.*

- *For* $a_{j_1j_1 0}(\overline{x}_{j_1} - a_{j_1j_1 1}) < 0$ *and* $a_{j_2j_2 0} > 0$,

$$(\overline{x}_{j_1}, a_{j_2j_2 1}) = \underbrace{(\mathrm{DI, US})}_{\text{decreasing-inflection upper-saddle}}. \tag{1.44}$$

The infinite-equilibrium of $x^*_{j_2} = a_{j_1j_2 1} = a_{j_2j_2 1}$ *is a* (DI,US)*-decreasing-inflection upper-saddle.*

- *For* $a_{j_1j_1 0}(\overline{x}_{j_1} - a_{j_1j_1 1}) > 0$ *and* $a_{j_2j_2 0} < 0$,

$$(\overline{x}_{j_1}, a_{j_2j_2 1}) = \underbrace{(\mathrm{DI, LS})}_{\text{decreasing-inflection lower-saddle}}. \tag{1.45}$$

The infinite-equilibrium of $x^*_{j_2} = a_{j_1j_2 1} = a_{j_2j_2 1}$ *is a* (DI,LS)*-decreasing-inflection lower-saddle.*

- *For* $a_{j_1j_1 0}(\overline{x}_{j_1} - a_{j_1j_1 1}) < 0$ *and* $a_{j_2j_2 0} < 0$,

$$(\overline{x}_{j_1}, a_{j_2j_2 1}) = \underbrace{(\mathrm{II, LS})}_{\text{increasing-inflection lower-saddle}}. \tag{1.46}$$

The infinite-equilibrium of $x^*_{j_2} = a_{j_1j_2 1} = a_{j_2j_2 1}$ *is an* (II,LS)*-increasing-inflection lower-saddle.*

(v_{2b}) The equilibrium of $(x^*_{j_1}, x^*_{j_2}) = (a_{j_1j_11}, a_{j_2j_21})$ *with* $a_{j_2j_21} = a_{j_1j_21}$ *has the following properties.*

- *For* $a_{j_1j_10} > 0$ *and* $a_{j_2j_20} > 0$,

$$(a_{j_1j_11}, a_{j_2j_21}) = \underbrace{(\mathrm{DI}:\mathrm{II}, \mathrm{US})}_{\text{inflection upper-saddle}}. \tag{1.47}$$

  The equilibrium of $(x^*_{j_1}, x^*_{j_2}) = (a_{j_1j_11}, a_{j_2j_21})$ *is a* (DI:II,US)-*inflection upper-saddle.*

- *For* $a_{j_1j_10} < 0$ *and* $a_{j_2j_20} > 0$,

$$(a_{j_1j_11}, a_{j_2j_21}) = \underbrace{(\mathrm{II}:\mathrm{DI}, \mathrm{US})}_{\text{inflection upper-saddle}}. \tag{1.48}$$

  The equilibrium of $(x^*_{j_1}, x^*_{j_2}) = (a_{j_1j_11}, a_{j_2j_21})$ *is an* (II:DI,US)-*inflection upper-saddle.*

- *For* $a_{j_1j_10} > 0$ *and* $a_{j_2j_20} < 0$,

$$(a_{j_1j_11}, a_{j_2j_21}) = \underbrace{(\mathrm{II}:\mathrm{DI}, \mathrm{LS})}_{\text{inflection lower-saddle}}. \tag{1.49}$$

  The equilibrium of $(x^*_{j_1}, x^*_{j_2}) = (a_{j_1j_11}, a_{j_2j_21})$ *is an* (II:DI,LS)-*inflection lower-saddle.*

- *For* $a_{j_1j_10} < 0$ *and* $a_{j_2j_20} < 0$,

$$(a_{j_1j_11}, a_{j_2j_21}) = \underbrace{(\mathrm{DI}:\mathrm{II}, \mathrm{LS})}_{\text{inflection lower-saddle}}. \tag{1.50}$$

  The equilibrium of $(x^*_{j_1}, x^*_{j_2}) = (a_{j_1j_11}, a_{j_2j_21})$ *is a* (DI:II,LS)-*inflection lower-saddle.*

(vi) For $\Delta_{j_1j_2} = 0$ *and* $\Delta_{j_2j_2} > 0$, *the standard form is*

$$\begin{aligned} \dot{x}_{j_2} &= a_{j_2j_20}(x_{j_2} - a_{j_2j_21})(x_{j_2} - a_{j_2j_22}), \\ \dot{x}_{j_1} &= a_{j_1j_10}(x_{j_1} - a_{j_1j_11})(x_{j_2} - a_{j_1j_21})^2 \end{aligned} \tag{1.51}$$

where

$$\begin{aligned} &a_{j_1j_21} = -\frac{1}{2}B_{j_1j_2}, \\ &b_{j_2j_21}, b_{j_2j_22} = -\frac{1}{2}(B_{j_2j_2} \pm \sqrt{\Delta_{j_2j_2}}), \\ &\{a_{j_2j_21}, a_{j_2j_22}\} = \mathrm{sort}\{b_{j_2j_21}, b_{j_2j_22}\}, a_{j_2j_21} < a_{j_2j_22}. \end{aligned} \tag{1.52}$$

(vi_1) The first integral manifold ($a_{j_2j_2s_1} \neq a_{j_1j_2 1}$) is

$$\ln\frac{|x_{j_1}-a_{j_1j_1 1}|}{|x_{j_1 0}-a_{j_1j_1 1}|}=\frac{a_{j_1j_1 0}}{a_{j_2j_2 0}}\Big\{\big[(x_{j_2}-a_{j_2j_2 1})-(x_{j_2 0}-a_{j_2j_2 1})\big] \\ +\sum\nolimits_{s_1=1}^{2}\frac{(a_{j_2j_2s_1}-a_{j_1j_2 1})^2}{(a_{j_2j_2s_1}-a_{j_2j_2s_2})}\ln\frac{|x_{j_2}-a_{j_2j_2s_1}|}{|x_{j_2 0}-a_{j_2j_2s_1}|}\Big\}. \tag{1.53}$$

(vi_{1a}) The equilibrium of $(x^*_{j_1},x^*_{j_2})=(a_{j_1j_1 1},a_{j_2j_2s_1})$ ($s_1,\ s_2\in\{1,2\},\ s_1\neq s_2$) has the following properties.

- *For $a_{j_1j_1 0}(a_{j_2j_2s_1}-a_{j_1j_2 1})^2>0$ and $a_{j_2j_2 0}(a_{j_2j_2s_1}-a_{j_2j_2s_2})>0$,*

$$(a_{j_1j_1 1},a_{j_2j_2s_1})=\underbrace{(\mathrm{SO},\mathrm{SO})}_{\text{source}}. \tag{1.54}$$

  The equilibrium of $(x^*_{j_1},x^*_{j_2})=(a_{j_1j_1 1},a_{j_2j_2s_1})$ is an (SO,SO)-source.

- *For $a_{j_1j_1 0}(a_{j_2j_2s_1}-a_{j_1j_2 1})^2<0$ and $a_{j_2j_2 0}(a_{j_2j_2s_1}-a_{j_2j_2s_2})>0$,*

$$(a_{j_1j_1 1},a_{j_2j_2s_1})=\underbrace{(\mathrm{SI},\mathrm{SO})}_{\text{saddle}}. \tag{1.55}$$

  The equilibrium of $(x^*_{j_1},x^*_{j_2})=(a_{j_1j_1 1},a_{j_2j_2s_1})$ is an (SI,SO)-saddle.

- *For $a_{j_1j_1 0}(a_{j_2j_2s_1}-a_{j_1j_2 1})^2>0$ and $a_{j_2j_2 0}(a_{j_2j_2s_1}-a_{j_2j_2s_2})<0$,*

$$(a_{j_1j_1 1},a_{j_2j_2s_1})=\underbrace{(\mathrm{SO},\mathrm{SI})}_{\text{saddle}}. \tag{1.56}$$

  The equilibrium of $(x^*_{j_1},x^*_{j_2})=(a_{j_1j_1 1},a_{j_2j_2s_1})$ is an (SO,SI)-saddle.

- *For $a_{j_1j_1 0}(a_{j_2j_2s_1}-a_{j_1j_2 1})^2<0$ and $a_{j_2j_2 0}(a_{j_2j_2s_1}-a_{j_2j_2s_2})<0$,*

$$(a_{j_1j_1 1},a_{j_2j_2s_1})=\underbrace{(\mathrm{SI},\mathrm{SI})}_{\text{sink}}. \tag{1.57}$$

  The equilibrium of $(x^*_{j_1},x^*_{j_2})=(a_{j_1j_1 1},a_{j_2j_2s_1})$ is an (SI,SI)-sink.

(vi_{1b}) The equilibrium of $(x^*_{j_1},x^*_{j_2})=(a_{j_1j_1 1},a_{j_1j_2 1})$ with $a_{j_1j_2 1}\neq a_{j_2j_2 1},a_{j_2j_2 2}$ has the following properties.

- *For $a_{j_1j_1 0}>0$ and $a_{j_2j_2 0}\prod_{s_1=1}^{2}(a_{j_1j_2 1}-a_{j_2j_2s_1})>0$,*

$$(a_{j_1j_1 1},a_{j_1j_2 1})=\underbrace{({}_{\mathrm{DI:II}}\mathrm{SO},\mathrm{pF})}_{\text{inflection source flow }(+)}. \tag{1.58}$$

  The equilibrium of $(x^*_{j_1},x^*_{j_2})=(a_{j_1j_1 1},a_{j_1j_2 1})$ is a (${}_{\mathrm{DI:II}}$SO, pF)-positive inflection source flow.

- *For* $a_{j_1j_10}<0$ *and* $a_{j_2j_20}\prod_{s_1=1}^{2}(a_{j_1j_21}-a_{j_2j_2s_1})>0$,

$$(a_{j_1j_11},a_{j_1j_21})=\underbrace{(_{\text{II:DI}}\text{SI, pF})}_{\text{inflection sink flow }(+)}. \tag{1.59}$$

The equilibrium of $(x_{j_1}^*,x_{j_2}^*)=(a_{j_1j_11},a_{j_1j_21})$ *is an* ($_{\text{II:DI}}$SI, pF)*-positive inflection sink flow.*

- *For* $a_{j_1j_10}>0$ *and* $a_{j_2j_20}\prod_{s_1=1}^{2}(a_{j_1j_21}-a_{j_2j_2s_1})<0$,

$$(a_{j_1j_11},a_{j_1j_21})=\underbrace{(_{\text{II:DI}}\text{SO, nF})}_{\text{inflection source flow }(-)}. \tag{1.60}$$

The equilibrium of $(x_{j_1}^*,x_{j_2}^*)=(a_{j_1j_11},a_{j_1j_21})$ *is an* ($_{\text{II:DI}}$SO, nF)*-negative inflection source flow.*

- *For* $a_{j_1j_10}<0$ *and* $a_{j_2j_20}\prod_{s_1=1}^{2}(a_{j_1j_21}-a_{j_2j_2s_1})<0$,

$$(a_{j_1j_11},a_{j_1j_21})=\underbrace{(_{\text{DI:II}}\text{SI, nF})}_{\text{inflection sink flow }(-)}. \tag{1.61}$$

The equilibrium of $(x_{j_1}^*,x_{j_2}^*)=(a_{j_1j_11},a_{j_1j_21})$ *is a* ($_{\text{DI:II}}$SI, nF)*-negative inflection sink flow.*

(vi$_2$) The first integral manifold $(a_{j_2j_2s_1}=a_{j_1j_21})$ *is*

$$\begin{aligned}\ln\frac{|x_{j_1}-a_{j_1j_11}|}{|x_{j_10}-a_{j_1j_11}|}=\frac{a_{j_1j_10}}{a_{j_2j_20}}\Big\{&\big[(x_{j_2}-a_{j_2j_2s_2})-(x_{j_20}-a_{j_2j_2s_2})\big]\\&+(a_{j_2j_2s_2}-a_{j_1j_21})\ln\frac{|x_{j_2}-a_{j_2j_2s_2}|}{|x_{j_20}-a_{j_2j_2s_2}|}\Big\}.\end{aligned} \tag{1.62}$$

(vi$_{2a}$) The infinite-equilibrium of $x_{j_2}^*=a_{j_1j_21}=a_{j_2j_2s_1}$ *with* $\overline{x}_{j_1}\neq a_{j_1j_11}$ *has the following properties.*

- *For* $a_{j_1j_10}(\overline{x}_{j_1}-a_{j_1j_11})>0$ *and* $a_{j_2j_20}(a_{j_2j_2s_1}-a_{j_2j_2s_2})>0$,

$$(\overline{x}_{j_1},a_{j_2j_2s_1})=\underbrace{(\text{UP, SO})}_{\text{up-parabola source}}. \tag{1.63}$$

The infinite-equilibrium of $x_{j_2}^*=a_{j_1j_21}=a_{j_2j_2s_1}$ *is a* (UP,SO)*-up-parabola source.*

- *For* $a_{j_1j_10}(\overline{x}_{j_1} - a_{j_1j_11}) < 0$ *and* $a_{j_2j_20}(a_{j_2j_2s_1} - a_{j_2j_2s_2}) > 0$,

$$(\overline{x}_{j_1}, a_{j_2j_2s_1}) = \underbrace{(\mathrm{DP}, \mathrm{SO})}_{\text{down-parabola source}}. \tag{1.64}$$

The infinite-equilibrium of $x^*_{j_2} = a_{j_1j_21} = a_{j_2j_2s_1}$ *is a* (DP,SO)*-down-parabola source.*

- *For* $a_{j_1j_10}(\overline{x}_{j_1} - a_{j_1j_11}) > 0$ *and* $a_{j_2j_20}(a_{j_2j_2s_1} - a_{j_2j_2s_2}) < 0$,

$$(\overline{x}_{j_1}, a_{j_2j_2s_1}) = \underbrace{(\mathrm{DP}, \mathrm{SI})}_{\text{down-parabola sink}}. \tag{1.65}$$

The infinite-equilibrium of $x^*_{j_2} = a_{j_1j_21} = a_{j_2j_2s_1}$ *is a* (DP,SI)*-down-parabola sink.*

- *For* $a_{j_1j_10}(\overline{x}_{j_1} - a_{j_1j_11}) < 0$ *and* $a_{j_2j_20}(a_{j_2j_2s_1} - a_{j_2j_2s_2}) < 0$,

$$(\overline{x}_{j_1}, a_{j_2j_2s_1}) = \underbrace{(\mathrm{UP}, \mathrm{SI})}_{\text{up-parabola sink}}. \tag{1.66}$$

The infinite-equilibrium of $x^*_{j_2} = a_{j_1j_21} = a_{j_2j_2s_1}$ *is a* (UP,SI)*-up-parabola sink.*

(vi$_{2b}$) The equilibrium of $(x^*_{j_1}, x^*_{j_2}) = (a_{j_1j_11}, a_{j_2j_2s_1})$ *with* $a_{j_2j_2s_1} = a_{j_1j_21}$ *has the following properties.*

- *For* $a_{j_1j_10} > 0$ *and* $a_{j_2j_20}(a_{j_2j_2s_1} - a_{j_2j_2s_2}) > 0$,

$$(a_{j_1j_11}, a_{j_2j_2s_1}) = \underbrace{(\mathrm{DP}:\mathrm{UP}, \mathrm{SO})}_{\text{hyperbolic source}}. \tag{1.67}$$

The equilibrium of $(x^*_{j_1}, x^*_{j_2}) = (a_{j_1j_11}, a_{j_2j_2s_1})$ *is a* (DP:UP,SO)*-hyperbolic source.*

- *For* $a_{j_1j_10} < 0$ *and* $a_{j_2j_20}(a_{j_2j_2s_1} - a_{j_2j_2s_2}) > 0$,

$$(a_{j_1j_11}, a_{j_2j_2s_1}) = \underbrace{(\mathrm{UP}:\mathrm{DP}, \mathrm{SO})}_{\text{hyperbolic-secant source}}. \tag{1.68}$$

The equilibrium of $(x^*_{j_1}, x^*_{j_2}) = (a_{j_1j_11}, a_{j_2j_2s_1})$ *is a* (UP:DP,SO)*-hyperbolic-secant source.*

- *For* $a_{j_1j_10} > 0$ *and* $a_{j_2j_20}(a_{j_2j_2s_1} - a_{j_2j_2s_2}) < 0$,

$$(a_{j_1j_11}, a_{j_2j_2s_1}) = \underbrace{(\mathrm{UP}:\mathrm{DP}, \mathrm{SI})}_{\text{hyperbolic-secant sink}}. \tag{1.69}$$

The equilibrium of $(x^*_{j_1}, x^*_{j_2}) = (a_{j_1j_1 1}, a_{j_2j_2s_1})$ *is a* (UP:DP,SI)-*hyperbolic-secant sink.*

- *For* $a_{j_1j_1 0} < 0$ *and* $a_{j_2j_2 0}(a_{j_2j_2s_1} - a_{j_2j_2s_2}) < 0$,

$$(a_{j_1j_1 1}, a_{j_2j_2s_1}) = \underbrace{(\mathrm{DP : UP, SI})}_{\text{hyperbolic sink}}. \tag{1.70}$$

The equilibrium of $(x^*_{j_1}, x^*_{j_2}) = (a_{j_1j_1 1}, a_{j_2j_2s_1})$ *is a* (DP:UP,SI)-*hyperbolic sink.*

(vi_{2c}) *The equilibrium of* $(x^*_{j_1}, x^*_{j_2}) = (a_{j_1j_1 1}, a_{j_2j_2s_2})$ *with* $a_{j_2j_2s_1} = a_{j_1j_2 1}$ *has the following properties.*

- *For* $a_{j_1j_1 0}(a_{j_2j_2s_1} - a_{j_1j_2 1})^2 > 0$ *and* $a_{j_2j_2 0}(a_{j_2j_2s_2} - a_{j_2j_2s_1}) > 0$,

$$(a_{j_1j_1 1}, a_{j_2j_2s_2}) = \underbrace{(\mathrm{SO, SO})}_{\text{source}}. \tag{1.71}$$

The equilibrium of $(x^*_{j_1}, x^*_{j_2}) = (a_{j_1j_1 1}, a_{j_2j_2s_2})$ *is an* (SO,SO)-*source.*

- *For* $a_{j_1j_1 0}(a_{j_2j_2s_1} - a_{j_1j_2 1})^2 < 0$ *and* $a_{j_2j_2 0}(a_{j_2j_2s_2} - a_{j_2j_2s_1}) > 0$,

$$(a_{j_1j_1 1}, a_{j_2j_2s_2}) = \underbrace{(\mathrm{SI, SO})}_{\text{saddle}}. \tag{1.72}$$

The equilibrium of $(x^*_{j_1}, x^*_{j_2}) = (a_{j_1j_1 1}, a_{j_2j_2s_2})$ *is an* (SI,SO)-*saddle.*

- *For* $a_{j_1j_1 0}(a_{j_2j_2s_1} - a_{j_1j_2 1})^2 > 0$ *and* $a_{j_2j_2 0}(a_{j_2j_2s_2} - a_{j_2j_2s_1}) < 0$,

$$(a_{j_1j_1 1}, a_{j_2j_2s_2}) = \underbrace{(\mathrm{SO, SI})}_{\text{saddle}}. \tag{1.73}$$

The equilibrium of $(x^*_{j_1}, x^*_{j_2}) = (a_{j_1j_1 1}, a_{j_2j_2s_2})$ *is an* (SO,SI)-*saddle.*

- *For* $a_{j_1j_1 0}(a_{j_2j_2s_1} - a_{j_1j_2 1})^2 < 0$ *and* $a_{j_2j_2 0}(a_{j_2j_2s_2} - a_{j_2j_2s_1}) < 0$,

$$(a_{j_1j_1 1}, a_{j_2j_2s_2}) = \underbrace{(\mathrm{SI, SI})}_{\text{sink}}. \tag{1.74}$$

The equilibrium of $(x^*_{j_1}, x^*_{j_2}) = (a_{j_1j_1 1}, a_{j_2j_2s_2})$ *is an* (SI,SI)-*sink.*

(*vii*) *For* $\Delta_{j_1j_2} > 0$ *and* $\Delta_{j_2j_2} < 0$, *the standard form is*

$$\begin{aligned} \dot{x}_{j_2} &= a_{j_2j_2 0}\left[(x_{j_2} - a_{j_2j_2})^2 + b_{j_2j_2}\right], \\ \dot{x}_{j_1} &= a_{j_1j_1 0}(x_{j_1} - a_{j_1j_1 1})(x_{j_2} - a_{j_1j_2 1})(x_{j_2} - a_{j_1j_2 2}) \end{aligned} \tag{1.75}$$

where

$$\begin{aligned} & b_{j_1j_21}, b_{j_1j_22} = -\frac{1}{2}(B_{j_1j_2} \pm \sqrt{\Delta_{j_1j_2}}), \\ & \{a_{j_1j_21}, a_{j_1j_22}\} = \text{sort}\{b_{j_1j_21}, b_{j_1j_22}\}, a_{j_1j_21} < a_{j_1j_22}; \\ & a_{j_2j_2} = -\frac{1}{2}B_{j_2j_2}, b_{j_2j_2} = \frac{1}{4}(-\Delta_{j_2j_2}). \end{aligned} \tag{1.76}$$

The first integral manifold is

$$\begin{aligned} \ln \frac{|x_{j_1} - a_{j_1j_11}|}{|x_{j_10} - a_{j_1j_11}|} = \frac{a_{j_1j_10}}{a_{j_2j_20}} \Big\{ & [(x_{j_2} - a_{j_2j_2}) - (x_{j_20} - a_{j_2j_2})] \\ & + \sum_{s_1=1}^{2} (a_{j_2j_2} - a_{j_1j_2s_1}) \ln \frac{|(x_{j_2} - a_{j_2j_2})^2 + b_{j_2j_2}|}{|(x_{j_20} - a_{j_2j_2})^2 + b_{j_2j_2}|} \\ & + \Big[\prod_{s_1=1}^{2} (a_{j_2j_2} - a_{j_1j_2s_1}) - b_{j_2j_2}\Big] \frac{1}{\sqrt{b_{j_2j_2}}} \\ & \times \Big(\arctan \frac{x_{j_2} - a_{j_2j_2}}{\sqrt{b_{j_2j_2}}} - \arctan \frac{x_{j_20} - a_{j_2j_2}}{\sqrt{b_{j_2j_2}}} \Big) \Big\}. \end{aligned} \tag{1.77}$$

The equilibrium of $(x^*_{j_1}, x^*_{j_2}) = (a_{j_1j_11}, a_{j_1j_2s_1})$ $(s_1, s_2 \in \{1,2\}, s_1 \neq s_2)$ *has the following properties.*

- *For* $a_{j_1j_10}(a_{j_1j_2s_1} - a_{j_1j_2s_2}) > 0$ *and* $a_{j_2j_20} > 0$,

$$(a_{j_1j_11}, a_{j_1j_2s_1}) = \underbrace{(\text{DP} : \text{UP}, \text{pF})}_{\text{hyperbolic flow } (+)}. \tag{1.78}$$

  The equilibrium of $(x^*_{j_1}, x^*_{j_2}) = (a_{j_1j_11}, a_{j_1j_2s_1})$ *is a* (DP:UP,pF)-*positive hyperbolic flow.*

- *For* $a_{j_1j_10}(a_{j_1j_2s_1} - a_{j_1j_2s_2}) < 0$ *and* $a_{j_2j_20} > 0$,

$$(a_{j_1j_11}, a_{j_1j_2s_1}) = \underbrace{(\text{UP} : \text{DP}, \text{pF})}_{\text{hyperbolic-secant flow } (+)}. \tag{1.79}$$

  The equilibrium of $(x^*_{j_1}, x^*_{j_2}) = (a_{j_1j_11}, a_{j_1j_2s_1})$ *is a* (UP:DP,pF)-*positive hyperbolic-secant flow.*

- *For* $a_{j_1j_10}(a_{j_1j_2s_1} - a_{j_1j_2s_2}) > 0$ *and* $a_{j_2j_20} < 0$,

$$(a_{j_1j_11}, a_{j_1j_2s_1}) = \underbrace{(\text{UP} : \text{DP}, \text{nF})}_{\text{hyperbolic-secant flow } (-)}. \tag{1.80}$$

  The equilibrium of $(x^*_{j_1}, x^*_{j_2}) = (a_{j_1j_11}, a_{j_1j_2s_1})$ *is a* (UP:DP,nF)-*negative hyperbolic-secant flow.*

- *For* $a_{j_1j_10}(a_{j_1j_2s_1}-a_{j_1j_2s_2})<0$ *and* $a_{j_2j_20}<0$,

$$(a_{j_1j_11},a_{j_1j_2s_1})=\underbrace{(\mathrm{DP:UP,nF})}_{\text{hyperbolic flow }(-)}. \tag{1.81}$$

The equilibrium of $(x^*_{j_1},x^*_{j_2})=(a_{j_1j_11},a_{j_1j_2s_1})$ *is a* (DP:UP,nF)-*negative hyperbolic flow.*

(viii) For $\Delta_{j_1j_2}>0$ *and* $\Delta_{j_2j_2}=0$, *the standard form is*

$$\begin{aligned}\dot{x}_{j_2}&=a_{j_2j_20}(x_{j_2}-a_{j_2j_21})^2,\\ \dot{x}_{j_1}&=a_{j_1j_10}(x_{j_1}-a_{j_1j_11})(x_{j_2}-a_{j_1j_21})(x_{j_2}-a_{j_1j_22})\end{aligned} \tag{1.82}$$

where

$$\begin{aligned}&b_{j_1j_21},b_{j_1j_22}=-\frac{1}{2}(B_{j_1j_2}\pm\sqrt{\Delta_{j_1j_2}}),\\ &\{a_{j_1j_21},a_{j_1j_22}\}=\mathrm{sort}\{b_{j_1j_21},b_{j_1j_22}\},a_{j_1j_21}<a_{j_1j_22};\\ &a_{j_2j_21}=-\frac{1}{2}B_{j_2j_2}.\end{aligned} \tag{1.83}$$

(viii$_1$) The first integral manifold for $a_{j_2j_21}\neq a_{j_1j_21},a_{j_1j_22}$ *is*

$$\begin{aligned}\ln\frac{|x_{j_1}-a_{j_1j_11}|}{|x_{j_10}-a_{j_1j_11}|}=&\frac{a_{j_1j_10}}{a_{j_2j_20}}\Big\{[(x_{j_2}-a_{j_2j_21})-(x_{j_20}-a_{j_2j_21})]\\ &+\textstyle\sum_{s_1=1}^{2}(a_{j_2j_21}-a_{j_1j_2s_1})\ln\frac{|x_{j_2}-a_{j_2j_21}|}{|x_{j_20}-a_{j_2j_21}|}\\ &-\textstyle\prod_{s_1=1}^{2}(a_{j_2j_21}-a_{j_1j_2s_1})(\frac{1}{x_{j_2}-a_{j_2j_21}}-\frac{1}{x_{j_20}-a_{j_2j_21}})\Big\}.\end{aligned} \tag{1.84}$$

(viii$_{1a}$) The equilibrium of $(x^*_{j_1},x^*_{j_2})=(a_{j_1j_11},a_{j_2j_21})$ *has the following properties.*

- *For* $a_{j_1j_10}\prod_{s_1=1}^{2}(a_{j_2j_21}-a_{j_1j_2s_1})>0$ *and* $a_{j_2j_20}>0$,

$$(a_{j_1j_11},a_{j_2j_21})=\underbrace{(\mathrm{SO,US})}_{\text{upper-saddle source}}. \tag{1.85}$$

The equilibrium of $(x^*_{j_1},x^*_{j_2})=(a_{j_1j_11},a_{j_2j_21})$ *is an* (SO,US)-*upper-saddle source.*

- *For* $a_{j_1j_10}\prod_{s_1=1}^{2}(a_{j_2j_21}-a_{j_1j_2s_1})<0$ *and* $a_{j_2j_20}>0$,

$$(a_{j_1j_11},a_{j_2j_21})=\underbrace{(\mathrm{SI,US})}_{\text{upper-saddle-sink}}. \tag{1.86}$$

The equilibrium of $(x^*_{j_1},x^*_{j_2})=(a_{j_1j_11},a_{j_2j_21})$ *is an* (SI,US)-*upper-saddle sink.*

- *For* $a_{j_1j_10}\prod_{s_1=1}^{2}(a_{j_2j_21}-a_{j_1j_2s_1})>0$ *and* $a_{j_2j_20}<0$,

$$(a_{j_1j_11},a_{j_2j_21})=\underbrace{(\text{SO},\text{LS})}_{\text{lower-saddle source}}. \tag{1.87}$$

The equilibrium of $(x_{j_1}^*,x_{j_2}^*)=(a_{j_1j_11},a_{j_2j_21})$ *is an* (SO,LS)-*lower-saddle source.*

- *For* $a_{j_1j_10}\prod_{s_1=1}^{2}(a_{j_2j_21}-a_{j_1j_2s_1})<0$ *and* $a_{j_2j_20}<0$,

$$(a_{j_1j_11},a_{j_2j_21})=\underbrace{(\text{SI},\text{LS})}_{\text{lower-saddle source}}. \tag{1.88}$$

The equilibrium of $(x_{j_1}^*,x_{j_2}^*)=(a_{j_1j_11},a_{j_2j_21})$ *is an* (SI,LS)-*lower-saddle sink.*

– *The saddle-source and saddle-sink are the appearing bifurcations of saddle-source and saddle-sink.*

(viii$_{1b}$) The equilibrium of $(x_{j_1}^*,x_{j_2}^*)=(a_{j_1j_11},a_{j_1j_2s_1})$ *with* $a_{j_1j_2s_1}\neq a_{j_2j_21}$ $(s_1, s_2\in\{1,2\}$, $s_1\neq s_2)$ *has the following properties.*

- *For* $a_{j_1j_10}(a_{j_1j_2s_1}-a_{j_1j_2s_2})>0$ *and* $a_{j_2j_20}(a_{j_1j_2s_1}-a_{j_2j_21})^2>0$,

$$(a_{j_1j_11},a_{j_1j_2s_1})=\underbrace{(\text{DP}:\text{UP},\text{pF})}_{\text{hyperbolic flow }(+)}. \tag{1.89}$$

The equilibrium of $(x_{j_1}^*,x_{j_2}^*)=(a_{j_1j_11},a_{j_1j_2s_1})$ *is a* (DP:UP, pF)-*positive hyperbolic flow.*

- *For* $a_{j_1j_10}(a_{j_1j_2s_1}-a_{j_1j_2s_2})<0$ *and* $a_{j_2j_20}(a_{j_1j_2s_1}-a_{j_2j_21})^2>0$,

$$(a_{j_1j_11},a_{j_1j_2s_1})=\underbrace{(\text{UP}:\text{DP},\text{pF})}_{\text{hyperbolic-secant flow }(+)}. \tag{1.90}$$

The equilibrium of $(x_{j_1}^*,x_{j_2}^*)=(a_{j_1j_11},a_{j_1j_2s_1})$ *is a* (UP:DP, pF)-*positive hyperbolic-secant flow.*

- *For* $a_{j_1j_10}(a_{j_1j_2s_1}-a_{j_1j_2s_2})>0$ *and* $a_{j_2j_20}(a_{j_1j_2s_1}-a_{j_2j_21})^2<0$,

$$(a_{j_1j_11},a_{j_1j_2s_1})=\underbrace{(\text{UP}:\text{DP},\text{nF})}_{\text{hyperbolic-secant flow }(-)}. \tag{1.91}$$

The equilibrium of $(x_{j_1}^*,x_{j_2}^*)=(a_{j_1j_11},a_{j_1j_2s_1})$ *is a* (UP:DP,nF)-*negative hyperbolic-secant flow.*

- *For* $a_{j_1j_10}(a_{j_1j_2s_1}-a_{j_1j_2s_2})<0$ *and* $a_{j_2j_20}(a_{j_1j_2s_1}-a_{j_2j_21})^2<0$,

$$(a_{j_1j_11},a_{j_1j_2s_1})=\underbrace{(\text{DP}:\text{UP},\text{nF})}_{\text{hyperbolic flow }(-)}. \tag{1.92}$$

The equilibrium of $(x_{j_1}^*,x_{j_2}^*)=(a_{j_1j_11},a_{j_1j_2s_1})$ *is a* (DP:UP,nF)*-negative hyperbolic flow.*

(viii₂) The first integral manifold for $a_{j_2j_21}=a_{j_1j_2s_1}$ $(s_1,\,s_2\in\{1,2\},\,s_1\neq s_2)$ *is*

$$\ln\frac{|x_{j_1}-a_{j_1j_11}|}{|x_{j_10}-a_{j_1j_11}|}=\frac{a_{j_1j_10}}{a_{j_2j_20}}\left\{[(x_{j_2}-a_{j_2j_21})-(x_{j_20}-a_{j_2j_21})]+(a_{j_2j_21}-a_{j_1j_2s_2})\ln\frac{|x_{j_2}-a_{j_2j_21}|}{|x_{j_20}-a_{j_2j_21}|}\right\}. \tag{1.93}$$

(viii$_{2a}$) The infinite-equilibrium of $x_{j_2}^*=a_{j_1j_2s_1}=a_{j_2j_21}$ *with* $\bar{x}_{j_1}\neq a_{j_1j_11}$ *has the following properties.*

- *For* $a_{j_1j_10}(\bar{x}_{j_1}-a_{j_1j_11})(a_{j_1j_2s_1}-a_{j_1j_2s_2})>0$ *and* $a_{j_2j_20}>0$,

$$(\bar{x}_{j_1},a_{j_2j_21})=\underbrace{(\text{DU},\text{US})}_{\text{down-up upper-saddle}}. \tag{1.94}$$

The infinite-equilibrium of $x_{j_2}^*=a_{j_1j_2s_1}=a_{j_2j_21}$ *is a* (DU,US)*-down-up upper-saddle.*

- *For* $a_{j_1j_10}(\bar{x}_{j_1}-a_{j_1j_11})(a_{j_1j_2s_1}-a_{j_1j_2s_2})<0$ *and* $a_{j_2j_20}>0$,

$$(\bar{x}_{j_1},a_{j_2j_21})=\underbrace{(\text{UD},\text{US})}_{\text{up-down upper-saddle}}. \tag{1.95}$$

The infinite-equilibrium of $x_{j_2}^*=a_{j_1j_2s_1}=a_{j_2j_21}$ *is a* (UD,US)*-up-down upper-saddle.*

- *For* $a_{j_1j_10}(\bar{x}_{j_1}-a_{j_1j_11})(a_{j_1j_2s_1}-a_{j_1j_2s_2})>0$ *and* $a_{j_2j_20}<0$,

$$(\bar{x}_{j_1},a_{j_2j_21})=\underbrace{(\text{UD},\text{LS})}_{\text{up-down lower-saddle}}. \tag{1.96}$$

The infinite-equilibrium of $x_{j_2}^*=a_{j_1j_2s_1}=a_{j_2j_21}$ *is a* (UD,LS)*-up-down lower-saddle.*

- *For* $a_{j_1j_10}(\bar{x}_{j_1}-a_{j_1j_11})(a_{j_1j_2s_1}-a_{j_1j_2s_2})<0$ *and* $a_{j_2j_20}<0$,

$$(\overline{x}_{j_1}, a_{j_2j_21}) = \underbrace{(\text{DU}, \text{LS})}_{\text{down-up lower-saddle}}. \tag{1.97}$$

The infinite-equilibrium of $x^*_{j_2} = a_{j_1j_2s_1} = a_{j_2j_21}$ *is a* (DU,LS)-*down-up lower-saddle.*

(viii$_{2b}$) The equilibrium of $(x^*_{j_1}, x^*_{j_2}) = (a_{j_1j_11}, a_{j_2j_21})$ *with* $a_{j_2j_21} = a_{j_1j_2s_1}$ *has the following properties.*

- *For* $a_{j_1j_10}(a_{j_1j_2s_1} - a_{j_1j_2s_2}) > 0$ *and* $a_{j_2j_20} > 0$,

$$(a_{j_1j_11}, a_{j_2j_21}) = \underbrace{(\text{DU} : \text{UD}, \text{US})}_{\text{down-up to up-down upper-saddle}}. \tag{1.98}$$

  The equilibrium of $(x^*_{j_1}, x^*_{j_2}) = (a_{j_1j_11}, a_{j_2j_21})$ *is a* (DU:UD,US)-*down-up to down-up upper-saddle.*
- *For* $a_{j_1j_10}(a_{j_1j_2s_1} - a_{j_1j_2s_2}) < 0$ *and* $a_{j_2j_20} > 0$,

$$(a_{j_1j_11}, a_{j_2j_21}) = \underbrace{(\text{UD} : \text{DU}, \text{US})}_{\text{up-down to down-up upper-saddle}}. \tag{1.99}$$

  The equilibrium of $(x^*_{j_1}, x^*_{j_2}) = (a_{j_1j_11}, a_{j_2j_21})$ *is a* (UD:DU,US)-*up-down to down-up upper-saddle.*
- *For* $a_{j_1j_10}(a_{j_1j_2s_1} - a_{j_1j_2s_2}) > 0$ *and* $a_{j_2j_20} < 0$,

$$(a_{j_1j_11}, a_{j_2j_21}) = \underbrace{(\text{UD} : \text{DU}, \text{LS})}_{\text{up-down to down-up lower-saddle}}. \tag{1.100}$$

  The equilibrium of $(x^*_{j_1}, x^*_{j_2}) = (a_{j_1j_11}, a_{j_2j_21})$ *is a* (UD:DU,LS)-*up-down to down-up lower-saddle.*
- *For* $a_{j_1j_10}(a_{j_1j_2s_1} - a_{j_1j_2s_2}) < 0$ *and* $a_{j_2j_20} < 0$,

$$(a_{j_1j_11}, a_{j_2j_21}) = \underbrace{(\text{DU} : \text{UD}, \text{LS})}_{\text{down-up to up-down lower-saddle}}. \tag{1.101}$$

  The equilibrium of $(x^*_{j_1}, x^*_{j_2}) = (a_{j_1j_11}, a_{j_2j_21})$ *is a* (DU:UD,LS)-*down-up to up-down lower-saddle.*

(ix) For $\Delta_{j_1j_2} > 0$ *and* $\Delta_{j_2j_2} > 0$, *the standard form is*

$$\begin{aligned} \dot{x}_{j_2} &= a_{j_2j_20}(x_{j_2} - a_{j_2j_21})(x_{j_2} - a_{j_2j_22}), \\ \dot{x}_{j_1} &= a_{j_1j_10}(x_{j_1} - a_{j_1j_11})(x_{j_2} - a_{j_1j_21})(x_{j_2} - a_{j_1j_22}) \end{aligned} \tag{1.102}$$

where

$$
\begin{aligned}
& b_{j_1j_2 1}, b_{j_1j_2 2} = -\frac{1}{2}(B_{j_1j_2} \pm \sqrt{\Delta_{j_1j_2}}), \\
& \{a_{j_1j_2 1}, a_{j_1j_2 2}\} = \mathrm{sort}\{b_{j_1j_2 1}, b_{j_1j_2 2}\}, a_{j_1j_2 1} < a_{j_1j_2 2}; \\
& b_{j_2j_2 1}, b_{j_2j_2 2} = -\frac{1}{2}(B_{j_2j_2} \pm \sqrt{\Delta_{j_2j_2}}), \\
& \{a_{j_2j_2 1}, a_{j_2j_2 2}\} = \mathrm{sort}\{b_{j_2j_2 1}, b_{j_2j_2 2}\}, a_{j_2j_2 1} < a_{j_2j_2 2}.
\end{aligned}
\tag{1.103}
$$

(ix_1) The first integral manifold ($a_{j_2j_2s_1} \neq a_{j_1j_2 1}$) is

$$
\ln \frac{|x_{j_1} - a_{j_1j_1 1}|}{|x_{j_1 0} - a_{j_1j_1 1}|} = \frac{a_{j_1j_1 0}}{a_{j_2j_2 0}} \left[(x_{j_2} - x_{j_2 0}) + \sum\nolimits_{l_1=1}^{2} \frac{\prod_{s_1=1}^{2}(a_{j_2j_2l_1} - a_{j_1j_2s_1})}{(a_{j_2j_2l_1} - a_{j_2j_2l_2})} \ln \frac{|x_{j_2} - a_{j_2j_2l_1}|}{|x_{j_2 0} - a_{j_2j_2l_1}|} \right].
\tag{1.104}
$$

(ix_{1a}) The equilibrium of $(x_{j_1}^*, x_{j_2}^*) = (a_{j_1j_1 1}, a_{j_2j_2s_1})$ (s_1, $s_2 \in \{1,2\}$, $s_1 \neq s_2$) *has the following properties.*

- *For* $a_{j_1j_1 0}\prod_{l_1=1}^{2}(a_{j_2j_2s_1} - a_{j_1j_2l_1}) > 0$ *and* $a_{j_2j_2 0}(a_{j_2j_2s_1} - a_{j_2j_2s_2}) > 0$,

$$
(a_{j_1j_1 1}, a_{j_2j_2s_1}) = \underbrace{(\mathrm{SO}, \mathrm{SO})}_{\text{source}}.
\tag{1.105}
$$

The equilibrium of $(x_{j_1}^*, x_{j_2}^*) = (a_{j_1j_1 1}, a_{j_2j_2s_1})$ *is an* (SO,SO)*-source.*

- *For* $a_{j_1j_1 0}\prod_{l_1=1}^{2}(a_{j_2j_2s_1} - a_{j_1j_2l_1}) < 0$ *and* $a_{j_2j_2 0}(a_{j_2j_2s_1} - a_{j_2j_2s_2}) > 0$,

$$
(a_{j_1j_1 1}, a_{j_2j_2s_1}) = \underbrace{(\mathrm{SI}, \mathrm{SO})}_{\text{saddle}}.
\tag{1.106}
$$

The equilibrium of $(x_{j_1}^*, x_{j_2}^*) = (a_{j_1j_1 1}, a_{j_2j_2s_1})$ *is an* (SI,SO)*-saddle.*

- *For* $a_{j_1j_1 0}\prod_{l_1=1}^{2}(a_{j_2j_2s_1} - a_{j_1j_2l_1}) > 0$ *and* $a_{j_2j_2 0}(a_{j_2j_2s_1} - a_{j_2j_2s_2}) < 0$,

$$
(a_{j_1j_1 1}, a_{j_2j_2s_1}) = \underbrace{(\mathrm{SO}, \mathrm{SI})}_{\text{saddle}}.
\tag{1.107}
$$

The equilibrium of $(x_{j_1}^*, x_{j_2}^*) = (a_{j_1j_1 1}, a_{j_2j_2s_1})$ *is an* (SO,SI)*-saddle.*

- *For* $a_{j_1j_1 0}\prod_{l_1=1}^{2}(a_{j_2j_2s_1} - a_{j_1j_2l_1}) < 0$ *and* $a_{j_2j_2 0}(a_{j_2j_2s_1} - a_{j_2j_2s_2}) < 0$,

$$
(a_{j_1j_1 1}, a_{j_2j_2s_1}) = \underbrace{(\mathrm{SI}, \mathrm{SI})}_{\text{sink}}.
\tag{1.108}
$$

The equilibrium of $(x_{j_1}^*, x_{j_2}^*) = (a_{j_1j_1 1}, a_{j_2j_2s_1})$ *is an* (SI,SI)*-sink.*

(ix_{1b}) The equilibrium of $(x^*_{j_1}, x^*_{j_2}) = (a_{j_1j_1 1}, a_{j_1j_2l_1})$ $(l_1,\ l_2 \in \{1,2\},\ l_1 \neq l_2)$ *with* $a_{j_1j_2l_1} \neq a_{j_2j_2 1}, a_{j_2j_2 2}$ *has the following properties.*

- *For* $a_{j_1j_1 0}(a_{j_1j_2l_1} - a_{j_1j_2l_2}) > 0$ *and* $a_{j_2j_2 0}\prod_{s_1=1}^{2}(a_{j_1j_2l_1} - a_{j_2j_2s_1}) > 0$,

$$(a_{j_1j_1 1}, a_{j_1j_2l_1}) = \underbrace{(\mathrm{DP:UP,pF})}_{\text{hyperbolic flow }(+)}. \tag{1.109}$$

The equilibrium of $(x^*_{j_1}, x^*_{j_2}) = (a_{j_1j_1 1}, a_{j_1j_2l_1})$ *is a* (DP:UP,pF)*-positive hyperbolic flow.*

- *For* $a_{j_1j_1 0}(a_{j_1j_2l_1} - a_{j_1j_2l_2}) < 0$ *and* $a_{j_2j_2 0}\prod_{s_1=1}^{2}(a_{j_1j_2l_1} - a_{j_2j_2s_1}) > 0$,

$$(a_{j_1j_1 1}, a_{j_1j_2l_1}) = \underbrace{(\mathrm{UP:DP,pF})}_{\text{hyperbolic-secant flow }(+)}. \tag{1.110}$$

The equilibrium of $(x^*_{j_1}, x^*_{j_2}) = (a_{j_1j_1 1}, a_{j_1j_2l_1})$ *is a* (UP:DP,pF)*-positive hyperbolic-secant flow.*

- *For* $a_{j_1j_1 0}(a_{j_1j_2l_1} - a_{j_1j_2l_2}) > 0$ *and* $a_{j_2j_2 0}\prod_{s_1=1}^{2}(a_{j_1j_2l_1} - a_{j_2j_2s_1}) < 0$,

$$(a_{j_1j_1 1}, a_{j_1j_2l_1}) = \underbrace{(\mathrm{UP:DP,nF})}_{\text{hyperbolic-secant flow }(-)}. \tag{1.111}$$

The equilibrium of $(x^*_{j_1}, x^*_{j_2}) = (a_{j_1j_1 1}, a_{j_1j_2l_1})$ *is a* (UP:DP,nF)*-negative hyperbolic-secant flow.*

- *For* $a_{j_1j_1 0}(a_{j_1j_2l_1} - a_{j_1j_2l_2}) < 0$ *and* $a_{j_2j_2 0}\prod_{s_1=1}^{2}(a_{j_1j_2l_1} - a_{j_2j_2s_1}) < 0$,

$$(a_{j_1j_1 1}, a_{j_1j_2l_1}) = \underbrace{(\mathrm{DP:UP,nF})}_{\text{hyperbolic flow }(-)}. \tag{1.112}$$

The equilibrium of $(x^*_{j_1}, x^*_{j_2}) = (a_{j_1j_1 1}, a_{j_1j_2l_1})$ *is a* (DP:UP,nF)*-negative hyperbolic flow.*

(ix_2) The first integral manifold ($a_{j_2j_2s_1} = a_{j_1j_2l_1}$ *and* $a_{j_2j_2s_2} \neq a_{j_1j_2l_2}$, $s_1, s_2, l_1, l_2 \in \{1,2\}$; $s_1 \neq s_2$, $l_1 \neq l_2$) *is*

$$\begin{aligned}\ln\frac{|x_{j_1} - a_{j_1j_1 1}|}{|x_{j_1 0} - a_{j_1j_1 1}|} = \frac{a_{j_1j_1 0}}{a_{j_2j_2 0}}\Big\{ & [(x_{j_2} - a_{j_2j_2s_2}) - (x_{j_2 0} - a_{j_2j_2s_2})] \\ & + (a_{j_2j_2s_2} - a_{j_1j_2l_2})\ln\frac{|x_{j_2} - a_{j_2j_2s_2}|}{|x_{j_2 0} - a_{j_2j_2s_2}|}\Big\}.\end{aligned} \tag{1.113}$$

(ix$_{2a}$) The infinite-equilibrium of $x^*_{j_2} = a_{j_1j_2l_1} = a_{j_2j_2s_1}$ with $\overline{x}_{j_1} \neq a_{j_1j_1 1}$ has the following properties.

- *For $a_{j_1j_1 0}(\overline{x}_{j_1} - a_{j_1j_1 1})(a_{j_1j_2l_1} - a_{j_1j_2l_2}) > 0$ and $a_{j_2j_2 0}(a_{j_2j_2s_1} - a_{j_2j_2s_2}) > 0$,*

$$(\overline{x}_{j_1}, a_{j_2j_2s_1}) = \underbrace{(\text{II, SO})}_{\text{increasing-inflection source}}. \tag{1.114}$$

  The infinite-equilibrium of $x^*_{j_2} = a_{j_1j_2l_1} = a_{j_2j_2s_1}$ is an (II,SO)-increasing-inflection source.
- *For $a_{j_1j_1 0}(\overline{x}_{j_1} - a_{j_1j_1 1})(a_{j_1j_2l_1} - a_{j_1j_2l_2}) < 0$ and $a_{j_2j_2 0}(a_{j_2j_2s_1} - a_{j_2j_2s_2}) > 0$,*

$$(\overline{x}_{j_1}, a_{j_2j_2s_1}) = \underbrace{(\text{DI, SO})}_{\text{decreasing-inflection source}}. \tag{1.115}$$

  The infinite-equilibrium of $x^*_{j_2} = a_{j_1j_2l_1} = a_{j_2j_2s_1}$ is a (DI,SO)-decreasing-inflection source.
- *For $a_{j_1j_1 0}(\overline{x}_{j_1} - a_{j_1j_1 1})(a_{j_1j_2l_1} - a_{j_1j_2l_2}) > 0$ and $a_{j_2j_2 0}(a_{j_2j_2s_1} - a_{j_2j_2s_2}) < 0$,*

$$(\overline{x}_{j_1}, a_{j_2j_2s_1}) = \underbrace{(\text{DI, SI})}_{\text{decreasing-inflection sink}}. \tag{1.116}$$

  The infinite-equilibrium of $x^*_{j_2} = a_{j_1j_2l_1} = a_{j_2j_2s_1}$ is a (DI,SI)-decreasing-inflection sink.
- *For $a_{j_1j_1 0}(\overline{x}_{j_1} - a_{j_1j_1 1})(a_{j_1j_2l_1} - a_{j_1j_2l_2}) < 0$ and $a_{j_2j_2 0}(a_{j_2j_2s_1} - a_{j_2j_2s_2}) < 0$,*

$$(\overline{x}_{j_1}, a_{j_2j_2s_1}) = \underbrace{(\text{II, SI})}_{\text{increasing-inflection sink}}. \tag{1.117}$$

  The infinite-equilibrium of $x^*_{j_2} = a_{j_1j_2l_1} = a_{j_2j_2s_1}$ is an (II,SI)-increasing-inflection sink.

(ix$_{2b}$) The equilibrium of $(x^*_{j_1}, x^*_{j_2}) = (a_{j_1j_1 1}, a_{j_2j_2s_1})$ with $a_{j_2j_2s_1} = a_{j_1j_2l_1}$ has the following properties.

- *For $a_{j_1j_1 0}(a_{j_1j_2l_1} - a_{j_1j_2l_2}) > 0$ and $a_{j_2j_2 0}(a_{j_2j_2s_1} - a_{j_2j_2s_2}) > 0$,*

$$(a_{j_1j_1 1}, a_{j_2j_2s_1}) = \underbrace{(\text{DI : II, SO})}_{\text{inflection source}}. \tag{1.118}$$

  The equilibrium of $(x^*_{j_1}, x^*_{j_2}) = (a_{j_1j_1 1}, a_{j_2j_2s_1})$ is a (DI:II,SO)-inflection source.

- *For* $a_{j_1j_10}(a_{j_1j_2l_1} - a_{j_1j_2l_2}) < 0$ *and* $a_{j_2j_20}(a_{j_2j_2s_1} - a_{j_2j_2s_2}) > 0$,

$$(a_{j_1j_1 1}, a_{j_2j_2s_1}) = \underbrace{(\text{II} : \text{DI}, \text{SO})}_{\text{inflection source}}. \quad (1.119)$$

The equilibrium of $(x^*_{j_1}, x^*_{j_2}) = (a_{j_1j_1 1}, a_{j_2j_2s_1})$ *is an* (II:DI,SO)-*inflection source.*

- *For* $a_{j_1j_10}(a_{j_1j_2l_1} - a_{j_1j_2l_2}) > 0$ *and* $a_{j_2j_20}(a_{j_2j_2s_1} - a_{j_2j_2s_2}) < 0$,

$$(a_{j_1j_1 1}, a_{j_2j_2s_1}) = \underbrace{(\text{II} : \text{DI}, \text{SI})}_{\text{inflection sink}}. \quad (1.120)$$

The equilibrium of $(x^*_{j_1}, x^*_{j_2}) = (a_{j_1j_1 1}, a_{j_2j_2s_1})$ *is an* (II:DI,SI)-*inflection sink.*

- *For* $a_{j_1j_10}(a_{j_1j_2l_1} - a_{j_1j_2l_2}) < 0$ *and* $a_{j_2j_20}(a_{j_2j_2s_1} - a_{j_2j_2s_2}) < 0$,

$$(a_{j_1j_1 1}, a_{j_2j_2s_1}) = \underbrace{(\text{DI} : \text{II}, \text{SI})}_{\text{inflection sink}}. \quad (1.121)$$

The equilibrium of $(x^*_{j_1}, x^*_{j_2}) = (a_{j_1j_1 1}, a_{j_2j_2s_1})$ *is a* (DI:II,SI)-*inflection sink.*

(ix$_{2c}$*) The equilibrium of* $(x^*_{j_1}, x^*_{j_2}) = (a_{j_1j_1 1}, a_{j_2j_2s_2})$ *with* $a_{j_2j_2s_1} = a_{j_1j_2l_1}$ *has the following properties.*

- *For* $a_{j_1j_10}\prod_{l_1=1}^{2}(a_{j_2j_2s_2} - a_{j_1j_2l_1}) > 0$ *and* $a_{j_2j_20}(a_{j_2j_2s_2} - a_{j_2j_2s_1}) > 0$,

$$(a_{j_1j_1 1}, a_{j_2j_2s_2}) = \underbrace{(\text{SO}, \text{SO})}_{\text{source}}. \quad (1.122)$$

The equilibrium of $(x^*_{j_1}, x^*_{j_2}) = (a_{j_1j_1 1}, a_{j_2j_2s_2})$ *is an* (SO,SO)-*source.*

- *For* $a_{j_1j_10}\prod_{l_1=1}^{2}(a_{j_2j_2s_2} - a_{j_1j_2l_1}) < 0$ *and* $a_{j_2j_20}(a_{j_2j_2s_2} - a_{j_2j_2s_1}) > 0$,

$$(a_{j_1j_1 1}, a_{j_2j_2s_2}) = \underbrace{(\text{SI}, \text{SO})}_{\text{saddle}}. \quad (1.123)$$

The equilibrium of $(x^*_{j_1}, x^*_{j_2}) = (a_{j_1j_1 1}, a_{j_2j_2s_2})$ *is an* (SI,SO)-*saddle.*

- *For* $a_{j_1j_10}\prod_{l_1=1}^{2}(a_{j_2j_2s_2} - a_{j_1j_2l_1}) > 0$ *and* $a_{j_2j_20}(a_{j_2j_2s_2} - a_{j_2j_2s_1}) < 0$,

$$(a_{j_1j_1 1}, a_{j_2j_2s_2}) = \underbrace{(\text{SO}, \text{SI})}_{\text{saddle}}. \quad (1.124)$$

The equilibrium of $(x^*_{j_1}, x^*_{j_2}) = (a_{j_1j_1 1}, a_{j_2j_2s_2})$ *is an* (SO,SI)-*saddle.*

- *For* $a_{j_1j_10}\prod_{l_1=1}^{2}(a_{j_2j_2s_2}-a_{j_1j_2l_1})<0$ *and* $a_{j_2j_20}(a_{j_2j_2s_2}-a_{j_2j_2s_1})<0$,

$$(a_{j_1j_11},a_{j_2j_2s_2})=\underbrace{(\text{SI},\text{SI})}_{\text{sink}}. \tag{1.125}$$

The equilibrium of $(x_{j_1}^*,x_{j_2}^*)=(a_{j_1j_11},a_{j_2j_2s_2})$ *is an* (SI,SI)*-sink.*

(ix_{2d}) The equilibrium of $(x_{j_1}^*,x_{j_2}^*)=(a_{j_1j_11},a_{j_1j_2l_2})$ $(l_1,\ l_2\in\{1,2\},\ l_1\neq l_2)$ *with* $a_{j_1j_2l_1}=a_{j_2j_2s_1}$ *has the following properties.*

- *For* $a_{j_1j_10}(a_{j_1j_2l_2}-a_{j_1j_2l_1})>0$ *and* $a_{j_2j_20}\prod_{s_1=1}^{2}(a_{j_1j_2l_2}-a_{j_2j_2s_1})>0$,

$$(a_{j_1j_11},a_{j_1j_2l_2})=\underbrace{(\text{DP}:\text{UP},\text{pF})}_{\text{hyperbolic flow }(+)}. \tag{1.126}$$

The equilibrium of $(x_{j_1}^*,x_{j_2}^*)=(a_{j_1j_11},a_{j_1j_2l_2})$ *is a* (DP:UP,pF)*-positive hyperbolic flow.*

- *For* $a_{j_1j_10}(a_{j_1j_2l_2}-a_{j_1j_2l_1})<0$ *and* $a_{j_2j_20}\prod_{s_1=1}^{2}(a_{j_1j_2l_2}-a_{j_2j_2s_1})>0$,

$$(a_{j_1j_11},a_{j_1j_2l_2})=\underbrace{(\text{UP}:\text{DP},\text{pF})}_{\text{hyperbolic-secant flow }(+)}. \tag{1.127}$$

The equilibrium of $(x_{j_1}^*,x_{j_2}^*)=(a_{j_1j_11},a_{j_1j_2l_2})$ *is a* (UP:DP,pF)*-positive hyperbolic-secant flow.*

- *For* $a_{j_1j_10}(a_{j_1j_2l_2}-a_{j_1j_2l_1})>0$ *and* $a_{j_2j_20}\prod_{s_1=1}^{2}(a_{j_1j_2l_2}-a_{j_2j_2s_1})<0$,

$$(a_{j_1j_11},a_{j_1j_2l_2})=\underbrace{(\text{UP}:\text{DP},\text{nF})}_{\text{hyperbolic-secant flow }(-)}. \tag{1.128}$$

The equilibrium of $(x_{j_1}^*,x_{j_2}^*)=(a_{j_1j_11},a_{j_1j_2l_2})$ *is a* (UP:DP,nF)*-negative hyperbolic-secant flow.*

- *For* $a_{j_1j_10}(a_{j_1j_2l_2}-a_{j_1j_2l_1})<0$ *and* $a_{j_2j_20}\prod_{s_1=1}^{2}(a_{j_1j_2l_2}-a_{j_2j_2s_1})<0$,

$$(a_{j_1j_11},a_{j_1j_2l_2})=\underbrace{(\text{DP}:\text{UP},\text{nF})}_{\text{hyperbolic flow }(-)}. \tag{1.129}$$

The equilibrium of $(x_{j_1}^*,x_{j_2}^*)=(a_{j_1j_11},a_{j_1j_2l_2})$ *is a* (DP:UP,nF)*-negative hyperbolic flow.*

(ix_3) The first integral manifold ($a_{j_2j_21}=a_{j_1j_22}$ *and* $a_{j_2j_22}=a_{j_1j_22}$) *is*

$$\ln\frac{|x_{j_1}-a_{j_1j_11}|}{|x_{j_10}-a_{j_1j_11}|}=\frac{a_{j_1j_10}}{a_{j_2j_20}}(x_{j_2}-x_{j_20}). \tag{1.130}$$

(ix_{3a}) The infinite-equilibrium of $x^*_{j_2} = a_{j_1 j_2 s_1} = a_{j_2 j_2 s_1}$ $(s_1 = 1, 2)$ *with* $\overline{x}_{j_1} \neq a_{j_1 j_1 1}$ *has the following properties.*

- *For* $a_{j_1 j_1 0}(\overline{x}_{j_1} - a_{j_1 j_1 1})(a_{j_1 j_2 s_1} - a_{j_1 j_2 s_2}) > 0$ *and* $a_{j_2 j_2 0}(a_{j_2 j_2 s_1} - a_{j_2 j_2 s_2}) > 0$,

$$(\overline{x}_{j_1}, a_{j_2 j_2 s_1}) = \underbrace{(\text{II}, \text{SO})}_{\text{increasing-inflection source}}. \tag{1.131}$$

  The infinite-equilibrium of $x^*_{j_2} = a_{j_1 j_2 s_1} = a_{j_2 j_2 s_1}$ *is an* (II,SO)*-increasing-inflection source.*

- *For* $a_{j_1 j_1 0}(\overline{x}_{j_1} - a_{j_1 j_1 1})(a_{j_1 j_2 s_1} - a_{j_1 j_2 s_2}) < 0$ *and* $a_{j_2 j_2 0}(a_{j_2 j_2 s_1} - a_{j_2 j_2 s_2}) > 0$,

$$(\overline{x}_{j_1}, a_{j_2 j_2 s_1}) = \underbrace{(\text{DI}, \text{SO})}_{\text{decreasing-inflection source}}. \tag{1.132}$$

  The infinite-equilibrium of $x^*_{j_2} = a_{j_1 j_2 s_1} = a_{j_2 j_2 s_1}$ *is a* (DI,SO)*-decreasing-inflection source.*

- *For* $a_{j_1 j_1 0}(\overline{x}_{j_1} - a_{j_1 j_1 1})(a_{j_1 j_2 s_1} - a_{j_1 j_2 s_2}) > 0$ *and* $a_{j_2 j_2 0}(a_{j_2 j_2 s_1} - a_{j_2 j_2 s_2}) < 0$,

$$(\overline{x}_{j_1}, a_{j_2 j_2 s_1}) = \underbrace{(\text{DI}, \text{SI})}_{\text{decreasing-inflection sink}}. \tag{1.133}$$

  The infinite-equilibrium of $x^*_{j_2} = a_{j_1 j_2 s_1} = a_{j_2 j_2 s_1}$ *is a* (DI,SI)*-decreasing-inflection sink.*

- *For* $a_{j_1 j_1 0}(\overline{x}_{j_1} - a_{j_1 j_1 1})(a_{j_1 j_2 s_1} - a_{j_1 j_2 s_2}) < 0$ *and* $a_{j_2 j_2 0}(a_{j_2 j_2 s_1} - a_{j_2 j_2 s_2}) < 0$,

$$(\overline{x}_{j_1}, a_{j_2 j_2 s_1}) = \underbrace{(\text{II}, \text{SI})}_{\text{increasing-inflection sink}}. \tag{1.134}$$

  The infinite-equilibrium of $x^*_{j_2} = a_{j_1 j_2 s_1} = a_{j_2 j_2 s_1}$ *is an* (II,SI)*-increasing-inflection sink.*

(ix_{3b}) The equilibrium of $(x^*_{j_1}, x^*_{j_2}) = (a_{j_1 j_1 1}, a_{j_2 j_2 s_1})$ *with* $a_{j_2 j_2 s_1} = a_{j_1 j_2 s_1}$ $(s_1 = 1, 2)$ *has the following properties.*

- *For* $a_{j_1 j_1 0}(a_{j_1 j_2 s_1} - a_{j_1 j_2 s_2}) > 0$ *and* $a_{j_2 j_2 0}(a_{j_2 j_2 s_1} - a_{j_2 j_2 s_2}) > 0$,

$$(a_{j_1 j_1 1}, a_{j_2 j_2 s_1}) = \underbrace{(\text{DI} : \text{II}, \text{SO})}_{\text{inflection source}}. \tag{1.135}$$

  The equilibrium of $(x^*_{j_1}, x^*_{j_2}) = (a_{j_1 j_1 1}, a_{j_2 j_2 s_1})$ *is a* (DI:II,SO)*-inflection source.*

- *For* $a_{j_1j_10}(a_{j_1j_2s_1} - a_{j_1j_2s_2}) < 0$ *and* $a_{j_2j_20}(a_{j_2j_2s_1} - a_{j_2j_2s_2}) > 0$,

$$(a_{j_1j_11}, a_{j_1j_2s_1}) = \underbrace{(\text{II} : \text{DI}, \text{SO})}_{\text{inflection source}}. \quad (1.136)$$

The equilibrium of $(x_{j_1}^*, x_{j_2}^*) = (a_{j_1j_11}, a_{j_1j_2s_1})$ *is an* (II:DI,SO)-*inflection source.*

- *For* $a_{j_1j_10}(a_{j_1j_2s_1} - a_{j_1j_2s_2}) > 0$ *and* $a_{j_2j_20}(a_{j_2j_2s_1} - a_{j_2j_2s_2}) < 0$,

$$(a_{j_1j_11}, a_{j_1j_2s_1}) = \underbrace{(\text{II} : \text{DI}, \text{SI})}_{\text{inflection sink}}. \quad (1.137)$$

The equilibrium of $(x_{j_1}^*, x_{j_2}^*) = (a_{j_1j_11}, a_{j_1j_2s_1})$ *is an* (II:DI,SI)-*inflection sink.*

- *For* $a_{j_1j_10}(a_{j_1j_2s_1} - a_{j_1j_2s_2}) < 0$ *and* $a_{j_2j_20}(a_{j_2j_2s_1} - a_{j_2j_2s_2}) < 0$,

$$(a_{j_1j_11}, a_{j_2j_2s_1}) = \underbrace{(\text{DI} : \text{II}, \text{SI})}_{\text{inflection sink}}. \quad (1.138)$$

The equilibrium of $(x_{j_1}^*, x_{j_2}^*) = (a_{j_1j_11}, a_{j_1j_2s_1})$ *is a* (DI:II,SI)-*inflection sink.*

1.2 Proof of Theorem 1.1

Consider a two-dimensional, nonlinear dynamical system as

$$\begin{aligned}\dot{x}_{j_2} &= a_{j_2j_20}(x_{j_2}^2 + B_{j_2j_2}x_{j_2} + C_{j_2j_2}),\\ \dot{x}_{j_1} &= a_{j_1j_10}(x_{j_1} - a_{j_1j_11})(x_{j_2}^2 + B_{j_1j_2}x_{j_2} + C_{j_1j_2})\end{aligned}$$

with

$$\begin{aligned}\Delta_{j_2j_2} &= B_{j_2j_2}^2 - 4C_{j_2j_2},\\ \Delta_{j_1j_2} &= B_{j_1j_2}^2 - 4C_{j_1j_2}.\end{aligned}$$

(i) For $\Delta_{j_1j_2} < 0$ and $\Delta_{j_2j_2} < 0$, the standard form is

$$\begin{aligned}\dot{x}_{j_2} &= a_{j_2j_20}\left[(x_{j_2} - a_{j_2j_2})^2 + b_{j_2j_2}\right],\\ \dot{x}_{j_1} &= a_{j_1j_10}(x_{j_1} - a_{j_1j_11})\left[(x_{j_2} - a_{j_1j_2})^2 + b_{j_1j_2}\right]\end{aligned}$$

where

$$a_{j_2j_2}=-\frac{1}{2}B_{j_2j_2},b_{j_2j_2}=\frac{1}{4}(-\Delta_{j_1j_2});$$
$$a_{j_1j_2}=-\frac{1}{2}B_{j_1j_2},b_{j_1j_2}=\frac{1}{4}(-\Delta_{j_1j_2}).$$

In phase space,

$$\frac{dx_{j_1}}{dx_{j_2}}=\frac{a_{j_1j_10}}{a_{j_2j_20}}\frac{(x_{j_1}-a_{j_1j_11})\left[(x_{j_2}-a_{j_1j_2})^2+b_{j_1j_2}\right]}{(x_{j_2}-a_{j_2j_2})^2+b_{j_2j_2}},$$

and the deformation of the foregoing equation is

$$\frac{1}{(x_{j_1}-a_{j_1j_11})}dx_{j_1}=\frac{a_{j_1j_10}}{a_{j_2j_20}}\left[1+\frac{2(a_{j_2j_2}-a_{j_1j_2})(x_{j_2}-a_{j_2j_2})}{(x_{j_2}-a_{j_2j_2})^2+b_{j_2j_2}}\right.$$
$$\left.+\frac{(a_{j_2j_2}-a_{j_1j_2})^2+b_{j_1j_2}-b_{j_2j_2}}{(x_{j_2}-a_{j_2j_2})^2+b_{j_2j_2}}\right]dx_{j_2}.$$

With an initial condition of (x_{j_10},x_{j_20}) at $t=t_0$, the integration of the above equation gives

$$\ln\frac{|x_{j_1}-a_{j_1j_11}|}{|x_{j_10}-a_{j_1j_11}|}=\frac{a_{j_1j_10}}{a_{j_2j_20}}\left\{\left[(x_{j_2}-a_{j_2j_2})-(x_{j_20}-a_{j_2j_2})\right]\right.$$
$$+(a_{j_2j_2}-a_{j_1j_2})\ln\frac{|(x_{j_2}-a_{j_2j_2})^2+b_{j_2j_2}|}{|(x_{j_20}-a_{j_2j_2})^2+b_{j_2j_2}|}$$
$$+\left[(a_{j_2j_2}-a_{j_1j_2})^2+b_{j_2j_1}-b_{j_2j_2}\right]$$
$$\left.\times\frac{1}{\sqrt{b_{j_2j_2}}}\left(\arctan\frac{x_{j_2}-a_{j_2j_2}}{\sqrt{b_{j_2j_2}}}-\arctan\frac{x_{j_20}-a_{j_2j_2}}{\sqrt{b_{j_2j_2}}}\right)\right\}.$$

At $x_{j_1}^*=a_{j_1j_11}$, the variational equation is

$$\Delta\dot{x}_{j_1}=a_{j_1j_10}\Delta x_{j_1}\left[(\overline{x}_{j_2}-a_{j_1j_2})^2+b_{j_1j_2}\right].$$

Because of

$$(\overline{x}_{j_2}-a_{j_1j_2})^2+b_{j_1j_2}>0,$$

the flows at $x_{j_1}^*=a_{j_1j_11}$ in the x_{j_1}-direction are source and sink for $a_{j_1j_10}>0$ and $a_{j_1j_10}<0$, respectively. Let

$$\dot{x}_{j_2}=a_{j_2j_20}\left[(\overline{x}_{j_2}-a_{j_2j_2})^2+b_{j_2j_2}\right],$$

and

$$(\bar{x}_{j_2} - a_{j_2j_2})^2 + b_{j_2j_2} > 0.$$

The flows at $x^*_{j_1} = a_{j_1j_1 1}$ in the x_{j_2}-direction are positive and negative for $a_{j_2j_2 0} > 0$ and $a_{j_2j_2 0} < 0$, respectively.

Thus, the equilibrium of $x^*_{j_1} = a_{j_1j_1 1}$ has the following properties as in Eqs. (1.6)–(1.9).

- For $a_{j_1j_1 0} > 0$ and $a_{j_2j_2 0} > 0$,

$$(a_{j_1j_1 1}, \dot{x}_{j_2}) = \underbrace{(\text{SO}, \text{pF})}_{\text{source flow } (+)}.$$

  The equilibrium of $x^*_{j_1} = a_{j_1j_1 1}$ is an (SO,pF)-positive source flow.
- For $a_{j_1j_1 0} < 0$ and $a_{j_2j_2 0} > 0$,

$$(a_{j_1j_1 1}, \dot{x}_{j_2}) = \underbrace{(\text{SI}, \text{pF})}_{\text{sink flow } (+)}.$$

  The equilibrium of $x^*_{j_1} = a_{j_1j_1 1}$ is an (SI,pF)-positive sink flow.
- For $a_{j_1j_1 0} > 0$ and $a_{j_2j_2 0} < 0$,

$$(a_{j_1j_1 1}, \dot{x}_{j_2}) = \underbrace{(\text{SO}, \text{nF})}_{\text{source flow } (-)}.$$

  The equilibrium of $x^*_{j_1} = a_{j_1j_1 1}$ is an (SO,nF)-negative source flow.
- For $a_{j_1j_1 0} < 0$ and $a_{j_2j_2 0} < 0$,

$$(a_{j_1j_1 1}, \dot{x}_{j_2}) = \underbrace{(\text{SI}, \text{nF})}_{\text{sink flow } (-)}.$$

  The equilibrium of $x^*_{j_1} = a_{j_1j_1 1}$ is an (SI,pF)-negative sink flow.

(ii) For $\Delta_{j_1j_2} < 0$ and $\Delta_{j_2j_2} = 0$, the standard form is

$$\begin{aligned}
\dot{x}_{j_2} &= a_{j_2j_2 0}(x_{j_2} - a_{j_2j_2 1})^2, \\
\dot{x}_{j_1} &= a_{j_1j_1 0}(x_{j_1} - a_{j_1j_1 1})\left[(x_{j_2} - a_{j_1j_2})^2 + b_{j_1j_2}\right]
\end{aligned}$$

where

$$a_{j_1j_2} = -\frac{1}{2}B_{j_1j_2}, b_{j_1j_2} = \frac{1}{4}(-\Delta_{j_1j_2}),$$
$$a_{j_2j_2 1} = -\frac{1}{2}B_{j_2j_2}.$$

In phase space,

$$\frac{dx_{j_1}}{dx_{j_2}} = \frac{a_{j_1j_1 0}}{a_{j_2j_2 0}} \frac{(x_{j_1} - a_{j_1j_1 1})\left[(x_{j_2} - a_{j_1j_2})^2 + b_{j_1j_2}\right]}{(x_{j_2} - a_{j_2j_2 1})^2},$$

and the deformation of the foregoing equation is

$$\frac{1}{(x_{j_1} - a_{j_1j_1 1})}dx_{j_1} = \frac{a_{j_1j_1 0}}{a_{j_2j_2 0}}\left[1 + \frac{2(a_{j_2j_2 1} - a_{j_1j_2})(x_{j_2} - a_{j_2j_2 1})}{(x_{j_2} - a_{j_2j_2 1})^2} + \frac{(a_{j_2j_2 1} - a_{j_1j_2})^2 + b_{j_1j_2}}{(x_{j_2} - a_{j_2j_2 1})^2}\right]dx_{j_2}.$$

With an initial condition of $(x_{j_1 0}, x_{j_2 0})$ at $t = t_0$, the integration of the above equation gives

$$\begin{aligned}\ln\frac{|x_{j_1} - a_{j_1j_1 1}|}{|x_{j_1 0} - a_{j_1j_1 1}|} = \frac{a_{j_1j_1 0}}{a_{j_2j_2 0}}\Big\{&\left[(x_{j_2} - a_{j_2j_2 1}) - (x_{j_2 0} - a_{j_2j_2 1})\right]\\ &+2(a_{j_2j_2 1} - a_{j_1j_2})\ln\frac{|x_{j_2} - a_{j_2j_2 1}|}{|x_{j_2 0} - a_{j_2j_2 1}|}\\ &-\left[(a_{j_2j_2 1} - a_{j_1j_2})^2 + b_{j_2j_1}\right]\left(\frac{1}{x_{j_2} - a_{j_2j_2 1}} - \frac{1}{x_{j_2 0} - a_{j_2j_2 1}}\right)\Big\}.\end{aligned}$$

At $(x^*_{j_1}, x^*_{j_2}) = (a_{j_1j_1 1}, a_{j_2j_2 1})$, the variational equations are

$$\begin{aligned}\Delta\dot{x}_{j_2} &= a_{j_2j_2 0}(\Delta x_{j_2})^2,\\ \Delta\dot{x}_{j_1} &= a_{j_1j_1 0}\left[(a_{j_2j_2 1} - a_{j_1j_2})^2 + b_{j_1j_2}\right]\Delta x_{j_1}.\end{aligned}$$

Because of

$$(a_{j_2j_2 1} - a_{j_1j_2})^2 + b_{j_1j_2} > 0,$$

the flows at $(x^*_{j_1}, x^*_{j_2}) = (a_{j_1j_1 1}, a_{j_2j_2 1})$ in the x_{j_1}-direction are source and sink for $a_{j_1j_1 0} > 0$ and $a_{j_1j_1 0} < 0$, respectively. The flows at $(x^*_{j_1}, x^*_{j_2}) = (a_{j_1j_1 1}, a_{j_2j_2 1})$ in the x_{j_2}-direction are upper-saddle and lower-saddle for $a_{j_2j_2 0} > 0$ and $a_{j_2j_2 0} < 0$, respectively.

Therefore, the equilibrium of $(x^*_{j_1}, x^*_{j_2}) = (a_{j_1j_1 1}, a_{j_2j_2 1})$ has the following properties as in Eqs. (1.13)–(1.16).

- For $a_{j_1j_10} > 0$ and $a_{j_2j_20} > 0$,

$$(a_{j_1j_11}, a_{j_2j_21}) = \underbrace{(\mathrm{SO}, \mathrm{US})}_{\text{upper-saddle source}}.$$

  The equilibrium of $(x^*_{j_1}, x^*_{j_2}) = (a_{j_1j_11}, a_{j_2j_21})$ is an (SO,US)-upper-saddle source.

- For $a_{j_1j_10} < 0$ and $a_{j_2j_20} > 0$,

$$(a_{j_1j_11}, a_{j_2j_21}) = \underbrace{(\mathrm{SI}, \mathrm{US})}_{\text{upper-saddle-sink}}.$$

  The equilibrium of $(x^*_{j_1}, x^*_{j_2}) = (a_{j_1j_11}, a_{j_2j_21})$ is an (SI,US)-upper-saddle sink.

- For $a_{j_1j_10} > 0$ and $a_{j_2j_20} < 0$,

$$(a_{j_1j_11}, a_{j_2j_21}) = \underbrace{(\mathrm{SO}, \mathrm{LS})}_{\text{lower-saddle source}}.$$

  The equilibrium of $(x^*_{j_1}, x^*_{j_2}) = (a_{j_1j_11}, a_{j_2j_21})$ is an (SO,LS)-lower-saddle source.

- For $a_{j_1j_10} < 0$ and $a_{j_2j_20} < 0$,

$$(a_{j_1j_11}, a_{j_2j_21}) = \underbrace{(\mathrm{SI}, \mathrm{LS})}_{\text{lower-saddle source}}.$$

  The equilibrium of $(x^*_{j_1}, x^*_{j_2}) = (a_{j_1j_11}, a_{j_2j_21})$ is an (SI,LS)-lower-saddle sink.

– *The saddle-source and saddle-sink are the appearing bifurcations of saddle-source and saddle-sink.*

(iii) For $\Delta_{j_1j_2} < 0$ and $\Delta_{j_2j_2} > 0$, the standard form is

$$\begin{aligned} \dot{x}_{j_2} &= a_{j_2j_20}(x_{j_2} - a_{j_2j_21})(x_{j_2} - a_{j_2j_22}), \\ \dot{x}_{j_1} &= a_{j_1j_10}(x_{j_1} - a_{j_1j_11})\left[(x_{j_2} - a_{j_1j_2})^2 + b_{j_1j_2}\right] \end{aligned}$$

where

$$\begin{aligned} & a_{j_1j_2} = -\frac{1}{2}B_{j_1j_2}, b_{j_1j_2} = \frac{1}{4}(-\Delta_{j_1j_2}), \\ & b_{j_2j_21}, b_{j_2j_22} = -\frac{1}{2}(B_{j_2j_2} \pm \sqrt{\Delta_{j_2j_2}}), \\ & \left\{a_{j_2j_21}, a_{j_2j_22}\right\} = \mathrm{sort}\left\{b_{j_2j_21}, b_{j_2j_22}\right\}, a_{j_2j_21} < a_{j_2j_22}. \end{aligned}$$

In phase space,

$$\frac{dx_{j_1}}{dx_{j_2}} = \frac{a_{j_1j_10}}{a_{j_2j_20}} \frac{(x_{j_1} - a_{j_1j_11})\left[(x_{j_2} - a_{j_1j_2})^2 + b_{j_1j_2}\right]}{(x_{j_2} - a_{j_2j_21})(x_{j_2} - a_{j_2j_22})},$$

and the deformation of the foregoing equation is

$$\begin{aligned} \frac{1}{(x_{j_1} - a_{j_1j_11})} dx_{j_1} = \frac{a_{j_1j_10}}{a_{j_2j_20}} \Bigg[1 &+ \frac{(a_{j_2j_21} - a_{j_1j_2})}{(x_{j_2} - a_{j_2j_21})} + \frac{(a_{j_2j_22} - a_{j_1j_2})}{(x_{j_2} - a_{j_2j_22})} \\ &+ \frac{(a_{j_2j_21} - a_{j_1j_2})(a_{j_2j_22} - a_{j_1j_2}) + b_{j_1j_2}}{(x_{j_2} - a_{j_2j_21})(x_{j_2} - a_{j_2j_22})} \Bigg] dx_{j_2}. \end{aligned}$$

With an initial condition of (x_{j_10}, x_{j_20}) at $t = t_0$, the integration of the above equation gives

$$\begin{aligned} \ln \frac{|x_{j_1} - a_{j_1j_11}|}{|x_{j_10} - a_{j_1j_11}|} = \frac{a_{j_1j_10}}{a_{j_2j_20}} \Bigg\{ & \left[(x_{j_2} - a_{j_2j_21}) - (x_{j_20} - a_{j_2j_21})\right] \\ &+ \textstyle\sum_{s_1=1}^{2} (a_{j_2j_2s_1} - a_{j_1j_2}) \ln \dfrac{|x_{j_2} - a_{j_2j_2s_1}|}{|x_{j_20} - a_{j_2j_2s_1}|} \\ &+ \left[\textstyle\prod_{s_1=1}^{2} (a_{j_2j_2s_1} - a_{j_1j_2}) + b_{j_2j_1}\right] \\ &\times \left(\textstyle\sum_{s_2=1}^{2} \dfrac{1}{a_{j_2j_2s_2} - a_{j_1j_2}} \ln \dfrac{|x_{j_2} - a_{j_2j_2s_2}|}{|x_{j_20} - a_{j_2j_2s_2}|}\right) \Bigg\}. \end{aligned}$$

At $(x_{j_1}^*, x_{j_2}^*) = (a_{j_1j_11}, a_{j_2j_2s_1})$, the variational equations are

$$\begin{aligned} \Delta \dot{x}_{j_2} &= a_{j_2j_20}(a_{j_2j_2s_1} - a_{j_2j_2s_2}) \Delta x_{j_2}, \\ \Delta \dot{x}_{j_1} &= a_{j_1j_10}\left[(a_{j_2j_2s_1} - a_{j_1j_2})^2 + b_{j_1j_2}\right] \Delta x_{j_1}. \end{aligned}$$

Because of

$$(a_{j_2j_2s_1} - a_{j_1j_2})^2 + b_{j_1j_2} > 0,$$

the flows at $(x_{j_1}^*, x_{j_2}^*) = (a_{j_1j_11}, a_{j_2j_2s_1})$ in the x_{j_1}-direction are source and sink for $a_{j_1j_10} > 0$ and $a_{j_1j_10} < 0$, respectively. The flows at $(x_{j_1}^*, x_{j_2}^*) = (a_{j_1j_11}, a_{j_2j_2s_1})$ in the x_{j_2}-direction are source and sink for $a_{j_2j_20}(a_{j_2j_2s_1} - a_{j_2j_2s_2}) > 0$ and $a_{j_2j_20}(a_{j_2j_2s_1} - a_{j_2j_2s_2}) < 0$, respectively.

Therefore, the equilibrium of $(x_{j_1}^*, x_{j_2}^*) = (a_{j_1j_11}, a_{j_2j_2s_1})$ $(s_1, s_2 \in \{1, 2\}, s_1 \neq s_2)$ has the following properties as in Eqs. (1.20)–(1.23).

- For $a_{j_1j_10} > 0$ and $a_{j_2j_20}(a_{j_2j_2s_1} - a_{j_2j_2s_2}) > 0$,

$$(a_{j_1j_11}, a_{j_2j_2s_1}) = \underbrace{(\text{SO}, \text{SO})}_{\text{source}}.$$

The equilibrium of $(x_{j_1}^*, x_{j_2}^*) = (a_{j_1j_11}, a_{j_2j_2s_1})$ is an (SO,SO)-source.

- For $a_{j_1j_10}<0$ and $a_{j_2j_20}(a_{j_2j_2s_1}-a_{j_2j_2s_2})>0$,

$$(a_{j_1j_11},a_{j_2j_2s_1})=\underbrace{(\text{SI, SO})}_{\text{saddle}}.$$

The equilibrium of $(x^*_{j_1},x^*_{j_2})=(a_{j_1j_11},a_{j_2j_2s_1})$ is an (SI,SO)-saddle.

- For $a_{j_1j_10}>0$ and $a_{j_2j_20}(a_{j_2j_2s_1}-a_{j_2j_2s_2})<0$,

$$(a_{j_1j_11},a_{j_2j_2s_1})=\underbrace{(\text{SO, SI})}_{\text{saddle}}.$$

The equilibrium of $(x^*_{j_1},x^*_{j_2})=(a_{j_1j_11},a_{j_2j_2s_1})$ is an (SO,SI)-saddle.

- For $a_{j_1j_10}<0$ and $a_{j_2j_20}(a_{j_2j_2s_1}-a_{j_2j_2s_2})<0$,

$$(a_{j_1j_11},a_{j_2j_2s_1})=\underbrace{(\text{SI, SI})}_{\text{sink}}.$$

The equilibrium of $(x^*_{j_1},x^*_{j_2})=(a_{j_1j_11},a_{j_2j_2s_1})$ is an (SI,SI)-sink.

(iv) For $\Delta_{j_1j_2}=0$ and $\Delta_{j_2j_2}<0$, the standard form is

$$\begin{aligned}\dot{x}_{j_2}&=a_{j_2j_20}\left[(x_{j_2}-a_{j_2j_2})^2+b_{j_2j_2}\right],\\ \dot{x}_{j_1}&=a_{j_1j_10}(x_{j_1}-a_{j_1j_11})(x_{j_2}-a_{j_1j_21})^2\end{aligned}$$

where

$$\begin{aligned}a_{j_1j_21}&=-\frac{1}{2}B_{j_1j_2},\\ a_{j_2j_2}&=-\frac{1}{2}B_{j_2j_2},b_{j_2j_2}=\frac{1}{4}(-\Delta_{j_2j_2}).\end{aligned}$$

In phase space,

$$\frac{dx_{j_1}}{dx_{j_2}}=\frac{a_{j_1j_10}}{a_{j_2j_20}}\frac{(x_{j_1}-a_{j_1j_11})(x_{j_2}-a_{j_1j_21})^2}{(x_{j_2}-a_{j_2j_2})^2+b_{j_2j_2}},$$

and the deformation of the foregoing equation is

$$\begin{aligned}\frac{1}{(x_{j_1}-a_{j_1j_11})}dx_{j_1}=\frac{a_{j_1j_10}}{a_{j_2j_20}}&\left[1+\frac{2(a_{j_2j_2}-a_{j_1j_21})(x_{j_2}-a_{j_2j_2})}{(x_{j_2}-a_{j_2j_2})^2+b_{j_2j_2}}\right.\\ &\left.+\frac{(a_{j_2j_2}-a_{j_1j_21})^2-b_{j_2j_2}}{(x_{j_2}-a_{j_2j_2})^2+b_{j_2j_2}}\right]dx_{j_2}.\end{aligned}$$

With an initial condition of (x_{j_10}, x_{j_20}) at $t = t_0$, the integration of the above equation gives

$$\begin{aligned}\ln\frac{|x_{j_1}-a_{j_1j_11}|}{|x_{j_10}-a_{j_1j_11}|} = \frac{a_{j_1j_10}}{a_{j_2j_20}}\Big\{&\big[(x_{j_2}-a_{j_2j_2})-(x_{j_20}-a_{j_2j_2})\big]\\ &+(a_{j_2j_2}-a_{j_1j_21})\ln\frac{|(x_{j_2}-a_{j_2j_2})^2+b_{j_2j_2}|}{|(x_{j_20}-a_{j_2j_2})^2+b_{j_2j_2}|}\\ &+\big[(a_{j_2j_2}-a_{j_1j_21})^2-b_{j_2j_2}\big]\\ &\times\frac{1}{\sqrt{b_{j_2j_2}}}\Big(\arctan\frac{x_{j_2}-a_{j_2j_2}}{\sqrt{b_{j_2j_2}}}-\arctan\frac{x_{j_20}-a_{j_2j_2}}{\sqrt{b_{j_2j_2}}}\Big)\Big\}.\end{aligned}$$

In phase space, at $x^*_{j_2} = a_{j_1j_21}$ with $\bar{x}_{j_1} \neq a_{j_1j_11}$,

$$\begin{aligned}\frac{dx_{j_1}}{dx_{j_2}}\Big|_{x^*_{j_2}=a_{j_1j_21}} &= \frac{a_{j_1j_10}}{a_{j_2j_20}}\frac{(\bar{x}_{j_1}-a_{j_1j_11})(x_{j_2}-a_{j_1j_21})^2}{(x_{j_2}-a_{j_2j_2})^2+b_{j_2j_2}}\Big|_{x^*_{j_2}=a_{j_1j_21}}=0,\\ \frac{d^2x_{j_1}}{dx^2_{j_2}}\Big|_{x^*_{j_2}=a_{j_1j_21}} &= 0.\end{aligned}$$

If

$$\frac{d^3x_{j_1}}{dx^3_{j_2}}\Big|_{x^*_{j_2}=a_{j_1j_21}} = \frac{a_{j_1j_10}}{a_{j_2j_20}}\frac{2(\bar{x}_{j_1}-a_{j_1j_11})}{(a_{j_1j_21}-a_{j_2j_2})^2+b_{j_2j_2}} > 0,$$

there is a third-order increasing-inflection flow at $x^*_{j_2} = a_{j_1j_21}$ with $\bar{x}_{j_1} \neq a_{j_1j_11}$. If

$$\frac{d^3x_{j_1}}{dx^3_{j_2}}\Big|_{x^*_{j_2}=a_{j_1j_21}} = \frac{a_{j_1j_10}}{a_{j_2j_20}}\frac{2(\bar{x}_{j_1}-a_{j_1j_11})}{(a_{j_1j_21}-a_{j_2j_2})^2+b_{j_2j_2}} < 0,$$

there is a third-order decreasing-inflection flow at $x^*_{j_2} = a_{j_1j_21}$ with $\bar{x}_{j_1} \neq a_{j_1j_11}$. Let

$$\dot{x}_{j_2} = a_{j_2j_20}\big[(a_{j_1j_21}-a_{j_2j_2})^2+b_{j_2j_2}\big],$$

with

$$(a_{j_1j_21}-a_{j_2j_2})^2+b_{j_2j_2} > 0.$$

The third-order inflection flows at $x^*_{j_2} = a_{j_1j_21}$ with $\bar{x}_{j_1} \neq a_{j_1j_11}$ are positive and negative for $a_{j_2j_20} > 0$ and $a_{j_2j_20} < 0$, respectively.

Thus, the equilibrium of $x^*_{j_2} = a_{j_1j_21}$ with $\bar{x}_{j_1} \neq a_{j_1j_11}$ has the following properties.

- For $a_{j_1j_10}(\bar{x}_{j_1} - a_{j_1j_11}) > 0$ and $a_{j_2j_20} > 0$,

$$(a_{j_1j_21}, \dot{x}_{j_2}) = \underbrace{(\mathrm{II}, \mathrm{pF})}_{\text{increasing-inflection flow } (+)}.$$

- For $a_{j_1j_10}(\bar{x}_{j_1} - a_{j_1j_11}) < 0$ and $a_{j_2j_20} > 0$,

$$(a_{j_1j_21}, \dot{x}_{j_2}) = \underbrace{(\mathrm{DI}, \mathrm{pF})}_{\text{decreasing-inflection flow } (+)}.$$

- For $a_{j_1j_10}(\bar{x}_{j_1} - a_{j_1j_11}) > 0$ and $a_{j_2j_20} < 0$,

$$(a_{j_1j_21}, \dot{x}_{j_2}) = \underbrace{(\mathrm{DI}, \mathrm{nF})}_{\text{decreasing-inflection flow } (-)}.$$

- For $a_{j_1j_10}(\bar{x}_{j_1} - a_{j_1j_11}) < 0$ and $a_{j_2j_20} < 0$,

$$(a_{j_1j_21}, \dot{x}_{j_2}) = \underbrace{(\mathrm{II}, \mathrm{nF})}_{\text{increasing-inflection flow } (-)}.$$

The variational equation at $x^*_{j_1} = a_{j_1j_11}$ with $\bar{x}_{j_2} \neq a_{j_1j_21}$ is

$$\Delta\dot{x}_{j_1} = a_{j_1j_10}(\bar{x}_{j_2} - a_{j_1j_21})^2 \Delta x_{j_1}$$

and

$$\dot{x}_{j_2} = a_{j_2j_20}\left[(\bar{x}_{j_2} - a_{j_2j_2})^2 + b_{j_2j_2}\right].$$

Thus, the equilibrium of $x^*_{j_2} = a_{j_1j_21}$ with $\bar{x}_{j_1} \neq a_{j_1j_11}$ has the following properties.

- For $a_{j_1j_10}(\bar{x}_{j_2} - a_{j_1j_21})^2 > 0$ and $a_{j_2j_20} > 0$,

$$(a_{j_1j_11}, \dot{x}_{j_2}) = \underbrace{(\mathrm{SO}, \mathrm{pF})}_{\text{source flow } (+)}.$$

- For $a_{j_1j_10}(\bar{x}_{j_2} - a_{j_1j_21})^2 < 0$ and $a_{j_2j_20} > 0$,

$$(a_{j_1j_11}, \dot{x}_{j_2}) = \underbrace{(\mathrm{SI}, \mathrm{pF})}_{\text{sink flow } (+)}.$$

- For $a_{j_1j_10}(\bar{x}_{j_2} - a_{j_1j_21})^2 > 0$ and $a_{j_2j_20} < 0$,

$$(a_{j_1j_1 1}, \dot{x}_{j_2}) = \underbrace{(\text{SO}, \text{nF})}_{\text{source flow } (-)}.$$

- For $a_{j_1j_10}(\bar{x}_{j_2} - a_{j_1j_21})^2 < 0$ and $a_{j_2j_20} < 0$,

$$(a_{j_1j_1 1}, \dot{x}_{j_2}) = \underbrace{(\text{SI}, \text{nF})}_{\text{sink flow } (-)}.$$

Therefore, the equilibrium of $(x^*_{j_1}, x^*_{j_2}) = (a_{j_1j_1}1, a_{j_1j_2}1)$ has the following properties as in Eqs. (1.27)–(1.30).

- For $a_{j_1j_10} > 0$ and $a_{j_2j_20} > 0$,

$$(a_{j_1j_1}1, a_{j_1j_2}1) = \underbrace{({}_{\text{DI:II}}\text{SO}, \text{pF})}_{\text{inflection source flow } (+)}.$$

  The equilibrium of $(x^*_{j_1}, x^*_{j_2}) = (a_{j_1j_1}1, a_{j_1j_2}1)$ is a $({}_{\text{DI:II}}\text{SO}$, pF)-positive inflection source flow.
- For $a_{j_1j_10} < 0$ and $a_{j_2j_20} > 0$,

$$(a_{j_1j_1}1, a_{j_1j_2}1) = \underbrace{({}_{\text{II:DI}}\text{SI}, \text{pF})}_{\text{inflection sink flow } (+)}.$$

  The equilibrium of $(x^*_{j_1}, x^*_{j_2}) = (a_{j_1j_1}1, a_{j_1j_2}1)$ is an $({}_{\text{II:DI}}\text{SI}$, pF)-positive inflection sink flow.
- For $a_{j_1j_10} > 0$ and $a_{j_2j_20} < 0$,

$$(a_{j_1j_1}1, a_{j_1j_2}1) = \underbrace{({}_{\text{II:DI}}\text{SO}, \text{nF})}_{\text{inflection source flow } (-)}.$$

  The equilibrium of $(x^*_{j_1}, x^*_{j_2}) = (a_{j_1j_1}1, a_{j_1j_2}1)$ is an $({}_{\text{II:DI}}\text{SO}$, nF)-negative inflection source flow.
- For $a_{j_1j_10} < 0$ and $a_{j_2j_20} < 0$,

$$(a_{j_1j_1}1, a_{j_1j_2}1) = \underbrace{({}_{\text{DI:II}}\text{SI}, \text{nF})}_{\text{inflection sink flow } (-)}.$$

  The equilibrium of $(x^*_{j_1}, x^*_{j_2}) = (a_{j_1j_1}1, a_{j_1j_2}1)$ is a $({}_{\text{DI:II}}\text{SI}$, nF)-negative inflection sink flow.

(v) For $\Delta_{j_1j_2}=0$ and $\Delta_{j_2j_2}=0$, the standard form is

$$\begin{aligned}\dot{x}_{j_2}&=a_{j_2j_20}(x_{j_2}-a_{j_2j_21})^2,\\ \dot{x}_{j_1}&=a_{j_1j_10}(x_{j_1}-a_{j_1j_11})(x_{j_2}-a_{j_1j_21})^2\end{aligned}$$

where

$$a_{j_1j_21}=-\frac{1}{2}B_{j_1j_2},\, a_{j_2j_21}=-\frac{1}{2}B_{j_2j_2}.$$

(v_1) For $a_{j_2j_21}\neq a_{j_1j_21}$, in phase space,

$$\frac{dx_{j_1}}{dx_{j_2}}=\frac{a_{j_1j_10}}{a_{j_2j_20}}\frac{(x_{j_1}-a_{j_1j_11})(x_{j_2}-a_{j_1j_21})^2}{(x_{j_2}-a_{j_2j_21})^2},$$

and the deformation of the foregoing equation is

$$\frac{1}{(x_{j_1}-a_{j_1j_11})}dx_{j_1}=\frac{a_{j_1j_10}}{a_{j_2j_20}}\left[1+\frac{2(a_{j_2j_21}-a_{j_1j_21})}{(x_{j_2}-a_{j_2j_21})}+\frac{(a_{j_2j_21}-a_{j_1j_21})^2}{(x_{j_2}-a_{j_2j_21})^2}\right]dx_{j_2}.$$

With an initial condition of (x_{j_10},x_{j_20}) at $t=t_0$, the integration of the above equation gives

$$\begin{aligned}\ln\frac{|x_{j_1}-a_{j_1j_11}|}{|x_{j_10}-a_{j_1j_11}|}=\frac{a_{j_1j_10}}{a_{j_2j_20}}\Big\{&[(x_{j_2}-a_{j_2j_21})-(x_{j_20}-a_{j_2j_21})]\\ &+2(a_{j_2j_21}-a_{j_1j_21})\ln\frac{|x_{j_2}-a_{j_2j_21}|}{|x_{j_20}-a_{j_2j_21}|}\\ &-(a_{j_2j_21}-a_{j_1j_21})^2(\frac{1}{x_{j_2}-a_{j_2j_21}}-\frac{1}{x_{j_20}-a_{j_2j_21}})\Big\}.\end{aligned}$$

(v_{1a}) For $(x_{j_1}^*,x_{j_2}^*)=(a_{j_1j_11},a_{j_2j_21})$, the variational equations are

$$\begin{aligned}\Delta\dot{x}_{j_2}&=a_{j_2j_20}(\Delta x_{j_2})^2,\\ \Delta\dot{x}_{j_1}&=a_{j_1j_10}(a_{j_2j_21}-a_{j_1j_21})^2\Delta x_{j_1}.\end{aligned}$$

The flows at $(x_{j_1}^*,x_{j_2}^*)=(a_{j_1j_11},a_{j_2j_21})$ in the x_{j_1}-direction are source and sink for $a_{j_1j_10}>0$ and $a_{j_1j_10}<0$, respectively. The flows at $(x_{j_1}^*,x_{j_2}^*)=(a_{j_1j_11},a_{j_2j_21})$ in the x_{j_2}-direction are upper-saddle and lower-saddle for $a_{j_2j_20}>0$ and $a_{j_2j_20}<0$, respectively.

Therefore, the equilibrium of $(x_{j_1}^*,x_{j_2}^*)=(a_{j_1j_11},a_{j_2j_21})$ with $a_{j_2j_21}\neq a_{j_1j_21}$ has the following properties as in Eqs. (1.34)–(1.37).

- For $a_{j_1j_10}(a_{j_2j_21}-a_{j_1j_21})^2>0$ and $a_{j_2j_20}>0$,

$$(a_{j_1j_11},a_{j_2j_21}) = \underbrace{(\text{SO},\text{US})}_{\text{upper-saddle source}}.$$

The equilibrium of $(x^*_{j_1},x^*_{j_2})=(a_{j_1j_11},a_{j_2j_21})$ is an (SO,US)-upper-saddle source.

- For $a_{j_1j_10}(a_{j_2j_21}-a_{j_1j_21})^2<0$ and $a_{j_2j_20}>0$,

$$(a_{j_1j_11},a_{j_2j_21}) = \underbrace{(\text{SI},\text{US})}_{\text{upper-saddle-sink}}.$$

The equilibrium of $(x^*_{j_1},x^*_{j_2})=(a_{j_1j_11},a_{j_2j_21})$ is an (SI,US)-upper-saddle sink.

- For $a_{j_1j_10}(a_{j_2j_21}-a_{j_1j_21})^2>0$ and $a_{j_2j_20}<0$,

$$(a_{j_1j_11},a_{j_2j_21}) = \underbrace{(\text{SO},\text{LS})}_{\text{lower-saddle source}}.$$

The equilibrium of $(x^*_{j_1},x^*_{j_2})=(a_{j_1j_11},a_{j_2j_21})$ is an (SO,LS)-lower-saddle source.

- For $a_{j_1j_10}(a_{j_2j_21}-a_{j_1j_21})^2<0$ and $a_{j_2j_20}<0$,

$$(a_{j_1j_11},a_{j_2j_21}) = \underbrace{(\text{SI},\text{LS})}_{\text{lower-saddle source}}.$$

The equilibrium of $(x^*_{j_1},x^*_{j_2})=(a_{j_1j_11},a_{j_2j_21})$ is an (SI,LS)-lower-saddle sink.

– The saddle-source and saddle-sink are the appearing bifurcations of saddle-source and saddle-sink.

($\text{v}_{1\text{b}}$) In phase space, at $x^*_{j_2}=a_{j_1j_21}$ with $\bar{x}_{j_1}\neq a_{j_1j_11}$,

$$\frac{dx_{j_1}}{dx_{j_2}}\bigg|_{x^*_{j_2}=a_{j_1j_21}} = \frac{a_{j_1j_10}}{a_{j_2j_20}}\frac{(\bar{x}_{j_1}-a_{j_1j_11})(x_{j_2}-a_{j_1j_21})^2}{(x_{j_2}-a_{j_2j_21})^2}\bigg|_{x^*_{j_2}=a_{j_1j_21}}=0,$$
$$\frac{d^2x_{j_1}}{dx^2_{j_2}}\bigg|_{x^*_{j_2}=a_{j_1j_21}} = 0.$$

If

$$\frac{d^3x_{j_1}}{dx^3_{j_2}}\bigg|_{x^*_{j_2}=a_{j_1j_21}} = \frac{a_{j_1j_10}}{a_{j_2j_20}}\frac{2(\bar{x}_{j_1}-a_{j_1j_11})}{(a_{j_1j_21}-a_{j_2j_21})^2}>0,$$

there is an increasing-inflection flow at $x^*_{j_2}=a_{j_1j_21}$ with $\bar{x}_{j_1}\neq a_{j_1j_11}$. If

$$\frac{d^3x_{j_1}}{dx_{j_2}^3}\Big|_{x_{j_2}^*=a_{j_1j_2 1}} = \frac{a_{j_1j_1 0}}{a_{j_2j_2 0}}\frac{2(\overline{x}_{j_1}-a_{j_1j_1 1})}{(a_{j_1j_2 1}-a_{j_2j_2 1})^2} < 0,$$

there is a decreasing-inflection flow at $x_{j_2}^*=a_{j_1j_2 1}$ with $\overline{x}_{j_1}\neq a_{j_1j_1 1}$. Let

$$\dot{x}_{j_2}=a_{j_2j_2 0}(a_{j_1j_2 1}-a_{j_2j_2 1})^2\neq 0.$$

The inflection flows at $x_{j_2}^*=a_{j_1j_2 1}$ with $\overline{x}_{j_1}\neq a_{j_1j_1 1}$ are positive and negative for $a_{j_2j_2 0}>0$ and $a_{j_2j_2 0}<0$, respectively.

Thus, the equilibrium of $x_{j_2}^*=a_{j_1j_2 1}$ with $\overline{x}_{j_1}\neq a_{j_1j_1 1}$ has the following properties.

- For $a_{j_1j_1 0}(\overline{x}_{j_1}-a_{j_1j_1 1})>0$ and $a_{j_2j_2 0}(a_{j_1j_2 1}-a_{j_2j_2 1})^2>0$,

$$(a_{j_1j_2 1},\dot{x}_{j_2}) = \underbrace{(\mathrm{II},\mathrm{pF})}_{\text{increasing-inflection flow }(+)}.$$

- For $a_{j_1j_1 0}(\overline{x}_{j_1}-a_{j_1j_1 1})<0$ and $a_{j_2j_2 0}(a_{j_1j_2 1}-a_{j_2j_2 1})^2>0$,

$$(a_{j_1j_2 1},\dot{x}_{j_2}) = \underbrace{(\mathrm{DI},\mathrm{pF})}_{\text{decreasing-inflection flow }(+)}.$$

- For $a_{j_1j_1 0}(\overline{x}_{j_1}-a_{j_1j_1 1})>0$ and $a_{j_2j_2 0}(a_{j_1j_2 1}-a_{j_2j_2 1})^2<0$,

$$(a_{j_1j_2 1},\dot{x}_{j_2}) = \underbrace{(\mathrm{DI},\mathrm{nF})}_{\text{decreasing-inflection flow }(-)}.$$

- For $a_{j_1j_1 0}(\overline{x}_{j_1}-a_{j_1j_1 1})<0$ and $a_{j_2j_2 0}(a_{j_1j_2 1}-a_{j_2j_2 1})^2<0$,

$$(a_{j_1j_2 1},\dot{x}_{j_2}) = \underbrace{(\mathrm{II},\mathrm{nF})}_{\text{increasing-inflection flow }(-)}.$$

The variational equation at $x_{j_1}^*=a_{j_1j_1 1}$ with $\overline{x}_{j_2}\neq a_{j_1j_2 1}$ is

$$\Delta\dot{x}_{j_1}=a_{j_1j_1 0}(\overline{x}_{j_2}-a_{j_1j_2 1})^2\Delta x_{j_1}$$

and

$$\dot{x}_{j_2}=a_{j_2j_2 0}(\overline{x}_{j_2}-a_{j_2j_2 1})^2.$$

Thus, the equilibrium of $x_{j_2}^*=a_{j_1j_2 1}$ with $\overline{x}_{j_1}\neq a_{j_1j_1 1}$ has the following properties.

- For $a_{j_1j_1 0}(\overline{x}_{j_2} - a_{j_1j_2 1})^2 > 0$ and $a_{j_2j_2 0}(\overline{x}_{j_2} - a_{j_2j_2 1})^2 > 0$,

$$(a_{j_1j_1 1}, \dot{x}_{j_2}) = \underbrace{(\mathrm{SO}, \mathrm{pF})}_{\text{source flow } (+)}.$$

- For $a_{j_1j_1 0}(\overline{x}_{j_2} - a_{j_1j_2 1})^2 < 0$ and $a_{j_2j_2 0}(\overline{x}_{j_2} - a_{j_2j_2 1})^2 > 0$,

$$(a_{j_1j_1 1}, \dot{x}_{j_2}) = \underbrace{(\mathrm{SI}, \mathrm{pF})}_{\text{sink flow } (+)}.$$

- For $a_{j_1j_1 0}(\overline{x}_{j_2} - a_{j_1j_2 1})^2 > 0$ and $a_{j_2j_2 0}(\overline{x}_{j_2} - a_{j_2j_2 1})^2 < 0$,

$$(a_{j_1j_1 1}, \dot{x}_{j_2}) = \underbrace{(\mathrm{SO}, \mathrm{nF})}_{\text{source flow } (-)}.$$

- For $a_{j_1j_1 0}(\overline{x}_{j_2} - a_{j_1j_2 1})^2 < 0$ and $a_{j_2j_2 0}(\overline{x}_{j_2} - a_{j_2j_2 1})^2 < 0$,

$$(a_{j_1j_1 1}, \dot{x}_{j_2}) = \underbrace{(\mathrm{SI}, \mathrm{nF})}_{\text{sink flow } (-)}.$$

From the foregoing analysis, the equilibrium of $(x^*_{j_1}, x^*_{j_2}) = (a_{j_1j_1 1}, a_{j_1j_2 1})$ with $a_{j_2j_2 1} \neq a_{j_1j_2 1}$ has the following properties as in Eqs (1.38)–(1.41).

- For $a_{j_1j_1 0} > 0$ and $a_{j_2j_2 0}(a_{j_1j_2 1} - a_{j_2j_2 1})^2 > 0$,

$$(a_{j_1j_1 1}, a_{j_1j_2 1}) = \underbrace{({}_{\mathrm{DI:II}}\mathrm{SO}, \mathrm{pF})}_{\text{inflection source flow}(+)}.$$

  The equilibrium of $(x^*_{j_1}, x^*_{j_2}) = (a_{j_1j_1 1}, a_{j_1j_2 1})$ is a $({}_{\mathrm{DI:II}}\mathrm{SO}, \mathrm{pF})$-positive inflection source flow.
- For $a_{j_1j_1 0} < 0$ and $a_{j_2j_2 0}(a_{j_1j_2 1} - a_{j_2j_2 1})^2 > 0$,

$$(a_{j_1j_1 1}, a_{j_1j_2 1}) = \underbrace{({}_{\mathrm{II:DI}}\mathrm{SI}, \mathrm{pF})}_{\text{inflection source flow } (+)}.$$

  The equilibrium of $(x^*_{j_1}, x^*_{j_2}) = (a_{j_1j_1 1}, a_{j_1j_2 1})$ is an $({}_{\mathrm{II:DI}}\mathrm{SI}, \mathrm{pF})$-positive inflection sink flow.

- For $a_{j_1j_10} > 0$ and $a_{j_2j_20}(a_{j_1j_21} - a_{j_2j_21})^2 < 0$,

$$(a_{j_1j_11}, a_{j_1j_21}) = \underbrace{({}_{\text{II:DI}}\text{SO, nF})}_{\text{inflection source flow }(-)}.$$

The equilibrium of $(x^*_{j_1}, x^*_{j_2}) = (a_{j_1j_11}, a_{j_1j_21})$ is an $({}_{\text{II:DI}}\text{SO, nF})$-negative inflection source flow.

- For $a_{j_1j_10} < 0$ and $a_{j_2j_20}(a_{j_1j_21} - a_{j_2j_21})^2 < 0$,

$$(a_{j_1j_11}, a_{j_1j_21}) = \underbrace{({}_{\text{DI:II}}\text{SI, nF})}_{\text{inflection sink flow }(-)}.$$

The equilibrium of $(x^*_{j_1}, x^*_{j_2}) = (a_{j_1j_11}, a_{j_1j_21})$ is a $({}_{\text{DI:II}}\text{SI, nF})$-negative inflection sink flow.

(v_2) For $a_{j_2j_21} = a_{j_1j_21}$, we have

$$\frac{dx_{j_1}}{dx_{j_2}} = \frac{a_{j_1j_10}}{a_{j_2j_20}}(x_{j_1} - a_{j_1j_11}),$$

and the deformation of the foregoing equation is

$$\frac{1}{(x_{j_1} - a_{j_1j_11})}dx_{j_1} = \frac{a_{j_1j_10}}{a_{j_2j_20}}dx_{j_2}.$$

With an initial condition of (x_{j_10}, x_{j_20}) at $t = t_0$, the integration of the above equation gives the first integral manifold for $a_{j_2j_21} = a_{j_1j_21}$ as

$$\ln\frac{|x_{j_1} - a_{j_1j_11}|}{|x_{j_10} - a_{j_1j_11}|} = \frac{a_{j_1j_10}}{a_{j_2j_20}}\left[(x_{j_2} - a_{j_2j_21}) - (x_{j_20} - a_{j_2j_21})\right].$$

($\text{v}_{2\text{a}}$) In phase space, at $x^*_{j_2} = a_{j_1j_21} = a_{j_2j_21}$ with $\overline{x}_{j_1} \neq a_{j_1j_11}$, if

$$\left.\frac{dx_{j_1}}{dx_{j_2}}\right|_{x^*_{j_2} = a_{j_1j_21}} = \frac{a_{j_1j_10}}{a_{j_2j_20}}(\overline{x}_{j_1} - a_{j_1j_11}) > 0,$$

there is an increasing-inflection flow at $x^*_{j_2} = a_{j_1j_21}$ with $\overline{x}_{j_1} \neq a_{j_1j_11}$. If

$$\left.\frac{dx_{j_1}}{dx_{j_2}}\right|_{x^*_{j_2} = a_{j_1j_21}} = \frac{a_{j_1j_10}}{a_{j_2j_20}}(\overline{x}_{j_1} - a_{j_1j_11}) < 0,$$

there is a decreasing-inflection flow at $x^*_{j_2} = a_{j_1j_21}$ with $\overline{x}_{j_1} \neq a_{j_1j_11}$. The variational equation is

$$\Delta\dot{x}_{j_2} = a_{j_2 j_2 0}(\Delta x_{j_2})^2.$$

The inflection flows at $x^*_{j_2} = a_{j_2 j_2 1}$ with $\bar{x}_{j_1} \neq a_{j_1 j_1 1}$ are upper-saddle and lower-saddle for $a_{j_2 j_2 0} > 0$ and $a_{j_2 j_2 0} < 0$, respectively.

Therefore, the infinite-equilibrium of $x^*_{j_2} = a_{j_1 j_2 1} = a_{j_2 j_2 1}$ with $\bar{x}_{j_1} \neq a_{j_1 j_1 1}$ has the following properties as in Eqs. (1.43)–(1.46).

- For $a_{j_1 j_1 0}(\bar{x}_{j_1} - a_{j_1 j_1 1}) > 0$ and $a_{j_2 j_2 0} > 0$,

$$(\bar{x}_{j_1}, a_{j_2 j_2 1}) = \underbrace{(\text{II}, \text{US})}_{\text{increasing-inflection upper-saddle}}.$$

  The infinite-equilibrium of $x^*_{j_2} = a_{j_1 j_2 1} = a_{j_2 j_2 1}$ is an (II,US)-increasing-inflection upper-saddle.
- For $a_{j_1 j_1 0}(\bar{x}_{j_1} - a_{j_1 j_1 1}) < 0$ and $a_{j_2 j_2 0} > 0$,

$$(\bar{x}_{j_1}, a_{j_2 j_2 1}) = \underbrace{(\text{DI}, \text{US})}_{\text{decreasing-inflection upper-saddle}}.$$

  The equilibrium of $x^*_{j_2} = a_{j_1 j_2 1} = a_{j_2 j_2 1}$ is a (DI,US)-decreasing-inflection upper-saddle.
- For $a_{j_1 j_1 0}(\bar{x}_{j_1} - a_{j_1 j_1 1}) > 0$ and $a_{j_2 j_2 0} < 0$,

$$(\bar{x}_{j_1}, a_{j_2 j_2 1}) = \underbrace{(\text{DI}, \text{LS})}_{\text{decreasing-inflection lower-saddle}}.$$

  The equilibrium of $x^*_{j_2} = a_{j_1 j_2 1} = a_{j_2 j_2 1}$ is a (DI,LS)-decreasing-inflection lower-saddle.
- For $a_{j_1 j_1 0}(\bar{x}_{j_1} - a_{j_1 j_1 1}) < 0$ and $a_{j_2 j_2 0} < 0$,

$$(\bar{x}_{j_1}, a_{j_2 j_2 1}) = \underbrace{(\text{II}, \text{LS})}_{\text{increasing-inflection lower-saddle}}.$$

  The equilibrium of $x^*_{j_2} = a_{j_1 j_2 1} = a_{j_2 j_2 1}$ is an (II,LS)-increasing-inflection lower-saddle.

($\text{v}_{2\text{b}}$) From ($\text{v}_{2\text{a}}$), let $\bar{x}_{j_1} = a_{j_1 j_1 1}$, the equilibrium of $(x^*_{j_1}, x^*_{j_2}) = (a_{j_1 j_1 1}, a_{j_2 j_2 1})$ with $a_{j_2 j_2 1} = a_{j_1 j_2 1}$ has the following properties as in Eqs. (1.47)–(1.50).

- For $a_{j_1 j_1 0} > 0$ and $a_{j_2 j_2 0} > 0$,

$$(a_{j_1j_11}, a_{j_2j_21}) = \underbrace{(\mathrm{DI}:\mathrm{II},\mathrm{US})}_{\text{inflection upper-saddle}}.$$

The equilibrium of $(x^*_{j_1}, x^*_{j_2}) = (a_{j_1j_11}, a_{j_2j_21})$ is a (DI:II,US)-inflection upper-saddle.

- For $a_{j_1j_10} < 0$ and $a_{j_2j_20} > 0$,

$$(a_{j_1j_11}, a_{j_2j_21}) = \underbrace{(\mathrm{II}:\mathrm{DI},\mathrm{US})}_{\text{inflection upper-saddle}}.$$

The equilibrium of $(x^*_{j_1}, x^*_{j_2}) = (a_{j_1j_11}, a_{j_2j_21})$ is an (II:DI,US)-inflection upper-saddle.

- For $a_{j_1j_10} > 0$ and $a_{j_2j_20} < 0$,

$$(a_{j_1j_11}, a_{j_2j_21}) = \underbrace{(\mathrm{II}:\mathrm{DI},\mathrm{LS})}_{\text{inflection lower-saddle}}.$$

The equilibrium of $(x^*_{j_1}, x^*_{j_2}) = (a_{j_1j_11}, a_{j_2j_21})$ is an (II:DI,LS)-inflection lower-saddle.

- For $a_{j_1j_10} < 0$ and $a_{j_2j_20} < 0$,

$$(a_{j_1j_11}, a_{j_2j_21}) = \underbrace{(\mathrm{DI}:\mathrm{II},\mathrm{LS})}_{\text{inflection lower-saddle}}.$$

The equilibrium of $(x^*_{j_1}, x^*_{j_2}) = (a_{j_1j_11}, a_{j_2j_21})$ is a (DI:II,LS)-inflection lower-saddle.

(vi) For $\Delta_{j_1j_2} = 0$ and $\Delta_{j_2j_2} > 0$, the standard form is

$$\begin{aligned}
\dot{x}_{j_2} &= a_{j_2j_20}(x_{j_2} - a_{j_2j_21})(x_{j_2} - a_{j_2j_22}),\\
\dot{x}_{j_1} &= a_{j_1j_10}(x_{j_1} - a_{j_1j_11})(x_{j_2} - a_{j_1j_21})^2
\end{aligned}$$

where

$$\begin{aligned}
&a_{j_1j_21} = -\frac{1}{2}B_{j_1j_2},\\
&b_{j_2j_21}, b_{j_2j_22} = -\frac{1}{2}(B_{j_2j_2} \pm \sqrt{\Delta_{j_2j_2}}),\\
&\{a_{j_2j_21}, a_{j_2j_22}\} = \mathrm{sort}\{b_{j_2j_21}, b_{j_2j_22}\}, a_{j_2j_21} < a_{j_2j_22}.
\end{aligned}$$

(vi$_1$) In phase space, $a_{j_2j_2s_1} \neq a_{j_1j_21}$, we have

$$\frac{dx_{j_1}}{dx_{j_2}} = \frac{a_{j_1j_10}}{a_{j_2j_20}} \frac{(x_{j_1} - a_{j_1j_11})(x_{j_2} - a_{j_1j_21})^2}{(x_{j_2} - a_{j_2j_21})(x_{j_2} - a_{j_2j_22})},$$

and the deformation of the foregoing equation is

$$\begin{aligned}\frac{1}{(x_{j_1} - a_{j_1j_11})} dx_{j_1} = \frac{a_{j_1j_10}}{a_{j_2j_20}} \Bigg[1 + &\frac{(a_{j_2j_21} - a_{j_1j_21})^2}{(a_{j_2j_21} - a_{j_2j_22})(x_{j_2} - a_{j_2j_21})} \\ + &\frac{(a_{j_2j_22} - a_{j_1j_21})^2}{(a_{j_2j_22} - a_{j_2j_21})(x_{j_2} - a_{j_2j_22})} \Bigg] dx_{j_2}.\end{aligned}$$

With an initial condition of (x_{j_10}, x_{j_20}) at $t = t_0$, the integration of the above equation gives the first integral manifold $(a_{j_2j_2s_1} \neq a_{j_1j_21})$ as

$$\begin{aligned}\ln \frac{|x_{j_1} - a_{j_1j_11}|}{|x_{j_10} - a_{j_1j_11}|} = \frac{a_{j_1j_10}}{a_{j_2j_20}} \Bigg\{ &\left[(x_{j_2} - a_{j_2j_21}) - (x_{j_20} - a_{j_2j_21}) \right] \\ &+ \sum\nolimits_{s_1=1}^{2} \frac{(a_{j_2j_2s_1} - a_{j_1j_21})^2}{(a_{j_2j_2s_1} - a_{j_2j_2s_2})} \ln \frac{|x_{j_2} - a_{j_2j_2s_1}|}{|x_{j_20} - a_{j_2j_2s_1}|} \Bigg\}.\end{aligned}$$

(vi_{1a}) At $(x_{j_1}^*, x_{j_2}^*) = (a_{j_1j_11}, a_{j_2j_2s_1})$, the variational equations are

$$\begin{aligned}\Delta \dot{x}_{j_2} &= a_{j_2j_20}(a_{j_2j_2s_1} - a_{j_2j_2s_2})\Delta x_{j_2}, \\ \Delta \dot{x}_{j_1} &= a_{j_1j_10}(a_{j_2j_2s_1} - a_{j_1j_21})^2 \Delta x_{j_1}.\end{aligned}$$

The flows $(x_{j_1}^*, x_{j_2}^*) = (a_{j_1j_11}, a_{j_2j_2s_1})$ in the x_{j_1}-direction are source and sink for $a_{j_1j_10}(a_{j_2j_2s_1} - a_{j_1j_21})^2 > 0$ and $a_{j_1j_10}(a_{j_2j_2s_1} - a_{j_1j_21})^2 < 0$, respectively. The flows $(x_{j_1}^*, x_{j_2}^*) = (a_{j_1j_11}, a_{j_2j_2s_1})$ are source and sink for $a_{j_2j_20}(a_{j_2j_2s_1} - a_{j_2j_2s_2}) > 0$ and $a_{j_2j_20}$ $(a_{j_2j_2s_1} - a_{j_2j_2s_2}) < 0$ in the x_{j_2}-direction, respectively.

Thus, the equilibrium of $(x_{j_1}^*, x_{j_2}^*) = (a_{j_1j_11}, a_{j_2j_2s_1})$ (s_1, $s_2 \in \{1, 2\}$, $s_1 \neq s_2$) has the following properties as in Eqs. (1.54)–(1.57).

- For $a_{j_1j_10}(a_{j_2j_2s_1} - a_{j_1j_21})^2 > 0$ and $a_{j_2j_20}(a_{j_2j_2s_1} - a_{j_2j_2s_2}) > 0$,

$$(a_{j_1j_11}, a_{j_2j_2s_1}) = \underbrace{(\mathrm{SO}, \mathrm{SO})}_{\text{source}}.$$

  The equilibrium of $(x_{j_1}^*, x_{j_2}^*) = (a_{j_1j_11}, a_{j_2j_2s_1})$ is an (SO,SO)-source.

- For $a_{j_1j_10}(a_{j_2j_2s_1} - a_{j_1j_21})^2 < 0$ and $a_{j_2j_20}(a_{j_2j_2s_1} - a_{j_2j_2s_2}) > 0$,

$$(a_{j_1j_11}, a_{j_2j_2s_1}) = \underbrace{(\mathrm{SI}, \mathrm{SO})}_{\text{saddle}}.$$

  The equilibrium of $(x_{j_1}^*, x_{j_2}^*) = (a_{j_1j_11}, a_{j_2j_2s_1})$ is an (SI,SO)-saddle.

- For $a_{j_1j_10}(a_{j_2j_2s_1}-a_{j_1j_21})^2>0$ and $a_{j_2j_20}(a_{j_2j_2s_1}-a_{j_2j_2s_2})<0$,

$$(a_{j_1j_11},a_{j_2j_2s_1})=\underbrace{(\text{SO},\text{SI})}_{\text{saddle}}.$$

The equilibrium of $(x_{j_1}^*,x_{j_2}^*)=(a_{j_1j_11},a_{j_2j_2s_1})$ is an (SO,SI)-saddle.

- For $a_{j_1j_10}(a_{j_2j_2s_1}-a_{j_1j_21})^2<0$ and $a_{j_2j_20}(a_{j_2j_2s_1}-a_{j_2j_2s_2})<0$,

$$(a_{j_1j_11},a_{j_2j_2s_1})=\underbrace{(\text{SI},\text{SI})}_{\text{sink}}.$$

The equilibrium of $(x_{j_1}^*,x_{j_2}^*)=(a_{j_1j_11},a_{j_2j_2s_1})$ is an (SI,SI)-sink.

(vi$_{1b}$) In phase space, at $x_{j_2}^*=a_{j_1j_21}$ with $\bar{x}_{j_1}\neq a_{j_1j_11}$,

$$\begin{aligned}&\frac{dx_{j_1}}{dx_{j_2}}\bigg|_{x_{j_2}^*=a_{j_1j_21}}=\frac{a_{j_1j_10}}{a_{j_2j_20}}\frac{(\bar{x}_{j_1}-a_{j_1j_11})(x_{j_2}-a_{j_1j_21})^2}{(x_{j_2}-a_{j_2j_21})(x_{j_2}-a_{j_2j_22})}\bigg|_{x_{j_2}^*=a_{j_1j_21}}=0,\\&\frac{d^2x_{j_1}}{dx_{j_2}^2}\bigg|_{x_{j_2}^*=a_{j_1j_21}}=0.\end{aligned}$$

If

$$\frac{d^3x_{j_1}}{dx_{j_2}^3}\bigg|_{x_{j_2}^*=a_{j_1j_21}}=\frac{a_{j_1j_10}}{a_{j_2j_20}}\frac{2(\bar{x}_{j_1}-a_{j_1j_11})}{(a_{j_1j_21}-a_{j_2j_21})(a_{j_1j_21}-a_{j_2j_22})}>0,$$

there is a third-order increasing-inflection flow at $x_{j_2}^*=a_{j_1j_21}$ with $\bar{x}_{j_1}\neq a_{j_1j_11}$. If

$$\frac{d^3x_{j_1}}{dx_{j_2}^3}\bigg|_{x_{j_2}^*=a_{j_1j_21}}=\frac{a_{j_1j_10}}{a_{j_2j_20}}\frac{2(\bar{x}_{j_1}-a_{j_1j_11})}{(a_{j_1j_21}-a_{j_2j_21})(a_{j_1j_21}-a_{j_2j_22})}<0,$$

there is a third-order decreasing-inflection flow at $x_{j_2}^*=a_{j_1j_21}$ with $\bar{x}_{j_1}\neq a_{j_1j_11}$. Let

$$\begin{aligned}\dot{x}_{j_2}&=a_{j_2j_20}(a_{j_1j_21}-a_{j_2j_21})(a_{j_1j_21}-a_{j_2j_22})\\&=a_{j_2j_20}\prod\nolimits_{s_1=1}^{2}(a_{j_1j_21}-a_{j_2j_2s_1})\neq 0.\end{aligned}$$

The third-order inflection flows at $x_{j_2}^*=a_{j_1j_21}$ with $\bar{x}_{j_1}\neq a_{j_1j_11}$ are positive and negative for $\dot{x}_{j_2}>0$ and $\dot{x}_{j_2}<0$, respectively.

Thus, the equilibrium of $x_{j_2}^*=a_{j_1j_21}$ with $\bar{x}_{j_1}\neq a_{j_1j_11}$ has the following properties.

- For $a_{j_1j_10}(\overline{x}_{j_1} - a_{j_1j_11}) > 0$ and $a_{j_2j_20}\prod_{s_1=1}^{2}(a_{j_1j_21} - a_{j_2j_2s_1}) > 0$,

$$(a_{j_1j_21}, \dot{x}_{j_2}) = \underbrace{(\mathrm{II}, \mathrm{pF})}_{\text{increasing-inflection flow } (+)}.$$

- For $a_{j_1j_10}(\overline{x}_{j_1} - a_{j_1j_11}) < 0$ and $a_{j_2j_20}\prod_{s_1=1}^{2}(a_{j_1j_21} - a_{j_2j_2s_1}) > 0$,

$$(a_{j_1j_21}, \dot{x}_{j_2}) = \underbrace{(\mathrm{DI}, \mathrm{pF})}_{\text{decreasing-inflection flow } (+)}.$$

- For $a_{j_1j_10}(\overline{x}_{j_1} - a_{j_1j_11}) > 0$ and $a_{j_2j_20}\prod_{s_1=1}^{2}(a_{j_1j_21} - a_{j_2j_2s_1}) < 0$,

$$(a_{j_1j_21}, \dot{x}_{j_2}) = \underbrace{(\mathrm{DI}, \mathrm{nF})}_{\text{decreasing-inflection flow } (-)}.$$

- For $a_{j_1j_10}(\overline{x}_{j_1} - a_{j_1j_11}) < 0$ and $a_{j_2j_20}\prod_{s_1=1}^{2}(a_{j_1j_21} - a_{j_2j_2s_1}) < 0$,

$$(a_{j_1j_21}, \dot{x}_{j_2}) = \underbrace{(\mathrm{II}, \mathrm{nF})}_{\text{increasing-inflection flow } (-)}.$$

The variational equation at $x_{j_1}^* = a_{j_1j_11}$ with $\overline{x}_{j_2} \neq a_{j_1j_21}$ is

$$\Delta\dot{x}_{j_1} = a_{j_1j_10}(\overline{x}_{j_2} - a_{j_1j_21})^2 \Delta x_{j_1}$$

and

$$\dot{x}_{j_2} = a_{j_2j_20}\prod_{s_1=1}^{2}(\overline{x}_{j_2} - a_{j_2j_2s_1}).$$

Thus, the equilibrium of $x_{j_1}^* = a_{j_1j_11}$ with $\overline{x}_{j_2} \neq a_{j_1j_21}, a_{j_1j_2s_1}$ ($s_1 = 1, 2$) has the following properties.

- For $a_{j_1j_10}(\overline{x}_{j_2} - a_{j_1j_21})^2 > 0$ and $a_{j_2j_20}\prod_{s_1=1}^{2}(\overline{x}_{j_2} - a_{j_2j_2s_1}) > 0$,

$$(a_{j_1j_11}, \dot{x}_{j_2}) = \underbrace{(\mathrm{SO}, \mathrm{pF})}_{\text{source flow } (+)}.$$

- For $a_{j_1j_10}(\overline{x}_{j_2} - a_{j_1j_21})^2 < 0$ and $a_{j_2j_20}\prod_{s_1=1}^{2}(\overline{x}_{j_2} - a_{j_2j_2s_1}) > 0$,

$$(a_{j_1j_1}1, \dot{x}_{j_2}) = \underbrace{(\text{SI, pF})}_{\text{sink flow }(+)}.$$

- For $a_{j_1j_1 0}(\bar{x}_{j_2} - a_{j_1j_2 1})^2 > 0$ and $a_{j_2j_2 0}\prod_{s_1=1}^{2}(\bar{x}_{j_2} - a_{j_2j_2 s_1}) < 0$,

$$(a_{j_1j_1}1, \dot{x}_{j_2}) = \underbrace{(\text{SO, nF})}_{\text{source flow }(-)}.$$

- For $a_{j_1j_1 0}(\bar{x}_{j_2} - a_{j_1j_2 1})^2 < 0$ and $a_{j_2j_2 0}\prod_{s_1=1}^{2}(\bar{x}_{j_2} - a_{j_2j_2 s_1}) < 0$,

$$(a_{j_1j_1}1, \dot{x}_{j_2}) = \underbrace{(\text{SI, nF})}_{\text{sink flow }(-)}.$$

From the above analysis, the equilibrium of $(x_{j_1}^*, x_{j_2}^*) = (a_{j_1j_1}1, a_{j_1j_2}1)$ with $a_{j_1j_2}1 \neq a_{j_2j_2}1, a_{j_2j_2}2$ has the following properties as in Eqs. (1.58)–(1.61).

- For $a_{j_1j_1 0} > 0$ and $a_{j_2j_2 0}\prod_{s_1=1}^{2}(a_{j_1j_2}1 - a_{j_2j_2 s_1}) > 0$,

$$(a_{j_1j_1}1, a_{j_1j_2}1) = \underbrace{({}_{\text{DI:II}}\text{SO, pF})}_{\text{inflection-source flow }(+)}.$$

  The equilibrium of $(x_{j_1}^*, x_{j_2}^*) = (a_{j_1j_1}1, a_{j_1j_2}1)$ is a $({}_{\text{DI:II}}\text{SO, pF})$-positive inflection-source flow.
- For $a_{j_1j_1 0} < 0$ and $a_{j_2j_2 0}\prod_{s_1=1}^{2}(a_{j_1j_2}1 - a_{j_2j_2 s_1}) > 0$,

$$(a_{j_1j_1}1, a_{j_1j_2}1) = \underbrace{({}_{\text{II:DI}}\text{SI, pF})}_{\text{inflection-sink flow }(+)}.$$

  The equilibrium of $(x_{j_1}^*, x_{j_2}^*) = (a_{j_1j_1}1, a_{j_1j_2}1)$ is an $({}_{\text{II:DI}}\text{SI, pF})$-positive inflection-sink flow.
- For $a_{j_1j_1 0} > 0$ and $a_{j_2j_2 0}\prod_{s_1=1}^{2}(a_{j_1j_2}1 - a_{j_2j_2 s_1}) < 0$,

$$(a_{j_1j_1}1, a_{j_1j_2}1) = \underbrace{({}_{\text{II:DI}}\text{SO, nF})}_{\text{inflection-source flow }(-)}.$$

  The equilibrium of $(x_{j_1}^*, x_{j_2}^*) = (a_{j_1j_1}1, a_{j_1j_2}1)$ is an $({}_{\text{II:DI}}\text{SO, nF})$-negative inflection-source flow.

- For $a_{j_1j_10}<0$ and $a_{j_2j_20}\prod_{s_1=1}^{2}(a_{j_1j_21}-a_{j_2j_2s_1})<0$,

$$(a_{j_1j_11}, a_{j_1j_21}) = \underbrace{({}_{\text{DI:II}}\text{SI, nF})}_{\text{inflection-sink flow }(-)}.$$

The equilibrium of $(x_{j_1}^*, x_{j_2}^*) = (a_{j_1j_11}, a_{j_1j_21})$ is a (${}_{\text{DI:II}}$SI, nF)-negative inflection-sink flow.

(vi$_2$) At $a_{j_2j_2s_1} = a_{j_1j_21}$ (s_1, $s_2 \in \{1,2\}$, $s_1 \neq s_2$), we have

$$\frac{dx_{j_1}}{dx_{j_2}} = \frac{a_{j_1j_10}}{a_{j_2j_20}} \frac{(x_{j_1}-a_{j_1j_11})(x_{j_2}-a_{j_1j_21})}{(x_{j_2}-a_{j_2j_2s_2})},$$

and the deformation of the foregoing equation is

$$\frac{1}{(x_{j_1}-a_{j_1j_11})}dx_{j_1} = \frac{a_{j_1j_10}}{a_{j_2j_20}}\left[1+\frac{(a_{j_2j_2s_2}-a_{j_1j_21})}{(x_{j_2}-a_{j_2j_2s_2})}\right]dx_{j_2}.$$

With an initial condition of (x_{j_10}, x_{j_20}) at $t = t_0$, the integration of the above equation gives the first integral manifold ($a_{j_2j_2s_1} = a_{j_1j_21}$) as

$$\begin{aligned}\ln\frac{|x_{j_1}-a_{j_1j_11}|}{|x_{j_10}-a_{j_1j_11}|} = \frac{a_{j_1j_10}}{a_{j_2j_20}}\Big\{&\left[(x_{j_2}-a_{j_2j_2s_2})-(x_{j_20}-a_{j_2j_2s_2})\right]\\ &+(a_{j_2j_2s_2}-a_{j_1j_21})\ln\frac{|x_{j_2}-a_{j_2j_2s_2}|}{|x_{j_2}-a_{j_2j_2s_2}|}\Big\}.\end{aligned}$$

(vi$_{2a}$) At $x_{j_2}^* = a_{j_1j_21} = a_{j_2j_2s_1}$ with $\bar{x}_{j_1} \neq a_{j_1j_11}$,

$$\left.\frac{dx_{j_1}}{dx_{j_2}}\right|_{x_{j_2}^*=a_{j_1j_21}} = \frac{a_{j_1j_10}}{a_{j_2j_20}}\left.\frac{(\bar{x}_{j_1}-a_{j_1j_11})(x_{j_2}-a_{j_1j_21})}{(x_{j_2}-a_{j_2j_2s_2})}\right|_{x_{j_2}^*=a_{j_1j_21}} = 0.$$

If

$$\left.\frac{d^2x_{j_1}}{dx_{j_2}^2}\right|_{x_{j_2}^*=a_{j_1j_21}} = \frac{a_{j_1j_10}}{a_{j_2j_20}}\frac{(\bar{x}_{j_1}-a_{j_1j_11})}{(a_{j_1j_21}-a_{j_2j_2s_2})} > 0,$$

there is an up-parabola flow at $x_{j_2}^* = a_{j_1j_21} = a_{j_2j_2s_1}$ with $\bar{x}_{j_1} \neq a_{j_1j_11}$. If

$$\left.\frac{d^2x_{j_1}}{dx_{j_2}^2}\right|_{x_{j_2}^*=a_{j_1j_21}} = \frac{a_{j_1j_10}}{a_{j_2j_20}}\frac{(\bar{x}_{j_1}-a_{j_1j_11})}{(a_{j_1j_21}-a_{j_2j_2s_2})} < 0,$$

there is a down-parabola flow at $x_{j_2}^* = a_{j_1 j_2 1} = a_{j_2 j_2 s_1}$ with $\bar{x}_{j_1} \neq a_{j_1 j_1 1}$. The corresponding variational equation is

$$\Delta \dot{x}_{j_2} = a_{j_2 j_2 0}(a_{j_2 j_2 s_1} - a_{j_2 j_2 s_2})\Delta x_{j_2}.$$

The parabola flow at $x_{j_2}^* = a_{j_1 j_2 1} = a_{j_2 j_2 s_1}$ with $\bar{x}_{j_1} \neq a_{j_1 j_1 1}$ are source and sink flows for $a_{j_2 j_2 0}(a_{j_2 j_2 s_1} - a_{j_2 j_2 s_2}) > 0$ and $a_{j_2 j_2 0}(a_{j_2 j_2 s_1} - a_{j_2 j_2 s_2}) < 0$, respectively.

Therefore, the infinite-equilibrium of $x_{j_2}^* = a_{j_1 j_2 1} = a_{j_2 j_2 s_1}$ with $\bar{x}_{j_1} \neq a_{j_1 j_1 1}$ has the following properties as in Eqs. (1.63)–(1.66).

- For $a_{j_1 j_1 0}(\bar{x}_{j_1} - a_{j_1 j_1 1}) > 0$ and $a_{j_2 j_2 0}(a_{j_2 j_2 s_1} - a_{j_2 j_2 s_2}) > 0$,

$$(\bar{x}_{j_1}, a_{j_2 j_2 s_2}) = \underbrace{(\text{UP, SO})}_{\text{up-parabola source}}.$$

The infinite-equilibrium of $x_{j_2}^* = a_{j_1 j_2 1} = a_{j_2 j_2 s_1}$ is a (UP,SO)-up-parabola source.

- For $a_{j_1 j_1 0}(\bar{x}_{j_1} - a_{j_1 j_1 1}) < 0$ and $a_{j_2 j_2 0}(a_{j_2 j_2 s_1} - a_{j_2 j_2 s_2}) > 0$,

$$(\bar{x}_{j_1}, a_{j_2 j_2 s_1}) = \underbrace{(\text{DP, SO})}_{\text{down-parabola source}}.$$

The infinite-equilibrium of $x_{j_2}^* = a_{j_1 j_2 1} = a_{j_2 j_2 s_1}$ is a (DP,SO)-down-parabola source.

- For $a_{j_1 j_1 0}(\bar{x}_{j_1} - a_{j_1 j_1 1}) > 0$ and $a_{j_2 j_2 0}(a_{j_2 j_2 s_1} - a_{j_2 j_2 s_2}) < 0$,

$$(\bar{x}_{j_1}, a_{j_2 j_2 s_1}) = \underbrace{(\text{DP, SI})}_{\text{down-parabola sink}}.$$

The infinite-equilibrium of $x_{j_2}^* = a_{j_1 j_2 1} = a_{j_2 j_2 s_1}$ is a (DP,SI)-down-parabola sink.

- For $a_{j_1 j_1 0}(\bar{x}_{j_1} - a_{j_1 j_1 1}) < 0$ and $a_{j_2 j_2 0}(a_{j_2 j_2 s_1} - a_{j_2 j_2 s_2}) < 0$,

$$(\bar{x}_{j_1}, a_{j_2 j_2 s_1}) = \underbrace{(\text{UP, SI})}_{\text{up-parabola sink}}.$$

The infinite-equilibrium of $x_{j_2}^* = a_{j_1 j_2 1} = a_{j_2 j_2 s_1}$ is a (UP,SI)-up-parabola sink.

(vi_{2b}) From case (vi_{2a}), for $\bar{x}_{j_1} = a_{j_1 j_1 1}$, thus, the equilibrium of $(x_{j_1}^*, x_{j_2}^*) = (a_{j_1 j_1 1}, a_{j_2 j_2 s_1})$ with $a_{j_2 j_2 s_1} = a_{j_1 j_2 1}$ has the following properties as in Eqs. (1.67)–(1.70).

- For $a_{j_1 j_1 0} > 0$ and $a_{j_2 j_2 0}(a_{j_2 j_2 s_1} - a_{j_2 j_2 s_2}) > 0$,

$$(a_{j_1j_11}, a_{j_2j_2s_1}) = \underbrace{(\mathrm{DP}:\mathrm{UP},\mathrm{SO})}_{\text{hyperbolic source}}.$$

The equilibrium of $(x^*_{j_1}, x^*_{j_2}) = (a_{j_1j_11}, a_{j_2j_2s_1})$ is a (DP:UP,SO)-hyperbolic source.

- For $a_{j_1j_10} < 0$ and $a_{j_2j_20}(a_{j_2j_2s_1} - a_{j_2j_2s_2}) > 0$,

$$(a_{j_1j_11}, a_{j_2j_2s_1}) = \underbrace{(\mathrm{UP}:\mathrm{DP},\mathrm{SO})}_{\text{hyperbolic-secant source}}.$$

The equilibrium of $(x^*_{j_1}, x^*_{j_2}) = (a_{j_1j_11}, a_{j_2j_2s_1})$ is a (UP:DP,SO)-hyperbolic-secant source.

- For $a_{j_1j_10} > 0$ and $a_{j_2j_20}(a_{j_2j_2s_1} - a_{j_2j_2s_2}) < 0$,

$$(a_{j_1j_11}, a_{j_2j_2s_1}) = \underbrace{(\mathrm{UP}:\mathrm{DP},\mathrm{SI})}_{\text{hyperbolic-secant sink}}.$$

The equilibrium of $(x^*_{j_1}, x^*_{j_2}) = (a_{j_1j_11}, a_{j_2j_2s_1})$ is a (UP:DP,SI)-hyperbolic-secant sink.

- For $a_{j_1j_10} < 0$ and $a_{j_2j_20}(a_{j_2j_2s_1} - a_{j_2j_2s_2}) < 0$,

$$(a_{j_1j_11}, a_{j_2j_2s_1}) = \underbrace{(\mathrm{DP}:\mathrm{UP},\mathrm{SI})}_{\text{hyperbolic sink}}.$$

The equilibrium of $(x^*_{j_1}, x^*_{j_2}) = (a_{j_1j_11}, a_{j_2j_2s_1})$ is a (DP:UP,SI)-hyperbolic sink.

(vi$_{2c}$) At $(x^*_{j_1}, x^*_{j_2}) = (a_{j_1j_11}, a_{j_2j_2s_2})$ with $a_{j_2j_2s_1} = a_{j_1j_21} \neq a_{j_2j_2s_2}$, the variational equation is

$$\begin{aligned}\Delta\dot{x}_{j_2} &= a_{j_2j_20}(a_{j_2j_2s_2} - a_{j_2j_2s_1})\Delta x_{j_2},\\ \Delta\dot{x}_{j_1} &= a_{j_1j_10}(a_{j_2j_2s_2} - a_{j_1j_21})^2\Delta x_{j_1}.\end{aligned}$$

Because of

$$(a_{j_2j_2s_2} - a_{j_1j_21})^2 > 0,$$

the flows at $(x^*_{j_1}, x^*_{j_2}) = (a_{j_1j_11}, a_{j_2j_2s_2})$ in the x_{j_1}-direction are source and sink for $a_{j_1j_10} > 0$ and $a_{j_1j_10} < 0$, respectively. The flows at $(x^*_{j_1}, x^*_{j_2}) = (a_{j_1j_11}, a_{j_2j_2s_2})$ are source and sink for $a_{j_2j_20}(a_{j_2j_2s_2} - a_{j_2j_2s_1}) > 0$ and $a_{j_2j_20}(a_{j_2j_2s_2} - a_{j_2j_2s_1}) < 0$ in the x_{j_2}-direction, respectively.

Therefore, the equilibrium of $(x^*_{j_1}, x^*_{j_2}) = (a_{j_1j_11}, a_{j_2j_2s_2})$ with $a_{j_2j_2s_1} = a_{j_1j_21}$ has the following properties as in Eqs. (1.71)–(1.74).

- For $a_{j_1j_10}(a_{j_2j_2s_2} - a_{j_1j_21})^2 > 0$ and $a_{j_2j_20}(a_{j_2j_2s_2} - a_{j_2j_2s_1}) > 0$,

$$(a_{j_1j_11}, a_{j_2j_2s_2}) = \underbrace{(\mathrm{SO}, \mathrm{SO})}_{\text{source}}.$$

The equilibrium of $(x^*_{j_1}, x^*_{j_2}) = (a_{j_1j_11}, a_{j_2j_2s_2})$ is an (SO,SO)-source.

- For $a_{j_1j_10}(a_{j_2j_2s_2} - a_{j_1j_21})^2 < 0$ and $a_{j_2j_20}(a_{j_2j_2s_2} - a_{j_2j_2s_1}) > 0$,

$$(a_{j_1j_11}, a_{j_2j_2s_2}) = \underbrace{(\mathrm{SI}, \mathrm{SO})}_{\text{saddle}}.$$

The equilibrium of $(x^*_{j_1}, x^*_{j_2}) = (a_{j_1j_11}, a_{j_2j_2s_2})$ is an (SI,SO)-saddle.

- For $a_{j_1j_10}(a_{j_2j_2s_2} - a_{j_1j_21})^2 > 0$ and $a_{j_2j_20}(a_{j_2j_2s_2} - a_{j_2j_2s_1}) < 0$,

$$(a_{j_1j_11}, a_{j_2j_2s_2}) = \underbrace{(\mathrm{SO}, \mathrm{SI})}_{\text{saddle}}.$$

The equilibrium of $(x^*_{j_1}, x^*_{j_2}) = (a_{j_1j_11}, a_{j_2j_2s_2})$ is an (SO,SI)-saddle.

- For $a_{j_1j_10}(a_{j_2j_2s_2} - a_{j_1j_21})^2 < 0$ and $a_{j_2j_20}(a_{j_2j_2s_2} - a_{j_2j_2s_1}) < 0$,

$$(a_{j_1j_11}, a_{j_2j_2s_2}) = \underbrace{(\mathrm{SI}, \mathrm{SI})}_{\text{sink}}.$$

The equilibrium of $(x^*_{j_1}, x^*_{j_2}) = (a_{j_1j_11}, a_{j_2j_2s_2})$ is an (SI,SI)-sink.

(vii) For $\Delta_{j_1j_2} > 0$ and $\Delta_{j_2j_2} < 0$, the standard form is

$$\begin{aligned}
\dot{x}_{j_2} &= a_{j_2j_20}\left[(x_{j_2} - a_{j_2j_2})^2 + b_{j_2j_2}\right],\\
\dot{x}_{j_1} &= a_{j_1j_10}(x_{j_1} - a_{j_1j_11})(x_{j_2} - a_{j_1j_21})(x_{j_2} - a_{j_1j_22})
\end{aligned}$$

where

$$\begin{aligned}
&b_{j_1j_21}, b_{j_1j_22} = -\frac{1}{2}(B_{j_1j_2} \pm \sqrt{\Delta_{j_1j_2}}),\\
&\{a_{j_1j_21}, a_{j_1j_22}\} = \mathrm{sort}\{b_{j_1j_21}, b_{j_1j_22}\}, a_{j_1j_21} < a_{j_1j_22};\\
&a_{j_2j_2} = -\frac{1}{2}B_{j_2j_2}, b_{j_2j_2} = \frac{1}{4}(-\Delta_{j_2j_2}).
\end{aligned}$$

In phase space,

$$\frac{dx_{j_1}}{dx_{j_2}} = \frac{a_{j_1j_10}}{a_{j_2j_20}} \frac{(x_{j_1} - a_{j_1j_11})(x_{j_2} - a_{j_1j_21})(x_{j_2} - a_{j_1j_22})}{(x_{j_2} - a_{j_2j_2})^2 + b_{j_2j_2}},$$

and the deformation of the foregoing equation is

$$
\begin{aligned}
\frac{1}{(x_{j_1}-a_{j_1j_1 1})}dx_{j_1} = \frac{a_{j_1j_1 0}}{a_{j_2j_2 0}}\Big[1 &+ \frac{(a_{j_2j_2}-a_{j_1j_2 1})(x_{j_2}-a_{j_2j_2})}{(x_{j_2}-a_{j_2j_2})^2+b_{j_2j_2}} \\
&+ \frac{(a_{j_2j_2}-a_{j_1j_2 2})(x_{j_2}-a_{j_2j_2})}{(x_{j_2}-a_{j_2j_2})^2+b_{j_2j_2}} \\
&+ \frac{(a_{j_2j_2}-a_{j_1j_2 1})(a_{j_2j_2}-a_{j_1j_2 2})-b_{j_2j_2}}{(x_{j_2}-a_{j_2j_2})^2+b_{j_2j_2}}\Big]dx_{j_2}.
\end{aligned}
$$

With an initial condition of $(x_{j_1 0}, x_{j_2 0})$ at $t = t_0$, the integration of the above equation gives the first integral manifold as

$$
\begin{aligned}
\ln\frac{|x_{j_1}-a_{j_1j_1 1}|}{|x_{j_1 0}-a_{j_1j_1 1}|} = \frac{a_{j_1j_1 0}}{a_{j_2j_2 0}}\Big\{ &[(x_{j_2}-u_{j_2j_2})-(x_{j_2 0}-a_{j_2j_2})] \\
&+\textstyle\sum_{s_1=1}^{2}(a_{j_2j_2}-a_{j_1j_2s_1})\ln\frac{|(x_{j_2}-a_{j_2j_2})^2+b_{j_2j_2}|}{|(x_{j_2 0}-a_{j_2j_2})^2+b_{j_2j_2}|} \\
&+\Big[\textstyle\prod_{s_1=1}^{2}(a_{j_2j_2}-a_{j_1j_2s_1})-b_{j_2j_2}\Big]\frac{1}{\sqrt{b_{j_2j_2}}} \\
&\times\Big(\arctan\frac{x_{j_2}-a_{j_2j_2}}{\sqrt{b_{j_2j_2}}} - \arctan\frac{x_{j_2 0}-a_{j_2j_2}}{\sqrt{b_{j_2j_2}}}\Big)\Big\}.
\end{aligned}
$$

For $x_{j_2}^* = a_{j_1j_2s_1}$ and $\bar{x}_{j_1} \neq a_{j_1j_1 1}$ ($s_1, s_2 \in \{1,2\}$, $s_1 \neq s_2$), in phase space,

$$
\frac{dx_{j_1}}{dx_{j_2}}\Big|_{x_{j_2}^*=a_{j_1j_2s_1}} = \frac{a_{j_1j_1 0}}{a_{j_2j_2 0}}\frac{(\bar{x}_{j_1}-a_{j_1j_1 1})(x_{j_2}-a_{j_1j_2s_1})(x_{j_2}-a_{j_1j_2s_2})}{(x_{j_2}-a_{j_2j_2})^2+b_{j_2j_2}}\Big|_{x_{j_2}^*=a_{j_1j_2s_1}} = 0.
$$

If

$$
\frac{d^2x_{j_1}}{dx_{j_2}^2}\Big|_{x_{j_2}^*=a_{j_1j_2s_1}} = \frac{a_{j_1j_1 0}}{a_{j_2j_2 0}}\frac{(\bar{x}_{j_1}-a_{j_1j_1 1})(a_{j_1j_2s_1}-a_{j_1j_2s_2})}{(a_{j_1j_2s_1}-a_{j_2j_2})^2+b_{j_2j_2}} > 0,
$$

there is an up-parabola flow at $x_{j_2}^* = a_{j_1j_2s_1}$ with $\bar{x}_{j_1} \neq a_{j_1j_1 1}$. If

$$
\frac{d^2x_{j_1}}{dx_{j_2}^2}\Big|_{x_{j_2}^*=a_{j_1j_2s_1}} = \frac{a_{j_1j_1 0}}{a_{j_2j_2 0}}\frac{(\bar{x}_{j_1}-a_{j_1j_1 1})(a_{j_1j_2s_1}-a_{j_1j_2s_2})}{(a_{j_1j_2s_1}-a_{j_2j_2})^2+b_{j_2j_2}} < 0,
$$

there is a down-parabola flow at $x_{j_2}^* = a_{j_1j_2s_1}$ with $\bar{x}_{j_1} \neq a_{j_1j_1 1}$. Let

$$
\dot{x}_{j_2} = a_{j_2j_2 0}\big[(a_{j_1j_2s_1}-a_{j_2j_2})^2+b_{j_2j_2}\big],
$$

with

$$(a_{j_1j_2s_1} - a_{j_2j_2})^2 + b_{j_2j_2} > 0.$$

The inflection flows at $x^*_{j_2} = a_{j_1j_2s_1}$ with $\bar{x}_{j_1} \neq a_{j_1j_1 1}$ are positive and negative for $a_{j_2j_20} > 0$ and $a_{j_2j_20} < 0$, respectively.

- For $a_{j_1j_10}(\bar{x}_{j_1} - a_{j_1j_11})(a_{j_1j_2s_1} - a_{j_1j_2s_2}) > 0$ and $a_{j_2j_20} > 0$,

$$(a_{j_1j_2s_1}, \dot{x}_{j_2}) = \underbrace{(\text{UP}, \text{pF})}_{\text{up-parabola flow }(+)}.$$

- For $a_{j_1j_10}(\bar{x}_{j_1} - a_{j_1j_11})(a_{j_1j_2s_1} - a_{j_1j_2s_2}) < 0$ and $a_{j_2j_20} > 0$,

$$(a_{j_1j_2s_1}, \dot{x}_{j_2}) = \underbrace{(\text{DP}, \text{pF})}_{\text{down-parabola flow }(+)}.$$

- For $a_{j_1j_10}(\bar{x}_{j_1} - a_{j_1j_11})(a_{j_1j_2s_1} - a_{j_1j_2s_2}) > 0$ and $a_{j_2j_20} < 0$,

$$(a_{j_1j_2s_1}, \dot{x}_{j_2}) = \underbrace{(\text{DP}, \text{nF})}_{\text{down-parabola flow }(-)}.$$

- For $a_{j_1j_10}(\bar{x}_{j_1} - a_{j_1j_11})(a_{j_1j_2s_1} - a_{j_1j_2s_2}) < 0$ and $a_{j_2j_20} < 0$,

$$(a_{j_1j_2s_1}, \dot{x}_{j_2}) = \underbrace{(\text{UP}, \text{nF})}_{\text{up-parabola flow }(-)}.$$

From the above analysis, thus, the equilibrium of $(x^*_{j_1}, x^*_{j_2}) = (a_{j_1j_11}, a_{j_1j_2s_1})$ (s_1, $s_2 \in \{1, 2\}$, $s_1 \neq s_2$) has the following properties as in Eqs. (1.78)–(1.81).

- For $a_{j_1j_10}(a_{j_1j_2s_1} - a_{j_1j_2s_2}) > 0$ and $a_{j_2j_20} > 0$,

$$(a_{j_1j_11}, a_{j_1j_2s_1}) = \underbrace{(\text{DP} : \text{UP}, \text{pF})}_{\text{hyperbolic flow }(+)}.$$

  The equilibrium of $(x^*_{j_1}, x^*_{j_2}) = (a_{j_1j_11}, a_{j_1j_2s_1})$ is a (DP:UP,pF)-positive hyperbolic flow.

- For $a_{j_1j_10}(a_{j_1j_2s_1} - a_{j_1j_2s_2}) < 0$ and $a_{j_2j_20} > 0$,

$$(a_{j_1j_11}, a_{j_1j_2s_1}) = \underbrace{(\text{UP} : \text{DP}, \text{pF})}_{\text{hyperbolic-secant flow }(+)}.$$

  The equilibrium of $(x^*_{j_1}, x^*_{j_2}) = (a_{j_1j_11}, a_{j_1j_2s_1})$ is a (UP:DP,pF)-positive hyperbolic-secant flow.

- For $a_{j_1j_10}(a_{j_1j_2s_1} - a_{j_1j_2s_2}) > 0$ and $a_{j_2j_20} < 0$,

$$(a_{j_1j_11}, a_{j_1j_2s_1}) = \underbrace{(\mathrm{UP} : \mathrm{DP}, \mathrm{nF})}_{\text{hyperbolic-secant flow }(-)} .$$

The equilibrium of $(x_{j_1}^*, x_{j_2}^*) = (a_{j_1j_11}, a_{j_1j_2s_1})$ is a (UP:DP,nF)-negative hyperbolic-secant flow.

- For $a_{j_1j_10}(a_{j_1j_2s_1} - a_{j_1j_2s_2}) < 0$ and $a_{j_2j_20} < 0$,

$$(a_{j_1j_11}, a_{j_1j_2s_1}) = \underbrace{(\mathrm{DP} : \mathrm{UP}, \mathrm{nF})}_{\text{hyperbolic flow }(\ )} .$$

The equilibrium of $(x_{j_1}^*, x_{j_2}^*) = (a_{j_1j_11}, a_{j_1j_2s_1})$ is a (DP:UP,nF)-negative hyperbolic flow.

(viii) For $\Delta_{j_1j_2} > 0$ and $\Delta_{j_2j_2} = 0$, the standard form is

$$\begin{aligned} \dot{x}_{j_2} &= a_{j_2j_20}(x_{j_2} - a_{j_2j_21})^2, \\ \dot{x}_{j_1} &= a_{j_1j_10}(x_{j_1} - a_{j_1j_11})(x_{j_2} - a_{j_1j_21})(x_{j_2} - a_{j_1j_22}) \end{aligned}$$

where

$$\begin{aligned} & b_{j_1j_21}, b_{j_1j_22} = -\frac{1}{2}(B_{j_1j_2} \pm \sqrt{\Delta_{j_1j_2}}), \\ & \{a_{j_1j_21}, a_{j_1j_22}\} = \mathrm{sort}\{b_{j_1j_21}, b_{j_1j_22}\}, a_{j_1j_21} < a_{j_1j_22}; \\ & a_{j_2j_21} = -\frac{1}{2}B_{j_2j_2}. \end{aligned}$$

(viii$_1$) For $a_{j_2j_21} \neq a_{j_1j_21}, a_{j_1j_22}$, in phase space,

$$\frac{dx_{j_1}}{dx_{j_2}} = \frac{a_{j_1j_10}}{a_{j_2j_20}} \frac{(x_{j_1} - a_{j_1j_11})(x_{j_2} - a_{j_1j_21})(x_{j_2} - a_{j_1j_22})}{(x_{j_2} - a_{j_2j_21})^2},$$

and the deformation of the foregoing equation is

$$\begin{aligned} \frac{1}{(x_{j_1} - a_{j_1j_11})} dx_{j_1} = \frac{a_{j_1j_10}}{a_{j_2j_20}} \Big[1 + & \sum\nolimits_{s_1=1}^{2} (a_{j_2j_21} - a_{j_1j_2s_1}) \frac{1}{(x_{j_2} - a_{j_2j_21})} \\ & + \prod\nolimits_{s_1=1}^{2} (a_{j_2j_21} - a_{j_1j_2s_1}) \frac{1}{(x_{j_2} - a_{j_2j_21})^2} \Big] dx_{j_2}. \end{aligned}$$

With an initial condition of $(x_{j_1 0}, x_{j_2 0})$ at $t = t_0$, the integration of the above equation gives the first integral manifold for $a_{j_2 j_2 1} \neq a_{j_1 j_2 1}, a_{j_1 j_2 2}$ as

$$\begin{aligned} \ln \frac{|x_{j_1} - a_{j_1 j_1 1}|}{|x_{j_1 0} - a_{j_1 j_1 1}|} &= \frac{a_{j_1 j_1 0}}{a_{j_2 j_2 0}} \Big\{ [(x_{j_2} - a_{j_2 j_2 1}) - (x_{j_2 0} - a_{j_2 j_2 1})] \\ &+ \textstyle\sum_{s_1=1}^{2} (a_{j_2 j_2 1} - a_{j_1 j_2 s_1}) \ln \frac{|x_{j_2} - a_{j_2 j_2 1}|}{|x_{j_2 0} - a_{j_2 j_2 1}|} \\ &- \textstyle\prod_{s_1=1}^{2} (a_{j_2 j_2 1} - a_{j_1 j_2 s_1}) (\frac{1}{x_{j_2} - a_{j_2 j_2 1}} - \frac{1}{x_{j_2 0} - a_{j_2 j_2 1}}) \Big\}. \end{aligned}$$

(viii$_{1a}$) At $(x_{j_1}^*, x_{j_2}^*) = (a_{j_1 j_1 1}, a_{j_2 j_2 1})$, the variational equations are

$$\begin{aligned} \Delta \dot{x}_{j_2} &= a_{j_2 j_2 0} (\Delta x_{j_2})^2, \\ \Delta \dot{x}_{j_1} &= a_{j_1 j_1 0} \textstyle\prod_{s_1=1}^{2} (a_{j_2 j_2 1} - a_{j_1 j_2 s_1}) \Delta x_{j_1}. \end{aligned}$$

The flows at $(x_{j_1}^*, x_{j_2}^*) = (a_{j_1 j_1 1}, a_{j_2 j_2 1})$ in the x_{j_1}-direction are source and sink for $a_{j_1 j_1 0} \prod_{s_1=1}^{2} (a_{j_2 j_2 1} - a_{j_1 j_2 s_1}) > 0$ and $a_{j_1 j_1 0} \prod_{s_1=1}^{2} (a_{j_2 j_2 1} - a_{j_1 j_2 s_1}) < 0$, respectively. The flows at $(x_{j_1}^*, x_{j_2}^*) = (a_{j_1 j_1 1}, a_{j_2 j_2 1})$ in the x_{j_2}-direction are upper-saddle and lower-saddle for $a_{j_2 j_2 0} > 0$ and $a_{j_2 j_2 0} < 0$, respectively.

Therefore, the equilibrium of $(x_{j_1}^*, x_{j_2}^*) = (a_{j_1 j_1 1}, a_{j_2 j_2 1})$ has the following properties as in Eqs. (1.85)–(1.87).

- For $a_{j_1 j_1 0} \prod_{s_1=1}^{2} (a_{j_2 j_2 1} - a_{j_1 j_2 s_1}) > 0$ and $a_{j_2 j_2 0} > 0$,

$$(a_{j_1 j_1 1}, a_{j_2 j_2 1}) = \underbrace{(\text{SO}, \text{US})}_{\text{upper-saddle source}}.$$

  The equilibrium of $(x_{j_1}^*, x_{j_2}^*) = (a_{j_1 j_1 1}, a_{j_2 j_2 1})$ is an (SO,US)-upper-saddle source.
- For $a_{j_1 j_1 0} \prod_{s_1=1}^{2} (a_{j_2 j_2 1} - a_{j_1 j_2 s_1}) < 0$ and $a_{j_2 j_2 0} > 0$,

$$(a_{j_1 j_1 1}, a_{j_2 j_2 1}) = \underbrace{(\text{SI}, \text{US})}_{\text{upper-saddle-sink}}.$$

  The equilibrium of $(x_{j_1}^*, x_{j_2}^*) = (a_{j_1 j_1 1}, a_{j_2 j_2 1})$ is an (SI,US)-upper-saddle sink.
- For $a_{j_1 j_1 0} \prod_{s_1=1}^{2} (a_{j_2 j_2 1} - a_{j_1 j_2 s_1}) > 0$ and $a_{j_2 j_2 0} < 0$,

$$(a_{j_1 j_1 1}, a_{j_2 j_2 1}) = \underbrace{(\text{SO}, \text{LS})}_{\text{lower-saddle source}}.$$

  The equilibrium of $(x_{j_1}^*, x_{j_2}^*) = (a_{j_1 j_1 1}, a_{j_2 j_2 1})$ is an (SO,LS)-lower-saddle source.

- For $a_{j_1j_10}\prod_{s_1=1}^{2}(a_{j_2j_21}-a_{j_1j_2s_1})<0$ and $a_{j_2j_20}<0$,

$$(a_{j_1j_11},a_{j_2j_21})=\underbrace{(\mathrm{SI},\mathrm{LS})}_{\text{lower-saddle source}}.$$

The equilibrium of $(x^*_{j_1},x^*_{j_2})=(a_{j_1j_11},a_{j_2j_21})$ is an (SI,LS)-lower-saddle sink.

(viii$_{1b}$) For $x^*_{j_2}=a_{j_1j_2s_1}\neq a_{j_2j_21}$ and $\bar{x}_{j_1}\neq a_{j_1j_11}$ (s_1, $s_2\in\{1,2\}$, $s_1\neq s_2$), in phase space,

$$\frac{dx_{j_1}}{dx_{j_2}}\bigg|_{x^*_{j_2}=a_{j_1j_2s_1}}=\frac{a_{j_1j_10}}{a_{j_2j_20}}\frac{(\bar{x}_{j_1}-a_{j_1j_11})(x_{j_2}-a_{j_1j_2s_1})(x_{j_2}-a_{j_1j_2s_2})}{(x_{j_2}-a_{j_2j_21})^2}\bigg|_{x^*_{j_2}=a_{j_1j_2s_1}}=0.$$

If

$$\frac{d^2x_{j_1}}{dx^2_{j_2}}\bigg|_{x^*_{j_2}=a_{j_1j_2s_1}}=\frac{a_{j_1j_10}}{a_{j_2j_20}}\frac{(\bar{x}_{j_1}-a_{j_1j_11})(a_{j_1j_2s_1}-a_{j_1j_2s_2})}{(a_{j_1j_2s_1}-a_{j_2j_21})^2}>0,$$

there is an up-parabola flow at $x^*_{j_2}=a_{j_1j_2s_1}\neq a_{j_2j_21}$ with $\bar{x}_{j_1}\neq a_{j_1j_11}$. If

$$\frac{d^2x_{j_1}}{dx^2_{j_2}}\bigg|_{x^*_{j_2}=a_{j_1j_2s_1}}=\frac{a_{j_1j_10}}{a_{j_2j_20}}\frac{(\bar{x}_{j_1}-a_{j_1j_11})(a_{j_1j_2s_1}-a_{j_1j_2s_2})}{(a_{j_1j_2s_1}-a_{j_2j_21})^2}<0,$$

there is a down-parabola flow at $x^*_{j_2}=a_{j_1j_2s_1}$ with $\bar{x}_{j_1}\neq a_{j_1j_11}$. Let

$$\dot{x}_{j_2}=a_{j_2j_20}(a_{j_1j_2s_1}-a_{j_2j_21})^2\neq 0.$$

The parabola flows at $x^*_{j_2}=a_{j_1j_2s_1}\neq a_{j_2j_21}$ with $\bar{x}_{j_1}\neq a_{j_1j_11}$ are positive and negative for $a_{j_2j_20}>0$ and $a_{j_2j_20}<0$, respectively.

- For $a_{j_1j_10}(\bar{x}_{j_1}-a_{j_1j_11})(a_{j_1j_2s_1}-a_{j_1j_2s_2})>0$ and $a_{j_2j_20}(a_{j_1j_2s_1}-a_{j_2j_21})^2>0$,

$$(a_{j_1j_2s_1},\dot{x}_{j_2})=\underbrace{(\mathrm{UP},\mathrm{pF})}_{\text{up-parabola flow }(+)}.$$

- For $a_{j_1j_10}(\bar{x}_{j_1}-a_{j_1j_11})(a_{j_1j_2s_1}-a_{j_1j_2s_2})<0$ and $a_{j_2j_20}(a_{j_1j_2s_1}-a_{j_2j_21})^2>0$,

$$(a_{j_1j_2s_1},\dot{x}_{j_2})=\underbrace{(\mathrm{DP},\mathrm{pF})}_{\text{down-parabola flow }(+)}.$$

- For $a_{j_1j_10}(\overline{x}_{j_1} - a_{j_1j_11})(a_{j_1j_2s_1} - a_{j_1j_2s_2}) > 0$ and $a_{j_2j_20}(a_{j_1j_2s_1} - a_{j_2j_21})^2 < 0$,

$$(a_{j_1j_2s_1}, \dot{x}_{j_2}) = \underbrace{(\text{DP}, \text{nF})}_{\text{down-parabola flow }(-)}.$$

- For $a_{j_1j_10}(\overline{x}_{j_1} - a_{j_1j_11})(a_{j_1j_2s_1} - a_{j_1j_2s_2}) < 0$ and $a_{j_2j_20}(a_{j_1j_2s_1} - a_{j_2j_21})^2 < 0$,

$$(a_{j_1j_2s_1}, \dot{x}_{j_2}) = \underbrace{(\text{UP}, \text{nF})}_{\text{up-parabola flow }(-)}.$$

Therefore, from the above analysis, for $\overline{x}_{j_1} = a_{j_1j_11}$, the equilibrium of $(x^*_{j_1}, x^*_{j_2}) = (a_{j_1j_11}, a_{j_1j_2s_1})$ with $a_{j_1j_2s_1} \neq a_{j_2j_21}$ (s_1, $s_2 \in \{1, 2\}$, $s_1 \neq s_2$) has the following properties as in Eqs. (1.89)–(1.92).

- For $a_{j_1j_10}(a_{j_1j_2s_1} - a_{j_1j_2s_2}) > 0$ and $a_{j_2j_20}(a_{j_1j_2s_1} - a_{j_2j_21})^2 > 0$,

$$(a_{j_1j_11}, a_{j_1j_2s_1}) = \underbrace{(\text{DP} : \text{UP}, \text{pF})}_{\text{hyperbolic flow }(+)}.$$

  The equilibrium of $(x^*_{j_1}, x^*_{j_2}) = (a_{j_1j_11}, a_{j_1j_2s_1})$ is a (DP:UP, pF)-positive hyperbolic flow.
- For $a_{j_1j_10}(a_{j_1j_2s_1} - a_{j_1j_2s_2}) < 0$ and $a_{j_2j_20}(a_{j_1j_2s_1} - a_{j_2j_21})^2 > 0$,

$$(a_{j_1j_11}, a_{j_1j_2s_1}) = \underbrace{(\text{UP} : \text{DP}, \text{pF})}_{\text{hyperbolic-secant flow }(+)}.$$

  The equilibrium of $(x^*_{j_1}, x^*_{j_2}) = (a_{j_1j_11}, a_{j_1j_2s_1})$ is a (UP:DP, pF)-positive hyperbolic-secant flow.
- For $a_{j_1j_10}(a_{j_1j_2s_1} - a_{j_1j_2s_2}) > 0$ and $a_{j_2j_20}(a_{j_1j_2s_1} - a_{j_2j_21})^2 < 0$,

$$(a_{j_1j_11}, a_{j_1j_2s_1}) = \underbrace{(\text{UP} : \text{DP}, \text{nF})}_{\text{hyperbolic-secant flow }(-)}.$$

  The equilibrium of $(x^*_{j_1}, x^*_{j_2}) = (a_{j_1j_11}, a_{j_1j_2s_1})$ is a (UP:DP,nF)-negative hyperbolic-secant flow.
- For $a_{j_1j_10}(a_{j_1j_2s_1} - a_{j_1j_2s_2}) < 0$ and $a_{j_2j_20}(a_{j_1j_2s_1} - a_{j_2j_21})^2 < 0$,

$$(a_{j_1j_11}, a_{j_1j_2s_1}) = \underbrace{(\text{DP} : \text{UP}, \text{nF})}_{\text{hyperbolic flow }(-)}.$$

The equilibrium of $(x^*_{j_1}, x^*_{j_2}) = (a_{j_1 j_1 1}, a_{j_1 j_2 s_1})$ is a (DP:UP,nF)-negative hyperbolic flow.

(viii$_2$) For $a_{j_2 j_2 1} = a_{j_1 j_2 s_1}$ (s_1, $s_2 \in \{1, 2\}$, $s_1 \neq s_2$),

$$\frac{dx_{j_1}}{dx_{j_2}} = \frac{a_{j_1 j_1 0}}{a_{j_2 j_2 0}} \frac{(x_{j_1} - a_{j_1 j_1 1})(x_{j_2} - a_{j_1 j_2 s_1})(x_{j_2} - a_{j_1 j_2 s_2})}{(x_{j_2} - a_{j_2 j_2 1})^2},$$

and the deformation of the foregoing equation is

$$\frac{1}{(x_{j_1} - a_{j_1 j_1 1})} dx_{j_1} = \frac{a_{j_1 j_1 0}}{a_{j_2 j_2 0}} \left[1 + (a_{j_2 j_2 1} - a_{j_1 j_2 s_2}) \frac{1}{(x_{j_2} - a_{j_2 j_2 1})} \right] dx_{j_2}.$$

With an initial condition of $(x_{j_1 0}, x_{j_2 0})$ at $t = t_0$, the integration of the above equation gives the first integral manifold for $a_{j_2 j_2 1} = a_{j_1 j_2 s_1}$ (s_1, $s_2 \in \{1, 2\}$, $s_1 \neq s_2$) as

$$\begin{aligned} \ln \frac{|x_{j_1} - a_{j_1 j_1 1}|}{|x_{j_1 0} - a_{j_1 j_1 1}|} = \frac{a_{j_1 j_1 0}}{a_{j_2 j_2 0}} \bigg\{ & \left[(x_{j_2} - a_{j_2 j_2 1}) - (x_{j_2 0} - a_{j_2 j_2 1}) \right] \\ & + (a_{j_2 j_2 1} - a_{j_1 j_2 s_2}) \ln \frac{|x_{j_2} - a_{j_2 j_2 1}|}{|x_{j_2 0} - a_{j_2 j_2 1}|} \bigg\}. \end{aligned}$$

(viii$_{2a}$) At $x^*_{j_2} = a_{j_1 j_2 s_1} = a_{j_2 j_2 1}$ with $\bar{x}_{j_1} \neq a_{j_1 j_1 1}$, from the first integral manifold, the flow is an asymptotic flow. In addition,

$$\left. \frac{dx_{j_1}}{dx_{j_2}} \right|_{x^*_{j_2} = a_{j_2 j_2 1} \pm \varepsilon} = \pm \frac{a_{j_1 j_1 0}}{a_{j_2 j_2 0}} \frac{(\bar{x}_{j_1} - a_{j_1 j_1 1})(a_{j_1 j_2 s_1} - a_{j_1 j_2 s_2})}{\varepsilon}.$$

If

$$\frac{a_{j_1 j_1 0}}{a_{j_2 j_2 0}} (\bar{x}_{j_1} - a_{j_1 j_1 1})(a_{j_1 j_2 s_1} - a_{j_1 j_2 s_2}) > 0,$$

the flow at $x^*_{j_2} = a_{j_1 j_2 s_1} = a_{j_2 j_2 1}$ is a down-up flow. If

$$\frac{a_{j_1 j_1 0}}{a_{j_2 j_2 0}} (\bar{x}_{j_1} - a_{j_1 j_1 1})(a_{j_1 j_2 s_1} - a_{j_1 j_2 s_2}) < 0,$$

the flow at $x^*_{j_2} = a_{j_1 j_2 s_1} = a_{j_2 j_2 1}$ is an up-down flow. The variational equation at $x^*_{j_2} = a_{j_1 j_2 s_1} = a_{j_2 j_2 1}$ is

$$\Delta \dot{x}_{j_2} = a_{j_2 j_2 0} (\Delta x_{j_2})^2.$$

The asymptotic flows in the x_{j_2}-direction are upper-saddle and lower-saddle for $a_{j_2j_20}>0$ and $a_{j_2j_20}<0$, respectively.

Therefore, the infinite-equilibrium of $x_{j_2}^{*}=a_{j_1j_2s_1}=a_{j_2j_21}$ with $\bar{x}_{j_1}\neq a_{j_1j_11}$ has the following properties as in Eqs. (1.94)–(1.97).

- For $a_{j_1j_10}(\bar{x}_{j_1}-a_{j_1j_11})(a_{j_1j_2s_1}-a_{j_1j_2s_2})>0$ and $a_{j_2j_20}>0$,

$$(\bar{x}_{j_1},a_{j_2j_21})=\underbrace{(\text{DU},\text{US})}_{\text{down-up upper-saddle}}.$$

  The infinite-equilibrium of $x_{j_2}^{*}=a_{j_1j_2s_1}=a_{j_2j_21}$ is an (DU,US)-down-up upper-saddle.
- For $a_{j_1j_10}(\bar{x}_{j_1}-a_{j_1j_11})(a_{j_1j_2s_1}-a_{j_1j_2s_2})<0$ and $a_{j_2j_20}>0$,

$$(\bar{x}_{j_1},a_{j_2j_21})=\underbrace{(\text{UD},\text{US})}_{\text{up-down upper-saddle}}.$$

  The infinite-equilibrium of $x_{j_2}^{*}=a_{j_1j_2s_1}=a_{j_2j_21}$ is a (UD,US)-up-down upper-saddle.
- For $a_{j_1j_10}(\bar{x}_{j_1}-a_{j_1j_11})(a_{j_1j_2s_1}-a_{j_1j_2s_2})>0$ and $a_{j_2j_20}<0$,

$$(\bar{x}_{j_1},a_{j_2j_21})=\underbrace{(\text{UD},\text{LS})}_{\text{up-down lower-saddle}}.$$

  The infinite-equilibrium of $x_{j_2}^{*}=a_{j_1j_2s_1}=a_{j_2j_21}$ is a (UD,LS)-up-down lower-saddle.
- For $a_{j_1j_10}(\bar{x}_{j_1}-a_{j_1j_11})(a_{j_1j_2s_1}-a_{j_1j_2s_2})<0$ and $a_{j_2j_20}<0$,

$$(\bar{x}_{j_1},a_{j_2j_21})=\underbrace{(\text{DU},\text{LS})}_{\text{down-up lower-saddle}}.$$

  The infinite-equilibrium of $x_{j_2}^{*}=a_{j_1j_2s_1}=a_{j_2j_21}$ is a (DU,LS)-down-up lower-saddle.

(viii$_{2b}$) From case (viii$_{2a}$), for $\bar{x}_{j_1}=a_{j_1j_11}$, the equilibrium of $(x_{j_1}^{*},x_{j_2}^{*})=(a_{j_1j_11},a_{j_2j_21})$ with $a_{j_2j_21}=a_{j_1j_2s_1}$ has the following properties as in Eqs. (1.98)–(1.101).

- For $a_{j_1j_10}(a_{j_1j_2s_1}-a_{j_1j_2s_2})>0$ and $a_{j_2j_20}>0$,

$$(a_{j_1j_11},a_{j_2j_21})=\underbrace{(\text{DU}:\text{UD},\text{US})}_{\text{down-up-to-up-down upper-saddle}}. \tag{1.98}$$

  The equilibrium of $(x_{j_1}^{*},x_{j_2}^{*})=(a_{j_1j_11},a_{j_2j_21})$ is a (DU:UD,US)-down-up-to-down-up upper-saddle.

- For $a_{j_1j_10}(a_{j_1j_2s_1}-a_{j_1j_2s_2})<0$ and $a_{j_2j_20}>0$,

$$(a_{j_1j_11},a_{j_2j_21})=\underbrace{(\mathrm{UD}:\mathrm{DU},\mathrm{US})}_{\text{up-down-to-down-up upper-saddle}}. \tag{1.99}$$

The equilibrium of $(x_{j_1}^*,x_{j_2}^*)=(a_{j_1j_11},a_{j_2j_21})$ is a (UD:DU,US)-up-down-to-down-up upper-saddle.

- For $a_{j_1j_10}(a_{j_1j_2s_1}-a_{j_1j_2s_2})>0$ and $a_{j_2j_20}<0$,

$$(a_{j_1j_11},a_{j_2j_21})=\underbrace{(\mathrm{UD}:\mathrm{DU},\mathrm{LS})}_{\text{up-down-to-down-up lower-saddle}}. \tag{1.100}$$

The equilibrium of $(x_{j_1}^*,x_{j_2}^*)=(a_{j_1j_11},a_{j_2j_21})$ is a (UD:DU,LS)-up-down-to-down-up lower-saddle.

- For $a_{j_1j_10}(a_{j_1j_2s_1}-a_{j_1j_2s_2})<0$ and $a_{j_2j_20}<0$,

$$(a_{j_1j_11},a_{j_2j_21})=\underbrace{(\mathrm{DU}:\mathrm{UD},\mathrm{LS})}_{\text{down-up-to-up-down lower-saddle}}. \tag{1.101}$$

The equilibrium of $(x_{j_1}^*,x_{j_2}^*)=(a_{j_1j_11},a_{j_2j_21})$ is a (DU:UD,LS)-down-up-to-up-down lower-saddle.

(ix) For $\Delta_{j_1j_2}>0$ and $\Delta_{j_2j_2}>0$, the standard form is

$$\begin{aligned}\dot{x}_{j_2}&=a_{j_2j_20}(x_{j_2}-a_{j_2j_21})(x_{j_2}-a_{j_2j_22}),\\ \dot{x}_{j_1}&=a_{j_1j_10}(x_{j_1}-a_{j_1j_11})(x_{j_2}-a_{j_1j_21})(x_{j_2}-a_{j_1j_22})\end{aligned}$$

where

$$\begin{aligned}b_{j_1j_21},b_{j_1j_22}&=-\frac{1}{2}(B_{j_1j_2}\pm\sqrt{\Delta_{j_1j_2}}),\\ \{a_{j_1j_21},a_{j_1j_22}\}&=\mathrm{sort}\{b_{j_1j_21},b_{j_1j_22}\},a_{j_1j_21}<a_{j_1j_22};\\ b_{j_2j_21},b_{j_2j_22}&=-\frac{1}{2}(B_{j_2j_2}\pm\sqrt{\Delta_{j_2j_2}}),\\ \{a_{j_2j_21},a_{j_2j_22}\}&=\mathrm{sort}\{b_{j_2j_21},b_{j_2j_22}\},a_{j_2j_21}<a_{j_2j_22}.\end{aligned}$$

(ix$_1$) For $a_{j_1j_2s_1}\neq a_{j_2j_21},a_{j_2j_22}$ ($s_1=1,2$), in phase space,

$$\frac{dx_{j_1}}{dx_{j_2}}=\frac{a_{j_1j_10}}{a_{j_2j_20}}\frac{(x_{j_1}-a_{j_1j_11})(x_{j_2}-a_{j_1j_21})(x_{j_2}-a_{j_1j_22})}{(x_{j_2}-a_{j_2j_21})(x_{j_2}-a_{j_2j_22})},$$

and the deformation of the foregoing equation is

$$\frac{1}{(x_{j_1}-a_{j_1j_11})}dx_{j_1}=\frac{a_{j_1j_10}}{a_{j_2j_20}}\left[1+\sum\nolimits_{l_1=1}^{2}\frac{\prod_{s_1=1}^{2}(a_{j_2j_2l_1}-a_{j_1j_2s_1})}{(a_{j_2j_2l_1}-a_{j_2j_2l_2})}\frac{1}{(x_{j_2}-a_{j_2j_2l_1})}\right]dx_{j_2}.$$

With an initial condition of $(x_{j_1 0}, x_{j_2 0})$ at $t = t_0$, the integration of the above equation gives the first integral manifold as

$$\ln \frac{|x_{j_1} - a_{j_1 j_1 1}|}{|x_{j_1 0} - a_{j_1 j_1 1}|} = \frac{a_{j_1 j_1 0}}{a_{j_2 j_2 0}} \Big\{ [(x_{j_2} - a_{j_2 j_2 1}) - (x_{j_2 0} - a_{j_2 j_2 1}) \\ + \textstyle\sum_{l_1=1}^{2} \frac{\prod_{s_1=1}^{2} (a_{j_2 j_2 l_1} - a_{j_1 j_2 s_1})}{(a_{j_2 j_2 l_1} - a_{j_2 j_2 l_2})} \ln \frac{|x_{j_2} - a_{j_2 j_2 l_1}|}{|x_{j_2 0} - a_{j_2 j_2 l_1}|} \Big\}.$$

(ix_{1a}) At $(x_{j_1}^*, x_{j_2}^*) = (a_{j_1 j_1 1}, a_{j_2 j_2 s_1})$ $(s_1, s_2 \in \{1,2\}, s_1 \neq s_2)$, the variational equations are

$$\Delta \dot{x}_{j_2} = a_{j_2 j_2 0} (a_{j_2 j_2 s_1} - a_{j_2 j_2 s_2}) \Delta x_{j_2},$$
$$\Delta \dot{x}_{j_1} = a_{j_1 j_1 0} \textstyle\prod_{l_1=1}^{2} (a_{j_2 j_2 s_1} - a_{j_1 j_2 l_1}) \Delta x_{j_1}.$$

The flows in the x_{j_1}-direction are source and sink for $a_{j_1 j_1 0} \prod_{l_1=1}^{2} (a_{j_2 j_2 s_1} - a_{j_1 j_2 l_1}) > 0$ and $a_{j_1 j_1 0} \prod_{l_1=1}^{2} (a_{j_2 j_2 s_1} - a_{j_1 j_2 l_1}) < 0$, respectively. The flows are source and sink for $a_{j_2 j_2 0}(a_{j_2 j_2 s_1} - a_{j_2 j_2 s_2}) > 0$ and $a_{j_2 j_2 0}(a_{j_2 j_2 s_1} - a_{j_2 j_2 s_2}) < 0$ in the x_{j_2}-direction, respectively.

Therefore, the equilibrium of $(x_{j_1}^*, x_{j_2}^*) = (a_{j_1 j_1 1}, a_{j_2 j_2 s_1})$ $(s_1, s_2 \in \{1,2\}, s_1 \neq s_2)$ has the following properties as in Eqs. (1.105)–(1.108).

- For $a_{j_1 j_1 0} \prod_{l_1=1}^{2} (a_{j_2 j_2 s_1} - a_{j_1 j_2 l_1}) > 0$ and $a_{j_2 j_2 0}(a_{j_2 j_2 s_1} - a_{j_2 j_2 s_2}) > 0$,

$$(a_{j_1 j_1 1}, a_{j_2 j_2 s_1}) = \underbrace{(\mathrm{SO}, \mathrm{SO})}_{\text{source}}.$$

  The equilibrium of $(x_{j_1}^*, x_{j_2}^*) = (a_{j_1 j_1 1}, a_{j_2 j_2 s_1})$ is an (SO,SO)-source.
- For $a_{j_1 j_1 0} \prod_{l_1=1}^{2} (a_{j_2 j_2 s_1} - a_{j_1 j_2 l_1}) < 0$ and $a_{j_2 j_2 0}(a_{j_2 j_2 s_1} - a_{j_2 j_2 s_2}) > 0$,

$$(a_{j_1 j_1 1}, a_{j_2 j_2 s_1}) = \underbrace{(\mathrm{SI}, \mathrm{SO})}_{\text{saddle}}.$$

  The equilibrium of $(x_{j_1}^*, x_{j_2}^*) = (a_{j_1 j_1 1}, a_{j_2 j_2 s_1})$ is an (SI,SO)-saddle.
- For $a_{j_1 j_1 0} \prod_{l_1=1}^{2} (a_{j_2 j_2 s_1} - a_{j_1 j_2 l_1}) > 0$ and $a_{j_2 j_2 0}(a_{j_2 j_2 s_1} - a_{j_2 j_2 s_2}) < 0$,

$$(a_{j_1 j_1 1}, a_{j_2 j_2 s_1}) = \underbrace{(\mathrm{SO}, \mathrm{SI})}_{\text{saddle}}.$$

  The equilibrium of $(x_{j_1}^*, x_{j_2}^*) = (a_{j_1 j_1 1}, a_{j_2 j_2 s_1})$ is an (SO,SI)-saddle.
- For $a_{j_1 j_1 0} \prod_{l_1=1}^{2} (a_{j_2 j_2 s_1} - a_{j_1 j_2 l_1}) < 0$ and $a_{j_2 j_2 0}(a_{j_2 j_2 s_1} - a_{j_2 j_2 s_2}) < 0$,

$$(a_{j_1j_11}, a_{j_2j_2s_1}) = \underbrace{(\mathrm{SI}, \mathrm{SI})}_{\text{sink}}.$$

The equilibrium of $(x^*_{j_1}, x^*_{j_2}) = (a_{j_1j_11}, a_{j_2j_2s_1})$ is an (SI,SI)-sink.

(ix$_{1b}$) For $x^*_{j_2} = a_{j_1j_2l_1} \neq a_{j_2j_21}, a_{j_2j_22}$ and $\bar{x}_{j_1} \neq a_{j_1j_11}$ (s_1, $s_2 \in \{1, 2\}$, $s_1 \neq s_2$), in phase space,

$$\frac{dx_{j_1}}{dx_{j_2}}\bigg|_{x^*_{j_2} = a_{j_1j_2l_1}} = \frac{a_{j_1j_10}}{a_{j_2j_20}} \frac{(\bar{x}_{j_1} - a_{j_1j_11})(x_{j_2} - a_{j_1j_2l_1})(x_{j_2} - a_{j_1j_2l_2})}{\prod_{s_1=1}^{2}(x_{j_2} - a_{j_2j_2s_1})}\bigg|_{x^*_{j_2} = a_{j_1j_2l_1}} = 0.$$

If

$$\frac{d^2x_{j_1}}{dx^2_{j_2}}\bigg|_{x^*_{j_2} = a_{j_1j_2l_1}} = \frac{a_{j_1j_10}}{a_{j_2j_20}} \frac{(\bar{x}_{j_1} - a_{j_1j_11})(a_{j_1j_2l_1} - a_{j_1j_2l_2})}{\prod_{s_1=1}^{2}(a_{j_1j_2l_1} - a_{j_2j_2s_1})} > 0,$$

there is an up-parabola flow at $x^*_{j_2} = a_{j_1j_2l_1} \neq a_{j_2j_21}, a_{j_2j_22}$ with $\bar{x}_{j_1} \neq a_{j_1j_11}$. If

$$\frac{d^2x_{j_1}}{dx^2_{j_2}}\bigg|_{x^*_{j_2} = a_{j_1j_2l_1}} = \frac{a_{j_1j_10}}{a_{j_2j_20}} \frac{(\bar{x}_{j_1} - a_{j_1j_11})(a_{j_1j_2l_1} - a_{j_1j_2l_2})}{\prod_{s_1=1}^{2}(a_{j_1j_2l_1} - a_{j_2j_2s_1})} < 0,$$

there is a down-parabola flow at $x^*_{j_2} = a_{j_1j_2l_1}$ with $\bar{x}_{j_1} \neq a_{j_1j_11}$. Let

$$\dot{x}_{j_2} = a_{j_2j_20}\prod_{s_1=1}^{2}(a_{j_1j_2l_1} - a_{j_2j_2s_1}) \neq 0.$$

The inflection flows at $x^*_{j_2} = a_{j_1j_2l_1} \neq a_{j_2j_21}$ with $\bar{x}_{j_1} \neq a_{j_1j_11}$ are positive and negative for $a_{j_2j_20} > 0$ and $a_{j_2j_20} < 0$, respectively.

- For $a_{j_1j_10}(\bar{x}_{j_1} - a_{j_1j_11})(a_{j_1j_2l_1} - a_{j_1j_2l_2}) > 0$ and $a_{j_2j_20}\prod_{s_1=1}^{2}(a_{j_1j_2l_1} - a_{j_2j_2s_1}) > 0$,

$$(a_{j_1j_2l_1}, \dot{x}_{j_2}) = \underbrace{(\mathrm{UP}, \mathrm{pF})}_{\text{up-parabola flow }(+)}.$$

- For $a_{j_1j_10}(\bar{x}_{j_1} - a_{j_1j_11})(a_{j_1j_2l_1} - a_{j_1j_2l_2}) < 0$ and $a_{j_2j_20}\prod_{s_1=1}^{2}(a_{j_1j_2l_1} - a_{j_2j_2s_1}) > 0$,

$$(a_{j_1j_2l_1}, \dot{x}_{j_2}) = \underbrace{(\mathrm{DP}, \mathrm{pF})}_{\text{down-parabola flow }(+)}.$$

- For $a_{j_1j_10}(\overline{x}_{j_1}-a_{j_1j_11})(a_{j_1j_2l_1}-a_{j_1j_2l_2})>0$ and $a_{j_2j_20}\prod_{s_1=1}^{2}(a_{j_1j_2l_1}-a_{j_2j_2s_1})<0$,

$$(a_{j_1j_2l_1},\dot{x}_{j_2})=\underbrace{(\text{DP},\text{nF})}_{\text{down-parabola flow }(-)}.$$

- For $a_{j_1j_10}(\overline{x}_{j_1}-a_{j_1j_11})(a_{j_1j_2l_1}-a_{j_1j_2l_2})<0$ and $a_{j_2j_20}\prod_{s_1=1}^{2}(a_{j_1j_2l_1}-a_{j_2j_2s_1})<0$,

$$(a_{j_1j_2l_1},\dot{x}_{j_2})=\underbrace{(\text{UP},\text{nF})}_{\text{up-parabola flow }(-)}.$$

From the above analysis, for $\overline{x}_{j_1}=a_{j_1j_11}$, therefore, the equilibrium of $(x_{j_1}^*,x_{j_2}^*)=(a_{j_1j_11},a_{j_1j_2l_1})$ $(l_1,\ l_2\in\{1,2\},\ l_1\neq l_2)$ with $a_{j_1j_2l_1}\neq a_{j_2j_21},a_{j_2j_22}$ has the following properties as in Eqs. (1.109)–(1.112).

- For $a_{j_1j_10}(a_{j_1j_2l_1}-a_{j_1j_2l_2})>0$ and $a_{j_2j_20}\prod_{s_1=1}^{2}(a_{j_1j_2l_1}-a_{j_2j_2s_1})>0$,

$$(a_{j_1j_11},a_{j_1j_2l_1})=\underbrace{(\text{DP}:\text{UP},\text{pF})}_{\text{hyperbolic flow }(+)}.$$

  The equilibrium of $(x_{j_1}^*,x_{j_2}^*)=(a_{j_1j_11},a_{j_1j_2l_1})$ is a (DP:UP,pF)-positive hyperbolic flow.

- For $a_{j_1j_10}(a_{j_1j_2l_1}-a_{j_1j_2l_2})<0$ and $a_{j_2j_20}\prod_{s_1=1}^{2}(a_{j_1j_2l_1}-a_{j_2j_2s_1})>0$,

$$(a_{j_1j_11},a_{j_1j_2l_1})=\underbrace{(\text{UP}:\text{DP},\text{pF})}_{\text{hyperbolic-secant flow }(+)}.$$

  The equilibrium of $(x_{j_1}^*,x_{j_2}^*)=(a_{j_1j_11},a_{j_1j_2l_1})$ is a (UP:DP,pF)-positive hyperbolic-secant flow.

- For $a_{j_1j_10}(a_{j_1j_2l_1}-a_{j_1j_2l_2})>0$ and $a_{j_2j_20}\prod_{s_1=1}^{2}(a_{j_1j_2l_1}-a_{j_2j_2s_1})<0$,

$$(a_{j_1j_11},a_{j_1j_2l_1})=\underbrace{(\text{UP}:\text{DP},\text{nF})}_{\text{hyperbolic-secant flow }(-)}.$$

  The equilibrium of $(x_{j_1}^*,x_{j_2}^*)=(a_{j_1j_11},a_{j_1j_2l_1})$ is a (UP:DP,nF)-negative hyperbolic-secant flow.

- For $a_{j_1j_10}(a_{j_1j_2l_1}-a_{j_1j_2l_2})<0$ and $a_{j_2j_20}\prod_{s_1=1}^{2}(a_{j_1j_2l_1}-a_{j_2j_2s_1})<0$,

$$(a_{j_1j_11},a_{j_1j_2l_1})=\underbrace{(\text{DP}:\text{UP},\text{nF})}_{\text{hyperbolic flow }(-)}.$$

The equilibrium of $(x_{j_1}^*, x_{j_2}^*) = (a_{j_1j_11}, a_{j_1j_2l_1})$ is a (DP:UP,nF)-negative hyperbolic flow.

(ix$_2$) For $a_{j_2j_2s_1} = a_{j_1j_2l_1}$ and $a_{j_2j_2s_2} \neq a_{j_1j_2l_2}$ (s_1, s_2, l_1, $l_2 \in \{1,2\}$; $s_1 \neq s_2$, $l_1 \neq l_2$), in phase space,

$$\begin{aligned}\frac{dx_{j_1}}{dx_{j_2}} &= \frac{a_{j_1j_10}}{a_{j_2j_20}} \frac{(x_{j_1} - a_{j_1j_11})(x_{j_2} - a_{j_1j_2l_1})(x_{j_2} - a_{j_1j_2l_2})}{(x_{j_2} - a_{j_2j_2s_1})(x_{j_2} - a_{j_2j_2s_2})}\\ &= \frac{a_{j_1j_10}}{a_{j_2j_20}} \frac{(x_{j_1} - a_{j_1j_11})(x_{j_2} - a_{j_1j_2l_2})}{(x_{j_2} - a_{j_2j_2s_2})},\end{aligned}$$

and the deformation of the foregoing equation is

$$\frac{dx_{j_1}}{(x_{j_1} - a_{j_1j_11})} = \frac{a_{j_1j_10}}{a_{j_2j_20}} \left[1 + \frac{(a_{j_2j_2s_2} - a_{j_1j_2l_2})}{(x_{j_2} - a_{j_2j_2s_2})}\right] dx_{j_2}.$$

With an initial condition of (x_{j_10}, x_{j_20}) at $t = t_0$, the integration of the above equation gives the first integral manifold as

$$\begin{aligned}\ln \frac{|x_{j_1} - a_{j_1j_11}|}{|x_{j_10} - a_{j_1j_11}|} &= \frac{a_{j_1j_10}}{a_{j_2j_20}} \Big\{ \left[(x_{j_2} - a_{j_2j_2s_2}) - (x_{j_20} - a_{j_2j_2s_2})\right]\\ &\quad + (a_{j_2j_2s_2} - a_{j_1j_2l_2}) \ln \frac{|x_{j_2} - a_{j_2j_2s_2}|}{|x_{j_20} - a_{j_2j_2s_2}|} \Big\}.\end{aligned}$$

(ix$_{2a}$) For $x_{j_2}^* = a_{j_1j_2l_1} = a_{j_2j_2s_1}$ with $\bar{x}_{j_1} \neq a_{j_1j_11}$, if

$$\left.\frac{dx_{j_1}}{dx_{j_2}}\right|_{x_{j_2}^* = a_{j_1j_2l_1}} = \frac{a_{j_1j_10}}{a_{j_2j_20}} \frac{(\bar{x}_{j_1} - a_{j_1j_11})(a_{j_1j_2l_1} - a_{j_1j_2l_2})}{(a_{j_2j_2s_1} - a_{j_2j_2s_2})} > 0,$$

there is an increasing-inflection flow, and if

$$\left.\frac{dx_{j_1}}{dx_{j_2}}\right|_{x_{j_2}^* = a_{j_1j_2l_1}} = \frac{a_{j_1j_10}}{a_{j_2j_20}} \frac{(\bar{x}_{j_1} - a_{j_1j_11})(a_{j_1j_2l_1} - a_{j_1j_2l_2})}{(a_{j_2j_2s_1} - a_{j_2j_2s_2})} < 0,$$

there is a decreasing-inflection flow. The variational equation is

$$\Delta \dot{x}_{j_2} = a_{j_2j_20}(a_{j_2j_2s_1} - a_{j_2j_2s_2}) \Delta x_{j_2}.$$

For $a_{j_2j_20}(a_{j_2j_2s_1} - a_{j_2j_2s_2}) > 0$ and $a_{j_2j_20}(a_{j_2j_2s_1} - a_{j_2j_2s_2}) < 0$, the inflection flows are source and sink, respectively.

Therefore, the infinite-equilibrium of $x_{j_2}^* = a_{j_1j_2l_1} = a_{j_2j_2s_1}$ with $\bar{x}_{j_1} \neq a_{j_1j_11}$ has the following properties as in Eqs. (1.114)–(1.117).

- For $a_{j_1j_10}(\bar{x}_{j_1} - a_{j_1j_11})(a_{j_1j_2l_1} - a_{j_1j_2l_2}) > 0$ and $a_{j_2j_20}(a_{j_2j_2s_1} - a_{j_2j_2s_2}) > 0$,

$$(\bar{x}_{j_1}, a_{j_2j_2s_1}) = \underbrace{(\text{II}, \text{SO})}_{\text{increasing-inflection source}}.$$

The infinite-equilibrium of $x^*_{j_2} = a_{j_1j_2l_1} = a_{j_2j_2s_1}$ is an (II,SO)-increasing-inflection source.

- For $a_{j_1j_10}(\bar{x}_{j_1} - a_{j_1j_11})(a_{j_1j_2l_1} - a_{j_1j_2l_2}) < 0$ and $a_{j_2j_20}(a_{j_2j_2s_1} - a_{j_2j_2s_2}) > 0$,

$$(\bar{x}_{j_1}, a_{j_2j_2s_1}) = \underbrace{(\text{DI}, \text{SO})}_{\text{decreasing-inflection source}}.$$

The infinite-equilibrium of $x^*_{j_2} = a_{j_1j_2l_1} = a_{j_2j_2s_1}$ is a (DI,SO)-decreasing-inflection source.

- For $a_{j_1j_10}(\bar{x}_{j_1} - a_{j_1j_11})(a_{j_1j_2l_1} - a_{j_1j_2l_2}) > 0$ and $a_{j_2j_20}(a_{j_2j_2s_1} - a_{j_2j_2s_2}) < 0$,

$$(\bar{x}_{j_1}, a_{j_2j_2s_1}) = \underbrace{(\text{DI}, \text{SI})}_{\text{decreasing-inflection sink}}.$$

The infinite-equilibrium of $x^*_{j_2} = a_{j_1j_2l_1} = a_{j_2j_2s_1}$ is a (DI,SI)-decreasing-inflection sink.

- For $a_{j_1j_10}(\bar{x}_{j_1} - a_{j_1j_11})(a_{j_1j_2l_1} - a_{j_1j_2l_2}) < 0$ and $a_{j_2j_20}(a_{j_2j_2s_1} - a_{j_2j_2s_2}) < 0$,

$$(\bar{x}_{j_1}, a_{j_2j_2s_1}) = \underbrace{(\text{II}, \text{SI})}_{\text{increasing-inflection sink}}.$$

The infinite-equilibrium of $x^*_{j_2} = a_{j_1j_2l_1} = a_{j_2j_2s_1}$ is an (II,SI)-increasing-inflection sink.

($\text{ix}_{2\text{b}}$) From case ($\text{ix}_{2\text{a}}$), for $\bar{x}_{j_1} = a_{j_1j_11}$, therefore, the equilibrium of $(x^*_{j_1}, x^*_{j_2}) = (a_{j_1j_11}, a_{j_2j_2s_1})$ with $a_{j_2j_2s_1} = a_{j_1j_2l_1}$ has the following properties as in Eqs. (1.118)–(1.121).

- For $a_{j_1j_10}(a_{j_1j_2l_1} - a_{j_1j_2l_2}) > 0$ and $a_{j_2j_20}(a_{j_2j_2s_1} - a_{j_2j_2s_2}) > 0$,

$$(a_{j_1j_11}, a_{j_2j_2s_1}) = \underbrace{(\text{DI} : \text{II}, \text{SO})}_{\text{inflection source}}.$$

The equilibrium of $(x^*_{j_1}, x^*_{j_2}) = (a_{j_1j_11}, a_{j_2j_2s_1})$ is a (DI:II,SO)-inflection source.

- For $a_{j_1j_10}(a_{j_1j_2l_1} - a_{j_1j_2l_2}) < 0$ and $a_{j_2j_20}(a_{j_2j_2s_1} - a_{j_2j_2s_2}) > 0$,

$$(a_{j_1j_11}, a_{j_2j_2s_1}) = \underbrace{(\text{II} : \text{DI}, \text{SO})}_{\text{inflection source}}.$$

The equilibrium of $(x^*_{j_1}, x^*_{j_2}) = (a_{j_1j_11}, a_{j_2j_2s_1})$ is an (II:DI,SO)-inflection source.

- For $a_{j_1j_1 0}(a_{j_1j_2l_1}-a_{j_1j_2l_2})>0$ and $a_{j_2j_2 0}(a_{j_2j_2s_1}-a_{j_2j_2s_2})<0$,

$$(a_{j_1j_1 1},a_{j_2j_2s_1})=\underbrace{(\text{II}:\text{DI},\text{SI})}_{\text{inflection sink}}.$$

The equilibrium of $(x_{j_1}^*,x_{j_2}^*)=(a_{j_1j_1 1},a_{j_2j_2s_1})$ is an (II:DI,SI)-inflection sink.

- For $a_{j_1j_1 0}(a_{j_1j_2l_1}-a_{j_1j_2l_2})<0$ and $a_{j_2j_2 0}(a_{j_2j_2s_1}-a_{j_2j_2s_2})<0$,

$$(a_{j_1j_1 1},a_{j_2j_2s_1})=\underbrace{(\text{DI}:\text{II},\text{SI})}_{\text{inflection sink}}.$$

The equilibrium of $(x_{j_1}^*,x_{j_2}^*)=(a_{j_1j_1 1},a_{j_2j_2s_1})$ is a (DI:II,SI)-inflection sink.

(ix$_{2c}$) At $(x_{j_1}^*,x_{j_2}^*)=(a_{j_1j_1 1},a_{j_2j_2s_2})$ with $a_{j_2j_2s_1}=a_{j_1j_2l_1}$, the variational equations are

$$\begin{aligned}\Delta\dot{x}_{j_2}&=a_{j_2j_2 0}(a_{j_2j_2s_2}-a_{j_2j_2s_1})\Delta x_{j_2},\\ \Delta\dot{x}_{j_1}&=a_{j_1j_1 0}\textstyle\prod_{l_1=1}^{2}(a_{j_2j_2s_2}-a_{j_1j_2l_1})\Delta x_{j_1}.\end{aligned}$$

The flows in the x_{j_1}-direction are source and sink for $a_{j_1j_1 0}\prod_{l_1=1}^{2}(a_{j_2j_2s_2}-a_{j_1j_2l_1})>0$ and $a_{j_1j_1 0}\prod_{l_1=1}^{2}(a_{j_2j_2s_2}-a_{j_1j_2l_1})<0$, respectively. The flows are source and sink for $a_{j_2j_2 0}(a_{j_2j_2s_2}-a_{j_2j_2s_1})>0$ and $a_{j_2j_2 0}(a_{j_2j_2s_2}-a_{j_2j_2s_1})<0$ in the x_{j_2}-direction, respectively.

Therefore, the equilibrium of $(x_{j_1}^*,x_{j_2}^*)=(a_{j_1j_1 1},a_{j_2j_2s_2})$ with $a_{j_2j_2s_1}=a_{j_1j_2l_1}$ has the following properties as in Eqs. (1.122)–(1.125).

- For $a_{j_1j_1 0}\prod_{l_1=1}^{2}(a_{j_2j_2s_2}-a_{j_1j_2l_1})>0$ and $a_{j_2j_2 0}(a_{j_2j_2s_2}-a_{j_2j_2s_1})>0$,

$$(a_{j_1j_1 1},a_{j_2j_2s_2})=\underbrace{(\text{SO},\text{SO})}_{\text{source}}.$$

The equilibrium of $(x_{j_1}^*,x_{j_2}^*)=(a_{j_1j_1 1},a_{j_2j_2s_2})$ is an (SO,SO)-source.

- For $a_{j_1j_1 0}\prod_{l_1=1}^{2}(a_{j_2j_2s_2}-a_{j_1j_2l_1})<0$ and $a_{j_2j_2 0}(a_{j_2j_2s_2}-a_{j_2j_2s_1})>0$,

$$(a_{j_1j_1 1},a_{j_2j_2s_2})=\underbrace{(\text{SI},\text{SO})}_{\text{saddle}}.$$

The equilibrium of $(x_{j_1}^*,x_{j_2}^*)=(a_{j_1j_1 1},a_{j_2j_2s_2})$ is an (SI,SO)-saddle.

- For $a_{j_1j_1 0}\prod_{l_1=1}^{2}(a_{j_2j_2s_2}-a_{j_1j_2l_1})>0$ and $a_{j_2j_2 0}(a_{j_2j_2s_2}-a_{j_2j_2s_1})<0$,

$$(a_{j_1j_1 1},a_{j_2j_2s_2})=\underbrace{(\text{SO},\text{SI})}_{\text{saddle}}.$$

The equilibrium of $(x_{j_1}^*,x_{j_2}^*)=(a_{j_1j_1 1},a_{j_2j_2s_2})$ is an (SO,SI)-saddle.

- For $a_{j_1j_1 0}\prod_{l_1=1}^{2}(a_{j_2j_2s_2}-a_{j_1j_2l_1})<0$ and $a_{j_2j_2 0}(a_{j_2j_2s_2}-a_{j_2j_2s_1})<0$,

$$(a_{j_1j_1 1},a_{j_2j_2s_2})=\underbrace{(\mathrm{SI},\mathrm{SI})}_{\text{sink}}.$$

The equilibrium of $(x_{j_1}^*,x_{j_2}^*)=(a_{j_1j_1 1},a_{j_2j_2s_2})$ is an (SI,SI)-sink.

(ix$_{2d}$) For $x_{j_2}^*=a_{j_1j_2l_2}\neq a_{j_2j_2 1},a_{j_2j_2 2}$ and $\bar{x}_{j_1}\neq a_{j_1j_1 1}$ (s_1, $s_2\in\{1,2\}$, $s_1\neq s_2$), in phase space,

$$\frac{dx_{j_1}}{dx_{j_2}}\bigg|_{x_{j_2}^*=a_{j_1j_2l_2}}=\frac{a_{j_1j_1 0}}{a_{j_2j_2 0}}\frac{(\bar{x}_{j_1}-a_{j_1j_1 1})(x_{j_2}-a_{j_1j_2l_1})(x_{j_2}-a_{j_1j_2l_2})}{(x_{j_2}-a_{j_2j_2 1})(x_{j_2}-a_{j_2j_2 2})}\bigg|_{x_{j_2}^*=a_{j_1j_2l_2}}=0.$$

If

$$\frac{d^2x_{j_1}}{dx_{j_2}^2}\bigg|_{x_{j_2}^*=a_{j_1j_2l_2}}=\frac{a_{j_1j_1 0}}{a_{j_2j_2 0}}\frac{(\bar{x}_{j_1}-a_{j_1j_1 1})(a_{j_1j_2l_2}-a_{j_1j_2l_1})}{(a_{j_1j_2l_2}-a_{j_2j_2 1})(a_{j_1j_2l_2}-a_{j_2j_2 2})}>0,$$

there is an up-parabola flow at $x_{j_2}^*=a_{j_1j_2l_2}\neq a_{j_2j_2 1}$ with $\bar{x}_{j_1}\neq a_{j_1j_1 1}$. If

$$\frac{d^2x_{j_1}}{dx_{j_2}^2}\bigg|_{x_{j_2}^*=a_{j_1j_2l_2}}=\frac{a_{j_1j_1 0}}{a_{j_2j_2 0}}\frac{(\bar{x}_{j_1}-a_{j_1j_1 1})(a_{j_1j_2l_2}-a_{j_1j_2l_1})}{(a_{j_1j_2l_2}-a_{j_2j_2 1})(a_{j_1j_2l_2}-a_{j_2j_2 2})}<0,$$

there is a down-parabola flow at $x_{j_2}^*=a_{j_1j_2l_2}$ with $\bar{x}_{j_1}\neq a_{j_1j_1 1}$. Let

$$\begin{aligned}\dot{x}_{j_2}&=a_{j_2j_2 0}(a_{j_1j_2l_2}-a_{j_2j_2 1})(a_{j_1j_2l_2}-a_{j_2j_2 2})\\&=a_{j_2j_2 0}\prod_{s_1=1}^{2}(a_{j_1j_2l_2}-a_{j_2j_2s_1})\neq 0.\end{aligned}$$

The parabola flows at $x_{j_2}^*=a_{j_1j_2l_2}\neq a_{j_2j_2 1}$ with $\bar{x}_{j_1}\neq a_{j_1j_1 1}$ are positive and negative for $a_{j_2j_2 0}>0$ and $a_{j_2j_2 0}<0$, respectively.

- For $a_{j_1j_1 0}(\bar{x}_{j_1}-a_{j_1j_1 1})(a_{j_1j_2l_2}-a_{j_1j_2l_1})>0$ and $a_{j_2j_2 0}\prod_{l_1=1}^{2}(a_{j_1j_2l_2}-a_{j_2j_2s_1})>0$,

$$(a_{j_1j_2l_2},\dot{x}_{j_2})=\underbrace{(\mathrm{UP},\mathrm{pF})}_{\text{up-parabola flow }(+)}.$$

- For $a_{j_1j_1 0}(\bar{x}_{j_1}-a_{j_1j_1 1})(a_{j_1j_2l_2}-a_{j_1j_2l_1})<0$ and $a_{j_2j_2 0}\prod_{s_1=1}^{2}(a_{j_1j_2l_2}-a_{j_2j_2s_1})>0$,

$$(a_{j_1j_2l_2},\dot{x}_{j_2})=\underbrace{(\mathrm{DP},\mathrm{pF})}_{\text{down-parabola flow }(+)}.$$

- For $a_{j_1j_10}(\overline{x}_{j_1} - a_{j_1j_11})(a_{j_1j_2l_2} - a_{j_1j_2l_1}) > 0$ and $a_{j_2j_20}\prod_{s_1=1}^{2}(a_{j_1j_2l_2} - a_{j_2j_2s_1}) < 0$,

$$(a_{j_1j_2l_2}, \dot{x}_{j_2}) = \underbrace{(\text{DP}, \text{nF})}_{\text{down-parabola flow }(-)}.$$

- For $a_{j_1j_10}(\overline{x}_{j_1} - a_{j_1j_11})(a_{j_1j_2l_2} - a_{j_1j_2l_1}) < 0$ and $a_{j_2j_20}\prod_{s_1=1}^{2}(a_{j_1j_2l_2} - a_{j_2j_2s_1}) < 0$,

$$(a_{j_1j_2l_2}, \dot{x}_{j_2}) = \underbrace{(\text{UP}, \text{nF})}_{\text{up-parabola flow }(-)}.$$

From the above analysis, for $\overline{x}_{j_1} = a_{j_1j_11}$, therefore, the equilibrium of $(x^*_{j_1}, x^*_{j_2}) = (a_{j_1j_11}, a_{j_1j_2l_2})$ ($l_1, l_2 \in \{1,2\}, l_1 \neq l_2$) with $a_{j_1j_2l_1} = a_{j_2j_2s_1}$ has the following properties as in Eqs. (1.126)–(1.129).

- For $a_{j_1j_10}(a_{j_1j_2l_2} - a_{j_1j_2l_1}) > 0$ and $a_{j_2j_20}\prod_{s_1=1}^{2}(a_{j_1j_2l_2} - a_{j_2j_2s_1}) > 0$,

$$(a_{j_1j_11}, a_{j_1j_2l_2}) = \underbrace{(\text{DP} : \text{UP}, \text{pF})}_{\text{hyperbolic flow }(+)}.$$

  The equilibrium of $(x^*_{j_1}, x^*_{j_2}) = (a_{j_1j_11}, a_{j_1j_2l_2})$ is a (DP:UP,pF)-positive hyperbolic flow.

- For $a_{j_1j_10}(a_{j_1j_2l_2} - a_{j_1j_2l_1}) < 0$ and $a_{j_2j_20}\prod_{s_1=1}^{2}(a_{j_1j_2l_2} - a_{j_2j_2s_1}) > 0$,

$$(a_{j_1j_11}, a_{j_1j_2l_2}) = \underbrace{(\text{UP} : \text{DP}, \text{pF})}_{\text{hyperbolic-secant flow }(+)}.$$

  The equilibrium of $(x^*_{j_1}, x^*_{j_2}) = (a_{j_1j_11}, a_{j_1j_2l_2})$ is a (UP:DP,pF)-positive hyperbolic-secant flow.

- For $a_{j_1j_10}(a_{j_1j_2l_2} - a_{j_1j_2l_1}) > 0$ and $a_{j_2j_20}\prod_{s_1=1}^{2}(a_{j_1j_2l_2} - a_{j_2j_2s_1}) < 0$,

$$(a_{j_1j_11}, a_{j_1j_2l_2}) = \underbrace{(\text{UP} : \text{DP}, \text{nF})}_{\text{hyperbolic-secant flow }(-)}.$$

  The equilibrium of $(x^*_{j_1}, x^*_{j_2}) = (a_{j_1j_11}, a_{j_1j_2l_2})$ is a (UP:DP,nF)-negative hyperbolic-secant flow.

- For $a_{j_1j_10}(a_{j_1j_2l_2} - a_{j_1j_2l_1}) < 0$ and $a_{j_2j_20}\prod_{s_1=1}^{2}(a_{j_1j_2l_2} - a_{j_2j_2s_1}) < 0$,

$$(a_{j_1j_1}1, a_{j_1j_2l_2}) = \underbrace{(\text{DP : UP, nF})}_{\text{hyperbolic flow } (-)}.$$

The equilibrium of $(x_{j_1}^*, x_{j_2}^*) = (a_{j_1j_1}1, a_{j_1j_2l_2})$ is a (DP:UP,nF)-negative hyperbolic flow.

(ix$_3$) For $a_{j_2j_2}1 = a_{j_1j_2}2$ and $a_{j_2j_2}2 = a_{j_1j_2}2$, in phase space,

$$\begin{aligned}\frac{dx_{j_1}}{dx_{j_2}} &= \frac{a_{j_1j_1}0}{a_{j_2j_2}0}\frac{(x_{j_1} - a_{j_1j_1}1)(x_{j_2} - a_{j_1j_2}1)(x_{j_2} - a_{j_1j_2}2)}{(x_{j_2} - a_{j_2j_2}1)(x_{j_2} - a_{j_2j_2}2)}\\ &= \frac{a_{j_1j_1}0}{a_{j_2j_2}0}(x_{j_1} - a_{j_1j_1}1),\end{aligned}$$

and the deformation of the foregoing equation is

$$\frac{dx_{j_1}}{(x_{j_1} - a_{j_1j_1}1)} = \frac{a_{j_1j_1}0}{a_{j_2j_2}0}dx_{j_2}.$$

With an initial condition of $(x_{j_1 0}, x_{j_2 0})$ at $t = t_0$, the integration of the above equation gives the first integral manifold ($a_{j_2j_2}1 = a_{j_1j_2}2$ and $a_{j_2j_2}2 = a_{j_1j_2}2$) as

$$\ln\frac{|x_{j_1} - a_{j_1j_1}1|}{|x_{j_1 0} - a_{j_1j_1}1|} = \frac{a_{j_1j_1}0}{a_{j_2j_2}0}(x_{j_2} - x_{j_2 0}).$$

(ix$_{3a}$) At $x_{j_2}^* = a_{j_1j_2s_1} = a_{j_2j_2s_1}$ with $\bar{x}_{j_1} \neq a_{j_1j_1}1$, in phase space, if

$$\left.\frac{dx_{j_1}}{dx_{j_2}}\right|_{x_{j_2}^* = a_{j_1j_2s_1} = a_{j_2j_2s_1}} = \frac{a_{j_1j_1}0}{a_{j_2j_2}0}(\bar{x}_{j_1} - a_{j_1j_1}1) > 0,$$

there is an increasing-inflection flow, and if

$$\left.\frac{dx_{j_1}}{dx_{j_2}}\right|_{x_{j_2}^* = a_{j_1j_2s_1} = a_{j_2j_2s_1}} = \frac{a_{j_1j_1}0}{a_{j_2j_2}0}(\bar{x}_{j_1} - a_{j_1j_1}1) < 0,$$

there is a decreasing-inflection flow. The variational equation is

$$\Delta\dot{x}_{j_2} = a_{j_2j_2}0(a_{j_2j_2s_1} - a_{j_2j_2s_2})\Delta x_{j_2}.$$

The inflection flows at $x_{j_2}^* = a_{j_1j_2s_1} = a_{j_2j_2s_1}$ with $\bar{x}_{j_1} \neq a_{j_1j_1}1$ are source and sink for $a_{j_2j_2}0(a_{j_2j_2s_1} - a_{j_2j_2s_2}) > 0$ and $a_{j_2j_2}0(a_{j_2j_2s_1} - a_{j_2j_2s_2}) < 0$, respectively.

Therefore, the infinite-equilibrium of $x_{j_2}^* = a_{j_1j_2s_1} = a_{j_2j_2s_1}$ ($s_1 = 1, 2$) with $\bar{x}_{j_1} \neq a_{j_1j_1}1$ has the following properties as in Eqs (1.131)–(1.134).

- For $a_{j_1j_10}(\overline{x}_{j_1}-a_{j_1j_11})(a_{j_1j_2s_1}-a_{j_1j_2s_2})>0$ and $a_{j_2j_20}(a_{j_2j_2s_1}-a_{j_2j_2s_2})>0$,

$$(\overline{x}_{j_1},a_{j_2j_2s_1})=\underbrace{(\mathrm{II},\mathrm{SO})}_{\text{increasing-inflection source}}.$$

The infinite-equilibrium of $x_{j_2}^*=a_{j_1j_2s_1}=a_{j_2j_2s_1}$ is an (II,SO)-increasing-inflection source.

- For $a_{j_1j_10}(\overline{x}_{j_1}-a_{j_1j_11})(a_{j_1j_2s_1}-a_{j_1j_2s_2})<0$ and $a_{j_2j_20}(a_{j_2j_2s_1}-a_{j_2j_2s_2})>0$,

$$(\overline{x}_{j_1},a_{j_2j_2s_1})=\underbrace{(\mathrm{DI},\mathrm{SO})}_{\text{decreasing-inflection source}}.$$

The infinite-equilibrium of $x_{j_2}^*=a_{j_1j_2s_1}=a_{j_2j_2s_1}$ is a (DI,SO)-decreasing-inflection source.

- For $a_{j_1j_10}(\overline{x}_{j_1}-a_{j_1j_11})(a_{j_1j_2s_1}-a_{j_1j_2s_2})>0$ and $a_{j_2j_20}(a_{j_2j_2s_1}-a_{j_2j_2s_2})<0$,

$$(\overline{x}_{j_1},a_{j_2j_2s_1})=\underbrace{(\mathrm{DI},\mathrm{SI})}_{\text{decreasing-inflection sink}}.$$

The infinite-equilibrium of $x_{j_2}^*=a_{j_1j_2s_1}=a_{j_2j_2s_1}$ is a (DI,SI)-decreasing-inflection sink.

- For $a_{j_1j_10}(\overline{x}_{j_1}-a_{j_1j_11})(a_{j_1j_2s_1}-a_{j_1j_2s_2})<0$ and $a_{j_2j_20}(a_{j_2j_2s_1}-a_{j_2j_2s_2})<0$,

$$(\overline{x}_{j_1},a_{j_2j_2s_1})=\underbrace{(\mathrm{II},\mathrm{SI})}_{\text{increasing-inflection sink}}.$$

The infinite-equilibrium of $x_{j_2}^*=a_{j_1j_2s_1}=a_{j_2j_2s_1}$ is an (II,SI)-increasing-inflection sink.

($\mathrm{ix}_{3\mathrm{b}}$) The equilibrium of $(x_{j_1}^*,x_{j_2}^*)=(a_{j_1j_11},a_{j_2j_2s_1})$ with $a_{j_2j_2s_1}=a_{j_1j_2s_1}(s_1=1,2)$ has the following properties as in Eqs. (1.135)–(1.138).

- For $a_{j_1j_10}(a_{j_1j_2s_1}-a_{j_1j_2s_2})>0$ and $a_{j_2j_20}(a_{j_2j_2s_1}-a_{j_2j_2s_2})>0$,

$$(a_{j_1j_11},a_{j_2j_2s_1})=\underbrace{(\mathrm{DI}:\mathrm{II},\mathrm{SO})}_{\text{inflection source}}.$$

The equilibrium of $(x_{j_1}^*,x_{j_2}^*)=(a_{j_1j_11},a_{j_1j_2s_1})$ is a (DI:II,SO)-inflection source.

- For $a_{j_1j_10}(a_{j_1j_2s_1}-a_{j_1j_2s_2})<0$ and $a_{j_2j_20}(a_{j_2j_2s_1}-a_{j_2j_2s_2})>0$,

$$(a_{j_1j_1}1, a_{j_1j_2s_1}) = \underbrace{(\text{II} : \text{DI}, \text{SO})}_{\text{inflection source}}.$$

The equilibrium of $(x^*_{j_1}, x^*_{j_2}) = (a_{j_1j_1}1, a_{j_1j_2s_1})$ is an (II:DI,SO)-inflection source.

- For $a_{j_1j_1}0(a_{j_1j_2s_1} - a_{j_1j_2s_2}) > 0$ and $a_{j_2j_2}0(a_{j_2j_2s_1} - a_{j_2j_2s_2}) < 0$,

$$(a_{j_1j_1}1, a_{j_1j_2s_1}) = \underbrace{(\text{II} : \text{DI}, \text{SI})}_{\text{inflection sink}}.$$

The equilibrium of $(x^*_{j_1}, x^*_{j_2}) = (a_{j_1j_1}1, a_{j_1j_2s_1})$ is an (II:DI,SI)-inflection sink.

- For $a_{j_1j_1}0(a_{j_1j_2s_1} - a_{j_1j_2s_2}) < 0$ and $a_{j_2j_2}0(a_{j_2j_2s_1} - a_{j_2j_2s_2}) < 0$,

$$(a_{j_1j_1}1, a_{j_2j_2s_1}) = \underbrace{(\text{DI} : \text{II}, \text{SI})}_{\text{inflection sink}}.$$

The equilibrium of $(x^*_{j_1}, x^*_{j_2}) = (a_{j_1j_1}1, a_{j_1j_2s_1})$ is a (DI:II,SI)-inflection sink.

In the end, this theorem is proved. ■

1.3 Crossing-Linear and Self-Quadratic Product Vector Fields

In a nonlinear cubic nonlinear system, consider the cubic nonlinear vector field to be a crossing-linear and self-quadratic product vector field, and the other vector field is a self-quadratic vector field.

Theorem 1.2 *Consider a two-dimensional, cubic dynamical system with a product cubic vector field and a self-univariate quadratic vector field as*

$$\begin{aligned}\dot{x}_{j_2} &= a_{j_2j_2}0(x^2_{j_2} + B_{j_2j_2}x_{j_2} + C_{j_2j_2}),\\ \dot{x}_{j_1} &= a_{j_1j_1}0(x_{j_2} - a_{j_1j_2}1)(x^2_{j_1} + B_{j_1j_1}x_{j_1} + C_{j_1j_1})\end{aligned} \tag{1.139}$$

with

$$\Delta_{j_1j_1} = B^2_{j_1j_1} - 4C_{j_1j_1}, \Delta_{j_2j_2} = B^2_{j_2j_2} - 4C_{j_2j_2}. \tag{1.140}$$

(i) For $\Delta_{j_1j_1} < 0$ and $\Delta_{j_2j_2} < 0$, the standard form is

$$\begin{aligned}\dot{x}_{j_2} &= a_{j_2j_20}\left[(x_{j_2} - a_{j_2j_2})^2 + b_{j_2j_2}\right],\\ \dot{x}_{j_1} &= a_{j_1j_10}(x_{j_2} - a_{j_1j_21})\left[(x_{j_1} - a_{j_1j_1})^2 + b_{j_1j_1}\right]\end{aligned} \tag{1.141}$$

where

$$\begin{aligned}a_{j_1j_1} &= -\frac{1}{2}B_{j_1j_1}, b_{j_1j_1} = \frac{1}{4}(-\Delta_{j_1j_1}),\\ a_{j_2j_2} &= -\frac{1}{2}B_{j_2j_2}, b_{j_2j_2} = \frac{1}{4}(-\Delta_{j_2j_2}).\end{aligned} \tag{1.142}$$

The first integral manifold is

$$\begin{aligned}&\frac{1}{\sqrt{b_{j_1j_1}}}\left(\arctan\frac{x_{j_1} - a_{j_1j_1}}{\sqrt{b_{j_1j_1}}} - \arctan\frac{x_{j_10} - a_{j_1j_1}}{\sqrt{b_{j_1j_1}}}\right)\\ &= \frac{a_{j_1j_10}}{a_{j_2j_20}}\left[\frac{1}{2}\ln\frac{\left|(x_{j_2} - a_{j_2j_2})^2 + b_{j_2j_2}\right|}{\left|(x_{j_20} - a_{j_2j_2})^2 + b_{j_2j_2}\right|}\right.\\ &\left.+\frac{(a_{j_2j_2} - a_{j_1j_21})}{\sqrt{b_{j_2j_2}}}\left(\arctan\frac{x_{j_2} - a_{j_2j_2}}{\sqrt{b_{j_2j_2}}} - \arctan\frac{x_{j_20} - a_{j_2j_2}}{\sqrt{b_{j_2j_2}}}\right)\right].\end{aligned} \tag{1.143}$$

The equilibrium of $x^*_{j_1} = a_{j_1j_21}$ *has the following properties.*

- *For* $a_{j_1j_10} > 0$ *and* $a_{j_2j_20} > 0$,

$$(a_{j_1j_21}, \dot{x}_{j_2}) = \underbrace{(\text{UP}, \text{pF})}_{\text{up-parabola flow }(+)}. \tag{1.144}$$

  The equilibrium of $x^*_{j_1} = a_{j_1j_21}$ *is a* (UP,pF)*-positive up-parabola flow.*
- *For* $a_{j_1j_10} < 0$ *and* $a_{j_2j_20} > 0$,

$$(a_{j_1j_21}, \dot{x}_{j_2}) = \underbrace{(\text{DP}, \text{pF})}_{\text{down-parabola flow }(+)}. \tag{1.145}$$

  The equilibrium of $x^*_{j_1} = a_{j_1j_11}$ *is a* (DP,pF)*-positive down-parabola flow.*
- *For* $a_{j_1j_10} > 0$ *and* $a_{j_2j_20} < 0$,

$$(a_{j_1j_21}, \dot{x}_{j_2}) = \underbrace{(\text{DP}, \text{nF})}_{\text{down-parabola flow }(-)}. \tag{1.146}$$

  The equilibrium of $x^*_{j_1} = a_{j_1j_21}$ *is a* (DP,nF)*-negative down-parabola flow.*
- *For* $a_{j_1j_10} < 0$ *and* $a_{j_2j_20} < 0$,

$$(a_{j_1j_2 1}, \dot{x}_{j_2}) = \underbrace{(\text{UP}, \text{nF})}_{\text{up-parabola flow }(-)} . \tag{1.147}$$

The equilibrium of $x^*_{j_1} = a_{j_1j_2 1}$ *is a* (UP,nF)-*negative up-parabola flow.*

(ii) For $\Delta_{j_1j_1} < 0$ *and* $\Delta_{j_2j_2} = 0$, *the standard form is*

$$\begin{aligned} \dot{x}_{j_2} &= a_{j_2j_2 0}(x_{j_2} - a_{j_2j_2 1})^2, \\ \dot{x}_{j_1} &= a_{j_1j_1 0}(x_{j_2} - a_{j_1j_2 1})\left[(x_{j_1} - a_{j_1j_1})^2 + b_{j_1j_1}\right] \end{aligned} \tag{1.148}$$

where

$$\begin{aligned} a_{j_1j_1} &= -\frac{1}{2}B_{j_1j_1}, b_{j_1j_1} = \frac{1}{4}(-\Delta_{j_1j_1}), \\ a_{j_2j_2 1} &= -\frac{1}{2}B_{j_2j_2}. \end{aligned} \tag{1.149}$$

(ii₁) The first integral manifold for $a_{j_1j_2 1} \neq a_{j_2j_2 1}$ *is*

$$\begin{aligned} &\frac{1}{\sqrt{b_{j_1j_1}}}(\arctan \frac{x_{j_1} - a_{j_1j_1}}{\sqrt{b_{j_1j_1}}} - \arctan \frac{x_{j_1 0} - a_{j_1j_1}}{\sqrt{b_{j_1j_1}}}) \\ &= \frac{a_{j_1j_1 0}}{a_{j_2j_2 0}}\left[\ln \frac{|x_{j_2} - a_{j_2j_2 1}|}{|x_{j_2 0} - a_{j_2j_2 1}|} - (a_{j_2j_2 1} - a_{j_1j_2 1})(\frac{1}{x_{j_2} - a_{j_2j_2 1}} - \frac{1}{x_{j_2 0} - a_{j_2j_2 1}})\right]. \end{aligned} \tag{1.150}$$

(ii₁ₐ) The equilibrium of $x^*_{j_2} = a_{j_1j_2 1} \neq a_{j_2j_2 1}$ *has the following properties.*

- *For* $a_{j_1j_1 0} > 0$ *and* $a_{j_2j_2 0}(a_{j_1j_2 1} - a_{j_2j_2 1})^2 > 0$,

$$(a_{j_1j_2 1}, \dot{x}_{j_2}) = \underbrace{(\text{UP}, \text{pF})}_{\text{up-parabola flow }(+)} . \tag{1.151}$$

  The equilibrium of $x^*_{j_1} = a_{j_1j_2 1}$ *is a* (UP,pF)-*positive up-parabola flow.*
- *For* $a_{j_1j_1 0} < 0$ *and* $a_{j_2j_2 0}(a_{j_1j_2 1} - a_{j_2j_2 1})^2 > 0$,

$$(a_{j_1j_2 1}, \dot{x}_{j_2}) = \underbrace{(\text{DP}, \text{pF})}_{\text{down-parabola flow }(+)} . \tag{1.152}$$

  The equilibrium of $x^*_{j_1} = a_{j_1j_2 1}$ *is a* (DP,pF)-*positive down-parabola flow.*
- *For* $a_{j_1j_1 0} > 0$ *and* $a_{j_2j_2 0}(a_{j_1j_2 1} - a_{j_2j_2 1})^2 < 0$,

$$(a_{j_1j_2 1}, \dot{x}_{j_2}) = \underbrace{(\text{DP}, \text{nF})}_{\text{down-parabola flow } (-)} . \tag{1.153}$$

The equilibrium of $x^*_{j_1} = a_{j_1j_2 1}$ *is a* (DP,nF)*-negative down-parabola flow.*

- *For* $a_{j_1j_1 0} < 0$ *and* $a_{j_2j_2 0}(a_{j_1j_2 1} - a_{j_2j_2 1})^2 < 0$,

$$(a_{j_1j_2 1}, \dot{x}_{j_2}) = \underbrace{(\text{UP}, \text{nF})}_{\text{up-parabola flow } (-)} . \tag{1.154}$$

The equilibrium of $x^*_{j_1} = a_{j_1j_2 1}$ *is a* (UP,nF)*-negative up-parabola flow.*

(ii_{1b}) *The equilibrium of* $x^*_{j_2} = a_{j_2j_2 1} \neq a_{j_1j_2 1}$ *has the following properties.*

- *For* $a_{j_1j_1 0}(a_{j_2j_2 1} - a_{j_1j_2 1}) > 0$ *and* $a_{j_2j_2 0} > 0$,

$$(\dot{x}_{j_1}, a_{j_2j_2 1}) = \underbrace{(\text{pF}, \text{US})}_{\text{upper-saddle flow } (+)} . \tag{1.155}$$

The equilibrium of $x^*_{j_2} = a_{j_2j_2 1} \neq a_{j_1j_2 1}$ *is a* (pF,US)*-positive upper-saddle flow.*

- *For* $a_{j_1j_1 0}(a_{j_2j_2 1} - a_{j_1j_2 1}) < 0$ *and* $a_{j_2j_2 0} > 0$,

$$(\dot{x}_{j_1}, a_{j_2j_2 1}) = \underbrace{(\text{nF}, \text{US})}_{\text{upper-saddle flow } (-)} . \tag{1.156}$$

The equilibrium of $x^*_{j_2} = a_{j_2j_2 1} \neq a_{j_1j_2 1}$ *is an* (nF,US)*-negative upper-saddle flow.*

- *For* $a_{j_1j_1 0}(a_{j_2j_2 1} - a_{j_1j_2 1}) > 0$ *and* $a_{j_2j_2 0} < 0$,

$$(\dot{x}_{j_1}, a_{j_2j_2 1}) = \underbrace{(\text{pF}, \text{LS})}_{\text{lower-saddle flow } (+)} . \tag{1.157}$$

The equilibrium of $x^*_{j_2} = a_{j_2j_2 1} \neq a_{j_1j_2 1}$ *is a* (pF,LS)*-positive lower-saddle flow.*

- *For* $a_{j_1j_1 0}(a_{j_2j_2 1} - a_{j_1j_2 1}) < 0$ *and* $a_{j_2j_2 0} < 0$,

$$(\dot{x}_{j_1}, a_{j_2j_2 1}) = \underbrace{(\text{nF}, \text{LS})}_{\text{lower-saddle flow } (-)} . \tag{1.158}$$

The equilibrium of $x^*_{j_2} = a_{j_2j_2 1} \neq a_{j_1j_2 1}$ *is an* (nF,LS)*-negative lower-saddle flow.*

– *The upper-saddle and lower-saddle flows are the appearing bifurcations of sink and source flows.*

(ii₂) The first integral manifold for $a_{j_1j_21} = a_{j_2j_21}$ *is*

$$\frac{1}{\sqrt{b_{j_1j_1}}}(\arctan\frac{x_{j_1}-a_{j_1j_1}}{\sqrt{b_{j_1j_1}}}-\arctan\frac{x_{j_10}-a_{j_1j_1}}{\sqrt{b_{j_1j_1}}})=\frac{a_{j_1j_10}}{a_{j_2j_20}}\ln\frac{|x_{j_2}-a_{j_2j_21}|}{|x_{j_20}-a_{j_2j_21}|}. \quad (1.159)$$

The infinite-equilibrium of $x^*_{j_2} = a_{j_1j_21} = a_{j_2j_21}$ *and* $\overline{x}_{j_1} \in (-\infty, \infty)$ *has the following properties.*

- *For* $a_{j_1j_10} > 0$ *and* $a_{j_2j_20} > 0$,

$$(\overline{x}_{j_1}, a_{j_2j_21}) = \underbrace{(\mathrm{DU},\mathrm{US})}_{\text{down-up upper-saddle}}. \quad (1.160)$$

The infinite-equilibrium of $x^*_{j_2} = a_{j_1j_21} = a_{j_2j_21}$ *is a* (DU,US)-*down-up upper-saddle.*

- *For* $a_{j_1j_10} < 0$ *and* $a_{j_2j_20} > 0$,

$$(\overline{x}_{j_1}, a_{j_2j_21}) = \underbrace{(\mathrm{UD},\mathrm{US})}_{\text{up-down upper-saddle}}. \quad (1.161)$$

The infinite-equilibrium of $x^*_{j_2} = a_{j_1j_21} = a_{j_2j_21}$ *is a* (UD,US)-*up-down upper-saddle.*

- *For* $a_{j_1j_10} > 0$ *and* $a_{j_2j_20} < 0$,

$$(\overline{x}_{j_1}, a_{j_2j_21}) = \underbrace{(\mathrm{UD},\mathrm{LS})}_{\text{up-down lower-saddle}}. \quad (1.162)$$

The infinite-equilibrium of $x^*_{j_2} = a_{j_1j_21} = a_{j_2j_21}$ *is a* (UD,LS)-*up-down-lower-saddle.*

- *For* $a_{j_1j_10} < 0$ *and* $a_{j_2j_20} < 0$,

$$(\overline{x}_{j_1} a_{j_2j_21}) = \underbrace{(\mathrm{DU},\mathrm{LS})}_{\text{down-up lower-saddle}}. \quad (1.163)$$

The infinite-equilibrium of $x^*_{j_2} = a_{j_1j_21} = a_{j_2j_21}$ *is a* (DU,LS)-*down-up lower-saddle.*

(iii) For $\Delta_{j_1j_1} < 0$ *and* $\Delta_{j_2j_2} > 0$*, the standard form is*

$$\begin{aligned} \dot{x}_{j_2} &= a_{j_2j_20}(x_{j_2}-a_{j_2j_21})(x_{j_2}-a_{j_2j_22}), \\ \dot{x}_{j_1} &= a_{j_1j_10}\left[(x_{j_1}-a_{j_1j_1})^2+b_{j_1j_1}\right](x_{j_2}-a_{j_1j_21}) \end{aligned} \quad (1.164)$$

where

$$\begin{aligned} & a_{j_1j_1} = -\frac{1}{2}B_{j_1j_1}, b_{j_1j_1} = \frac{1}{4}(-\Delta_{j_1j_1}), \\ & b_{j_2j_2 1}, b_{j_2j_2 2} = -\frac{1}{2}(B_{j_2j_2} \pm \sqrt{\Delta_{j_2j_2}}), \\ & \{a_{j_2j_2 1}, a_{j_2j_2 2}\} = \mathrm{sort}\{b_{j_2j_2 1}, b_{j_2j_2 2}\}, a_{j_2j_2 1} < a_{j_2j_2 2}. \end{aligned} \tag{1.165}$$

(iii_1) The first integral manifold ($a_{j_1j_2 1} \neq a_{j_2j_2s_1}, a_{j_2j_2s_2}$, $s_1, s_2 \in \{1,2\}$, $s_1 \neq s_2$) is

$$\begin{aligned} & \frac{1}{\sqrt{b_{j_1j_1}}}(\arctan\frac{x_{j_1} - a_{j_1j_1}}{\sqrt{b_{j_1j_1}}} - \arctan\frac{x_{j_1 0} - a_{j_1j_1}}{\sqrt{b_{j_1j_1}}}) \\ & = \frac{a_{j_1j_1 0}}{a_{j_2j_2 0}}\sum\nolimits_{s_1=1}^{2}\frac{(a_{j_2j_2s_1} - a_{j_1j_2 1})}{(a_{j_2j_2s_1} - a_{j_2j_2s_2})}\ln\frac{|x_{j_2} - a_{j_2j_2s_1}|}{|x_{j_2 0} - a_{j_2j_2s_1}|}. \end{aligned} \tag{1.166}$$

(iii_{1a}) The equilibrium of $x^*_{j_2} = a_{j_1j_2 1} \neq a_{j_2j_2s_1}, a_{j_2j_2s_2}$ ($s_1, s_2 \in \{1,2\}$, $s_1 \neq s_2$) has the following properties.

- *For $a_{j_1j_1 0} > 0$ and $a_{j_2j_2 0}\prod_{s_1=1}^{2}(a_{j_1j_2 1} - a_{j_2j_2s_1}) > 0$,*

$$(a_{j_1j_2 1}, \dot{x}_{j_2}) = \underbrace{(\mathrm{UP}, \mathrm{pF})}_{\text{up-parabola flow } (+)}. \tag{1.167}$$

  The equilibrium of $x^*_{j_1} = a_{j_1j_2 1}$ is a (UP,pF)-positive up-parabola flow.

- *For $a_{j_1j_1 0} < 0$ and $a_{j_2j_2 0}\prod_{s_1=1}^{2}(a_{j_1j_2 1} - a_{j_2j_2s_1}) > 0$,*

$$(a_{j_1j_2 1}, \dot{x}_{j_2}) = \underbrace{(\mathrm{DP}, \mathrm{pF})}_{\text{down-parabola flow } (+)}. \tag{1.168}$$

  The equilibrium of $x^*_{j_1} = a_{j_1j_2 1}$ is a (DP,pF)-positive down-parabola flow.

- *For $a_{j_1j_1 0} > 0$ and $a_{j_2j_2 0}\prod_{s_1=1}^{2}(a_{j_1j_2 1} - a_{j_2j_2s_1}) < 0$,*

$$(a_{j_1j_2 1}, \dot{x}_{j_2}) = \underbrace{(\mathrm{DP}, \mathrm{nF})}_{\text{down-parabola flow } (-)}. \tag{1.169}$$

  The equilibrium of $x^*_{j_1} = a_{j_1j_2 1}$ is a (DP,nF)-negative down-parabola flow.

- *For $a_{j_1j_1 0} < 0$ and $a_{j_2j_2 0}\prod_{s_1=1}^{2}(a_{j_1j_2 1} - a_{j_2j_2s_1}) < 0$,*

$$(a_{j_1j_2 1}, \dot{x}_{j_2}) = \underbrace{(\mathrm{UP}, \mathrm{nF})}_{\text{up-parabola flow } (-)}. \tag{1.170}$$

  The equilibrium of $x^*_{j_1} = a_{j_1j_2 1}$ is a (UP,nF)-negative up-parabola flow.

(iii_{1b}) The equilibrium of $x_{j_2}^{*} = a_{j_2 j_2 s_1} \neq a_{j_1 j_2 1}$ $(s_1, s_2 \in \{1,2\}, s_1 \neq s_2)$ *has the following properties.*

- *For* $a_{j_1 j_1 0}(a_{j_2 j_2 s_1} - a_{j_1 j_2 1}) > 0$ *and* $a_{j_2 j_2 0}(a_{j_2 j_2 s_1} - a_{j_2 j_2 s_2}) > 0$,

$$(\dot{x}_{j_1}, a_{j_2 j_2 s_1}) = \underbrace{(\mathrm{pF}, \mathrm{SO})}_{\text{source flow } (+)}. \tag{1.171}$$

The equilibrium of $x_{j_2}^{*} = a_{j_2 j_2 s_1} \neq a_{j_1 j_2 1}$ *is a* (pF,SO)*-positive source flow.*

- *For* $a_{j_1 j_1 0}(a_{j_2 j_2 s_1} - a_{j_1 j_2 1}) < 0$ *and* $a_{j_2 j_2 0}(a_{j_2 j_2 s_1} - a_{j_2 j_2 s_2}) > 0$,

$$(\dot{x}_{j_1}, a_{j_2 j_2 s_1}) = \underbrace{(\mathrm{nF}, \mathrm{SO})}_{\text{source flow } (-)}. \tag{1.172}$$

The equilibrium of $x_{j_2}^{*} = a_{j_2 j_2 s_1} \neq a_{j_1 j_2 1}$ *is an* (nF,SO)*-negative source flow.*

- *For* $a_{j_1 j_1 0}(a_{j_2 j_2 s_1} - a_{j_1 j_2 1}) > 0$ *and* $a_{j_2 j_2 0}(a_{j_2 j_2 s_1} - a_{j_2 j_2 s_2}) < 0$,

$$(\dot{x}_{j_1}, a_{j_2 j_2 s_1}) = \underbrace{(\mathrm{pF}, \mathrm{SI})}_{\text{sink flow } (+)}. \tag{1.173}$$

The equilibrium of $x_{j_2}^{*} = a_{j_2 j_2 s_1} \neq a_{j_1 j_2 1}$ *is a* (pF,SI)*-positive sink flow.*

- *For* $a_{j_1 j_1 0}(a_{j_2 j_2 s_1} - a_{j_1 j_2 1}) < 0$ *and* $a_{j_2 j_2 0}(a_{j_2 j_2 s_1} - a_{j_2 j_2 s_2}) < 0$,

$$(\dot{x}_{j_1}, a_{j_2 j_2 s_1}) = \underbrace{(\mathrm{nF}, \mathrm{SI})}_{\text{sink flow } (-)}. \tag{1.174}$$

The equilibrium of $x_{j_2}^{*} = a_{j_2 j_2 s_1} \neq a_{j_1 j_2 1}$ *is an* (nF,SI)*-negative sink flow.*

(iii_2) The first integral manifold for $a_{j_1 j_2 1} = a_{j_2 j_2 s_1}$ $(s_1, s_2 \in \{1,2\}, s_1 \neq s_2)$ *is*

$$\frac{1}{\sqrt{b_{j_1 j_1}}}\left(\arctan\frac{x_{j_1} - a_{j_1 j_1}}{\sqrt{b_{j_1 j_1}}} - \arctan\frac{x_{j_1 0} - a_{j_1 j_1}}{\sqrt{b_{j_1 j_1}}}\right) = \frac{a_{j_1 j_1 0}}{a_{j_2 j_2 0}} \ln\frac{|x_{j_2} - a_{j_2 j_2 s_2}|}{|x_{j_2 0} - a_{j_2 j_2 s_2}|}. \tag{1.175}$$

The infinite-equilibrium of $x_{j_2}^{*} = a_{j_1 j_2 1} = a_{j_2 j_2 s_1}$ *with* $\overline{x}_{j_1} \in (-\infty, \infty)$ *has the following properties.*

- *For* $a_{j_1 j_1 0} > 0$ *and* $a_{j_2 j_2 0}(a_{j_2 j_2 s_1} - a_{j_2 j_2 s_2}) > 0$,

$$(\overline{x}_{j_1}, a_{j_2 j_2 s_1}) = \underbrace{(\mathrm{II}, \mathrm{SO})}_{\text{increasing-inflection source}}. \tag{1.176}$$

The infinite-equilibrium of $x_{j_2}^{*} = a_{j_1 j_2 1} = a_{j_2 j_2 s_1}$ *is an* (II,SO)*-increasing-inflection source.*

- *For* $a_{j_1j_10} < 0$ *and* $a_{j_2j_20}(a_{j_2j_2s_1} - a_{j_2j_2s_2}) > 0$,

$$(\overline{x}_{j_1}, a_{j_2j_2s_1}) = \underbrace{(\text{DI, SO})}_{\text{decreasing-inflection source}}. \tag{1.177}$$

The infinite-equilibrium of $x^*_{j_2} = a_{j_1j_21} = a_{j_2j_2s_1}$ *is a* (DI,SO)-*down-parabola source.*

- *For* $a_{j_1j_10} > 0$ *and* $a_{j_2j_20}(a_{j_2j_2s_1} - a_{j_2j_2s_2}) < 0$,

$$(\overline{x}_{j_1}, a_{j_2j_2s_1}) = \underbrace{(\text{DI, SI})}_{\text{decreasing-inflection sink}}. \tag{1.178}$$

The infinite-equilibrium of $x^*_{j_2} = a_{j_1j_21} = a_{j_2j_2s_1}$ *is a* (DI,SI)-*decreasing-inflection sink.*

- *For* $a_{j_1j_10} < 0$ *and* $a_{j_2j_20}(a_{j_1j_21} - a_{j_2j_21})^2 < 0$,

$$(\overline{x}_{j_1}, a_{j_2j_2s_1}) = \underbrace{(\text{II, SI})}_{\text{increasing-inflection sink}}. \tag{1.179}$$

The infinite-equilibrium of $x^*_{j_2} = a_{j_1j_21} = a_{j_2j_2s_1}$ *is an* (II,SI)-*increasing-infection sink.*

(iv) For $\Delta_{j_1j_1} = 0$ and $\Delta_{j_2j_2} < 0$, *the standard form is*

$$\begin{aligned} \dot{x}_{j_2} &= a_{j_2j_20}\left[(x_{j_2} - a_{j_2j_2})^2 + b_{j_2j_2}\right], \\ \dot{x}_{j_1} &= a_{j_1j_10}(x_{j_1} - a_{j_1j_11})^2(x_{j_2} - a_{j_1j_21}) \end{aligned} \tag{1.180}$$

where

$$\begin{aligned} a_{j_1j_11} &= -\frac{1}{2}B_{j_1j_1}, \\ a_{j_2j_2} &= -\frac{1}{2}B_{j_2j_2}, b_{j_2j_2} = \frac{1}{4}(-\Delta_{j_2j_2}). \end{aligned} \tag{1.181}$$

The first integral manifold is

$$\begin{aligned} -\left(\frac{1}{x_{j_1} - a_{j_1j_11}} - \frac{1}{x_{j_1}1 - a_{j_1j_11}}\right) &= \frac{a_{j_1j_10}}{a_{j_2j_20}}\left[\frac{1}{2}\ln\frac{|(x_{j_2} - a_{j_2j_2})^2 + b_{j_2j_2}|}{|(x_{j_20} - a_{j_2j_2})^2 + b_{j_2j_2}|}\right. \\ &\left. + \frac{(a_{j_2j_2} - a_{j_1j_21})}{\sqrt{b_{j_2j_2}}}(\arctan\frac{x_{j_2} - a_{j_2j_2}}{\sqrt{b_{j_2j_2}}} - \arctan\frac{x_{j_20} - a_{j_2j_2}}{\sqrt{b_{j_2j_2}}})\right]. \end{aligned} \tag{1.182}$$

The equilibrium of $(x^*_{j_1}, x^*_{j_2}) = (a_{j_1j_11}, a_{j_1j_21})$ *has the following properties.*

- *For* $a_{j_1j_10} > 0$ *and* $a_{j_2j_20} > 0$,

$$(a_{j_1j_11}, a_{j_1j_21}) = \underbrace{(\mathrm{UP}:\mathrm{UP},\mathrm{pF})}_{\text{hyperbolic-secant-to-hyperbolic flow }(+)}. \tag{1.183}$$

The equilibrium of $(x^*_{j_1}, x^*_{j_2}) = (a_{j_1j_11}, a_{j_1j_21})$ *is* a (UP:UP, pF)-*positive hyperbolic-secant to hyperbolic flow.*

- *For* $a_{j_1j_10} < 0$ *and* $a_{j_2j_20} > 0$,

$$(a_{j_1j_11}, a_{j_1j_21}) = \underbrace{(\mathrm{DP}:\mathrm{DP},\mathrm{pF})}_{\text{hyperbolic-to-hyperbolic-secant flow }(+)}. \tag{1.184}$$

The equilibrium of $(x^*_{j_1}, x^*_{j_2}) = (a_{j_1j_11}, a_{j_1j_21})$ *is a* (DP:DP,pF)-*positive hyperbolic-to-hyperbolic-secant flow.*

- *For* $a_{j_1j_10} > 0$ *and* $a_{j_2j_20} < 0$,

$$(a_{j_1j_11}, a_{j_1j_21}) = \underbrace{(\mathrm{DP}:\mathrm{DP},\mathrm{nF})}_{\text{hyperbolic to hyperbolic-secant flow }(-)}. \tag{1.185}$$

The equilibrium of $(x^*_{j_1}, x^*_{j_2}) = (a_{j_1j_11}, a_{j_1j_21})$ *is a* (DP:DP, nF)-*negative hyperbolic-to-hyperbolic-secant flow.*

- *For* $a_{j_1j_10} < 0$ *and* $a_{j_2j_20} < 0$,

$$(a_{j_1j_11}, a_{j_1j_21}) = \underbrace{(\mathrm{UP}:\mathrm{UP},\mathrm{nF})}_{\text{hyperbolic-secant-to-hyperbolic flow }(-)}. \tag{1.186}$$

The equilibrium of $(x^*_{j_1}, x^*_{j_2}) = (a_{j_1j_11}, a_{j_1j_21})$ *is a* (UP:UP,nF)-*negative hyperbolic-secant-to-hyperbolic flow.*

(v) For $\Delta_{j_1j_1} = 0$ *and* $\Delta_{j_2j_2} = 0$, *the standard form is*

$$\begin{aligned} \dot{x}_{j_2} &= a_{j_2j_20}(x_{j_2} - a_{j_2j_21})^2, \\ \dot{x}_{j_1} &= a_{j_1j_10}(x_{j_1} - a_{j_1j_11})^2(x_{j_2} - a_{j_1j_21}) \end{aligned} \tag{1.187}$$

where

$$a_{j_1j_11} = -\frac{1}{2}B_{j_1j_1}, a_{j_2j_21} = -\frac{1}{2}B_{j_2j_2}. \tag{1.188}$$

(v_1) The first integral manifold for $a_{j_2j_21} \neq a_{j_1j_21}$ *is*

$$-\left(\frac{1}{x_{j_1}-a_{j_1j_1 1}}-\frac{1}{x_{j_1 1}-a_{j_1j_1 1}}\right)=\frac{a_{j_1j_1 0}}{a_{j_2j_2 0}}\Big[\ln\frac{|x_{j_2}-a_{j_2j_2 1}|}{|x_{j_2 0}-a_{j_2j_2 1}|}$$
$$-(a_{j_2j_2 1}-a_{j_1j_2 1})\left(\frac{1}{x_{j_2}-a_{j_2j_2 1}}-\frac{1}{x_{j_2 0}-a_{j_2j_2 1}}\right)\Big]. \tag{1.189}$$

(v_{1a}) The equilibrium of $(x^*_{j_1},x^*_{j_2})=(a_{j_1j_1 1},a_{j_2j_2 1})$ *with* $a_{j_2j_2 1}\neq a_{j_1j_2 1}$ *has the following properties.*

- *For* $a_{j_1j_1 0}(a_{j_2j_2 1}-a_{j_1j_2 1})>0$ *and* $a_{j_2j_2 0}>0$,

$$(a_{j_1j_1 1},a_{j_2j_2 1})=\underbrace{(\mathrm{US},\mathrm{US})}_{\text{double-saddle}}. \tag{1.190}$$

  The equilibrium of $(x^*_{j_1},x^*_{j_2})=(a_{j_1j_1 1},a_{j_2j_2 1})$ *is a* (US,US)*-double-saddle.*
- *For* $a_{j_1j_1 0}(a_{j_2j_2 1}-a_{j_1j_2 1})<0$ *and* $a_{j_2j_2 0}>0$,

$$(a_{j_1j_1 1},a_{j_2j_2 1})=\underbrace{(\mathrm{LS},\mathrm{US})}_{\text{double-saddle}}. \tag{1.191}$$

  The equilibrium of $(x^*_{j_1},x^*_{j_2})=(a_{j_1j_1 1},a_{j_2j_2 1})$ *is an* (LS,US)*-double-saddle.*
- *For* $a_{j_1j_1 0}(a_{j_2j_2 1}-a_{j_1j_2 1})>0$ *and* $a_{j_2j_2 0}<0$,

$$(a_{j_1j_1 1},a_{j_2j_2 1})=\underbrace{(\mathrm{US},\mathrm{LS})}_{\text{double-saddle}}. \tag{1.192}$$

  The equilibrium of $(x^*_{j_1},x^*_{j_2})=(a_{j_1j_1 1},a_{j_2j_2 1})$ *is a* (US,LS)*-double-saddle.*
- *For* $a_{j_1j_1 0}(a_{j_2j_2 1}-a_{j_1j_2 1})<0$ *and* $a_{j_2j_2 0}<0$,

$$(a_{j_1j_1 1},a_{j_2j_2 1})=\underbrace{(\mathrm{LS},\mathrm{LS})}_{\text{double-saddle}}. \tag{1.193}$$

  The equilibrium of $(x^*_{j_1},x^*_{j_2})=(a_{j_1j_1 1},a_{j_2j_2 1})$ *is an* (LS,LS)*-double saddle.*

– *The double-saddles are the appearing bifurcations of saddle-source and saddle-sink, and such bifurcations are also the appearing bifurcations for networks of saddles, source and sink.*

(v_{1b}) The equilibrium of $(x^*_{j_1},x^*_{j_2})=(a_{j_1j_1 1},a_{j_1j_2 1})$ *with* $a_{j_2j_2 1}\neq a_{j_1j_2 1}$ *has the following properties.*

- *For* $a_{j_1j_1 0}>0$ *and* $a_{j_2j_2 0}(a_{j_1j_2 1}-a_{j_2j_2 1})^2>0$,

$$(a_{j_1j_1 1},a_{j_1j_2 1})=\underbrace{(\mathrm{UP}:\mathrm{UP},\mathrm{pF})}_{\text{hyperbolic-secant to hyperbolic flow }(+)}. \tag{1.194}$$

The equilibrium of $(x^*_{j_1}, x^*_{j_2}) = (a_{j_1j_1}1, a_{j_1j_2}1)$ *is a* (UP:UP, pF)-*positive hyperbolic-secant to hyperbolic flow.*

- *For* $a_{j_1j_1}0 < 0$ *and* $a_{j_2j_2}0(a_{j_1j_2}1 - a_{j_2j_2}1)^2 > 0$,

$$(a_{j_1j_1}1, a_{j_1j_2}1) = \underbrace{(\mathrm{DP}:\mathrm{DP},\mathrm{pF})}_{\text{hyperbolic to hyperbolic-secant flow }(+)}. \tag{1.195}$$

The equilibrium of $(x^*_{j_1}, x^*_{j_2}) = (a_{j_1j_1}1, a_{j_1j_2}1)$ *is a* (DP:DP, pF)-*positive hyperbolic to hyperbolic-secant flow.*

- *For* $a_{j_1j_1}0 > 0$ *and* $a_{j_2j_2}0(a_{j_1j_2}1 - a_{j_2j_2}1)^2 < 0$,

$$(a_{j_1j_1}1, a_{j_1j_2}1) = \underbrace{(\mathrm{DP}:\mathrm{DP},\mathrm{nF})}_{\text{hyperbolic to hyperbolic-secant flow }(-)}. \tag{1.196}$$

The equilibrium of $(x^*_{j_1}, x^*_{j_2}) = (a_{j_1j_1}1, a_{j_1j_2}1)$ *is a* (DP:DP,nF)-*negative hyperbolic to hyperbolic-secant flow.*

- *For* $a_{j_1j_1}0 < 0$ *and* $a_{j_2j_2}0(a_{j_1j_2}1 - a_{j_2j_2}1)^2 < 0$,

$$(a_{j_1j_1}1, a_{j_1j_2}1) = \underbrace{(\mathrm{UP}:\mathrm{UP},\mathrm{nF})}_{\text{hyperbolic-secant to hyperbolic flow }(-)}. \tag{1.197}$$

The equilibrium of $(x^*_{j_1}, x^*_{j_2}) = (a_{j_1j_1}1, a_{j_1j_2}1)$ *is a* (UP:UP, nF)-*negative hyperbolic-secant to hyperbolic flow.*

(v₂) The first integral manifold for $a_{j_2j_2}1 = a_{j_1j_2}1$ *is*

$$-\left(\frac{1}{x_{j_1} - a_{j_1j_1}1} - \frac{1}{x_{j_1}1 - a_{j_1j_1}1}\right) = \frac{a_{j_1j_1}0}{a_{j_2j_2}0} \ln \frac{|x_{j_2} - a_{j_2j_2}1|}{|x_{j_2}0 - a_{j_2j_2}1|}. \tag{1.198}$$

(v_{2a}) The infinite-equilibrium of $x^*_{j_2} = a_{j_1j_2}1 = a_{j_2j_2}1$ *with* $\bar{x}_{j_1} \neq a_{j_1j_1}1$ *has the following properties.*

- *For* $a_{j_1j_1}0(\bar{x}_{j_1} - a_{j_1j_1}1)^2 > 0$ *and* $a_{j_2j_2}0 > 0$,

$$(\bar{x}_{j_1}, a_{j_1j_2}1) = \underbrace{(\mathrm{DU},\mathrm{US})}_{\text{down-up upper-saddle}}. \tag{1.199}$$

The infinite-equilibrium of $x^*_{j_2} = a_{j_1j_2}1 = a_{j_2j_2}1$ *is a* (DU,US)-*down-up upper-saddle.*

- *For* $a_{j_1j_1}0(\bar{x}_{j_1} - a_{j_1j_1}1)^2 < 0$ *and* $a_{j_2j_2}0 > 0$,

$$(\bar{x}_{j_1}, a_{j_1j_2 1}) = \underbrace{(\text{UD}, \text{US})}_{\text{up-down upper-saddle}}. \tag{1.200}$$

The infinite-equilibrium of $x^*_{j_2} = a_{j_1j_2 1} = a_{j_2j_2 1}$ *is a* (UD,US)-*up-down upper-saddle.*

- *For* $a_{j_1j_1 0}(\bar{x}_{j_1} - a_{j_1j_1 1})^2 > 0$ *and* $a_{j_2j_2 0} < 0$,

$$(\bar{x}_{j_1}, a_{j_1j_2 1}) = \underbrace{(\text{UD}, \text{LS})}_{\text{up-down lower-saddle}}. \tag{1.201}$$

The infinite-equilibrium of $x^*_{j_2} = a_{j_1j_2 1} = a_{j_2j_2 1}$ *is a* (UD,LS)-*up-down lower-saddle.*

- *For* $a_{j_1j_1 0}(\bar{x}_{j_1} - a_{j_1j_1 1})^2 < 0$ *and* $a_{j_2j_2 0} < 0$,

$$(\bar{x}_{j_1}, a_{j_1j_2 1}) = \underbrace{(\text{DU}, \text{LS})}_{\text{down-up lower-saddle}}. \tag{1.202}$$

The infinite-equilibrium of $x^*_{j_2} = a_{j_1j_2 1} = a_{j_2j_2 1}$ *is a* (DU,LS)-*down-up lower-saddle.*

(v_{2b}) *The equilibrium of* $(x^*_{j_1}, x^*_{j_2}) = (a_{j_1j_1 1}, a_{j_1j_2 1})$ *with* $a_{j_2j_2 1} = a_{j_1j_2 1}$ *has the following properties.*

- *For* $a_{j_1j_1 0} > 0$ *and* $a_{j_2j_2 0} > 0$,

$$(a_{j_1j_1 1}, a_{j_1j_2 1}) = \underbrace{(\text{DU} : \text{DU}, \text{US})}_{\text{down-up upper-saddle}}. \tag{1.203}$$

The equilibrium of $(x^*_{j_1}, x^*_{j_2}) = (a_{j_1j_1 1}, a_{j_1j_2 1})$ *is a* (DU:DU,US)-*down-up upper-saddle.*

- *For* $a_{j_1j_1 0} < 0$ *and* $a_{j_2j_2 0} > 0$,

$$(a_{j_1j_1 1}, a_{j_1j_2 1}) = \underbrace{(\text{UD} : \text{UD}, \text{US})}_{\text{up-down upper-saddle}}. \tag{1.204}$$

The equilibrium of $(x^*_{j_1}, x^*_{j_2}) = (a_{j_1j_1 1}, a_{j_1j_2 1})$ *is a* (UD:UD,US)-*up-down upper-saddle.*

- *For* $a_{j_1j_1 0} > 0$ *and* $a_{j_2j_2 0} < 0$,

$$(a_{j_1j_1 1}, a_{j_1j_2 1}) = \underbrace{(\text{UD} : \text{UD}, \text{LS})}_{\text{up-down lower-saddle}}. \tag{1.205}$$

The equilibrium of $(x^*_{j_1}, x^*_{j_2}) = (a_{j_1j_1 1}, a_{j_1j_2 1})$ *is a* (UD:UD,LS)-*up-down lower-saddle.*

- *For $a_{j_1j_10}<0$ and $a_{j_2j_20}<0$,*

$$(a_{j_1j_11}, a_{j_1j_21}) = \underbrace{(\text{DU}:\text{DU},\text{LS})}_{\text{down-up upper-saddle}}. \tag{1.206}$$

The equilibrium of $(x^*_{j_1}, x^*_{j_2}) = (a_{j_1j_11}, a_{j_1j_21})$ is a (DU:DU,LS)*-down-up lower-saddle.*

(vi) For $\Delta_{j_1j_1}=0$ and $\Delta_{j_2j_2}>0$, the standard form is

$$\begin{aligned}\dot{x}_{j_2} &= a_{j_2j_20}(x_{j_2} - a_{j_2j_21})(x_{j_2} - a_{j_2j_22}),\\ \dot{x}_{j_1} &= a_{j_1j_10}(x_{j_1} - a_{j_1j_11})^2(x_{j_2} - a_{j_1j_21})\end{aligned} \tag{1.207}$$

where

$$\begin{aligned}&a_{j_1j_11} = -\frac{1}{2}B_{j_1j_1},\\ &b_{j_2j_21}, b_{j_2j_22} = -\frac{1}{2}(B_{j_2j_2} \pm \sqrt{\Delta_{j_2j_2}}),\\ &\{a_{j_2j_21}, a_{j_2j_22}\} = \text{sort}\{b_{j_2j_21}, b_{j_2j_22}\}, a_{j_2j_21} < a_{j_2j_22}.\end{aligned} \tag{1.208}$$

(vi_1) The first integral manifold ($a_{j_2j_2s_1} \neq a_{j_1j_21}$) is

$$-\left(\frac{1}{x_{j_1} - a_{j_1j_11}} - \frac{1}{x_{j_1}1 - a_{j_1j_11}}\right) = \frac{a_{j_1j_10}}{a_{j_2j_20}}\sum\nolimits_{s_1=1}^{2}\frac{(a_{j_2j_2s_1} - a_{j_1j_21})}{(a_{j_2j_2s_1} - a_{j_2j_2s_2})}\ln\frac{|x_{j_2} - a_{j_2j_2s_1}|}{|x_{j_20} - a_{j_2j_2s_1}|}. \tag{1.209}$$

(vi_{1a}) The equilibrium of $(x^*_{j_1}, x^*_{j_2}) = (a_{j_1j_11}, a_{j_2j_2s_1})$ (s_1, $s_2 \in \{1,2\}$, $s_1 \neq s_2$) has the following properties.

- *For $a_{j_1j_10}(a_{j_2j_2s_1} - a_{j_1j_21})>0$ and $a_{j_2j_20}(a_{j_2j_2s_1} - a_{j_2j_2s_2})>0$,*

$$(a_{j_1j_11}, a_{j_2j_2s_1}) = \underbrace{(\text{US},\text{SO})}_{\text{upper-saddle source}}. \tag{1.210}$$

The equilibrium of $(x^*_{j_1}, x^*_{j_2}) = (a_{j_1j_11}, a_{j_2j_2s_1})$ is a (US,SO)*-upper-saddle source.*

- *For $a_{j_1j_10}(a_{j_2j_2s_1} - a_{j_1j_21})<0$ and $a_{j_2j_20}(a_{j_2j_2s_1} - a_{j_2j_2s_2})>0$,*

$$(a_{j_1j_11}, a_{j_2j_2s_1}) = \underbrace{(\text{LS},\text{SO})}_{\text{lower-saddle source}}. \tag{1.211}$$

The equilibrium of $(x^*_{j_1}, x^*_{j_2}) = (a_{j_1j_11}, a_{j_2j_2s_1})$ is an (LS,SO)*-lower-saddle source.*

- *For* $a_{j_1j_10}(a_{j_2j_2s_1} - a_{j_1j_21}) > 0$ *and* $a_{j_2j_20}(a_{j_2j_2s_1} - a_{j_2j_2s_2}) < 0$,

$$(a_{j_1j_11}, a_{j_2j_2s_1}) = \underbrace{(\text{US}, \text{SI})}_{\text{upper-saddle sink}}. \tag{1.212}$$

  The equilibrium of $(x_{j_1}^*, x_{j_2}^*) = (a_{j_1j_11}, a_{j_2j_2s_1})$ *is a* (US,SI)-*upper-saddle sink.*

- *For* $a_{j_1j_10}(a_{j_2j_2s_1} - a_{j_1j_21}) < 0$ *and* $a_{j_2j_20}(a_{j_2j_2s_1} - a_{j_2j_2s_2}) < 0$,

$$(a_{j_1j_11}, a_{j_2j_2s_1}) = \underbrace{(\text{LS}, \text{SI})}_{\text{lower-saddle sink}}. \tag{1.213}$$

  The equilibrium of $(x_{j_1}^*, x_{j_2}^*) = (a_{j_1j_11}, a_{j_2j_2s_1})$ *is an* (LS,SI)-*lower-saddle sink.*

(vi$_{1b}$) The equilibrium of $(x_{j_1}^*, x_{j_2}^*) = (a_{j_1j_11}, a_{j_1j_21})$ *with* $a_{j_1j_21} \neq a_{j_2j_21}, a_{j_2j_22}$ *has the following properties.*

- *For* $a_{j_1j_10} > 0$ *and* $a_{j_2j_20}\prod_{s_1=1}^{2}(a_{j_1j_21} - a_{j_2j_2s_1}) > 0$,

$$(a_{j_1j_11}, a_{j_1j_21}) = \underbrace{(\text{UP} : \text{UP}, \text{pF})}_{\text{hyperbolic-secant to hyperbolic flow } (+)}. \tag{1.214}$$

  The equilibrium of $(x_{j_1}^*, x_{j_2}^*) = (a_{j_1j_11}, a_{j_1j_21})$ *is a* (UP:UP, pF)-*positive hyperbolic-secant-to-hyperbolic flow.*

- *For* $a_{j_1j_10} < 0$ *and* $a_{j_2j_20}\prod_{s_1=1}^{2}(a_{j_1j_21} - a_{j_2j_2s_1}) > 0$,

$$(a_{j_1j_11}, a_{j_1j_21}) = \underbrace{(\text{DP} : \text{DP}, \text{pF})}_{\text{hyperbolic-to-hyperbolic-secant flow } (+)}. \tag{1.215}$$

  The equilibrium of $(x_{j_1}^*, x_{j_2}^*) = (a_{j_1j_11}, a_{j_1j_21})$ *is a* (DP:DP, pF)-*positive hyperbolic-to-hyperbolic-secant flow.*

- *For* $a_{j_1j_10} > 0$ *and* $a_{j_2j_20}\prod_{s_1=1}^{2}(a_{j_1j_21} - a_{j_2j_2s_1}) < 0$,

$$(a_{j_1j_11}, a_{j_1j_21}) = \underbrace{(\text{DP} : \text{DP}, \text{nF})}_{\text{hyperbolic-to-hyperbolic-secant flow } (-)}. \tag{1.216}$$

  The equilibrium of $(x_{j_1}^*, x_{j_2}^*) = (a_{j_1j_11}, a_{j_1j_21})$ *is a* (DP:DP, nF)-*negative hyperbolic-to-hyperbolic-secant flow.*

- *For* $a_{j_1j_10} < 0$ *and* $a_{j_2j_20}\prod_{s_1=1}^{2}(a_{j_1j_21} - a_{j_2j_2s_1}) < 0$,

$$(a_{j_1 j_1 1}, a_{j_1 j_2 1}) = \underbrace{(\text{UP}:\text{UP}, \text{nF})}_{\text{hyperbolic-secant-to-hyperbolic flow } (-)}. \tag{1.217}$$

The equilibrium of $(x^*_{j_1}, x^*_{j_2}) = (a_{j_1 j_1 1}, a_{j_1 j_2 1})$ *is a* (UP:UP, nF)*-negative hyperbolic-secant-to-hyperbolic flow.*

(vi$_2$*) The first integral manifold* $(a_{j_2 j_2 s_1} = a_{j_1 j_2 1})$ *is*

$$-\left(\frac{1}{x_{j_1} - a_{j_1 j_1 1}} - \frac{1}{x_{j_1 1} - a_{j_1 j_1 1}}\right) = \frac{a_{j_1 j_1 0}}{a_{j_2 j_2 0}} \ln \frac{|x_{j_2} - a_{j_2 j_2 s_2}|}{|x_{j_2} - a_{j_2 j_2 s_2}|}. \tag{1.218}$$

(vi$_{2a}$*) The infinite-equilibrium of* $x^*_{j_2} = a_{j_1 j_2 s_1} = a_{j_2 j_2 s_1}$ *with* $\overline{x}_{j_1} \neq a_{j_1 j_1 1}$ *has the following properties.*

- *For* $a_{j_1 j_1 0}(\overline{x}_{j_1} - a_{j_1 j_1 1})^2 > 0$ *and* $a_{j_2 j_2 0}(a_{j_2 j_2 s_1} - a_{j_2 j_2 s_2}) > 0$,

$$(\overline{x}_{j_1}, a_{j_2 j_2 s_1}) = \underbrace{(\text{II}, \text{SO})}_{\text{increasing-inflection source}}. \tag{1.219}$$

The infinite-equilibrium of $x^*_{j_2} = a_{j_1 j_2 1} = a_{j_2 j_2 s_1}$ *is an* (II,SO)*-increasing-inflection source.*

- *For* $a_{j_1 j_1 0}(\overline{x}_{j_1} - a_{j_1 j_1 1})^2 < 0$ *and* $a_{j_2 j_2 0}(a_{j_2 j_2 s_1} - a_{j_2 j_2 s_2}) > 0$,

$$(\overline{x}_{j_1}, a_{j_2 j_2 s_1}) = \underbrace{(\text{DI}, \text{SO})}_{\text{decreasing-inflection source}}. \tag{1.220}$$

The infinite-equilibrium of $x^*_{j_2} = a_{j_1 j_2 1} = a_{j_2 j_2 s_1}$ *is a* (DI,SO)*-decreasing-inflection source.*

- *For* $a_{j_1 j_1 0}(\overline{x}_{j_1} - a_{j_1 j_1 1})^2 > 0$ *and* $a_{j_2 j_2 0}(a_{j_2 j_2 s_1} - a_{j_2 j_2 s_2}) < 0$,

$$(\overline{x}_{j_1}, a_{j_2 j_2 s_1}) = \underbrace{(\text{DI}, \text{SI})}_{\text{decreasing-inflection sink}}. \tag{1.221}$$

The infinite-equilibrium of $x^*_{j_2} = a_{j_1 j_2 1} = a_{j_2 j_2 s_1}$ *is a* (DI,SI)*-decreasing-inflection sink.*

- *For* $a_{j_1 j_1 0}(\overline{x}_{j_1} - a_{j_1 j_1 1})^2 < 0$ *and* $a_{j_2 j_2 0}(a_{j_2 j_2 s_1} - a_{j_2 j_2 s_2}) < 0$,

$$(\overline{x}_{j_1}, a_{j_2 j_2 s_1}) = \underbrace{(\text{II}, \text{SI})}_{\text{increasing-inflection sink}}. \tag{1.222}$$

The infinite-equilibrium of $x^*_{j_2} = a_{j_1 j_2 1} = a_{j_2 j_2 s_1}$ *is an* (II,SI)*-increasing-inflection sink.*

(vi$_{2b}$) The equilibrium of $(x^*_{j_1}, x^*_{j_2}) = (a_{j_1j_1}1, a_{j_2j_2s_1})$ *with* $a_{j_2j_2s_1} = a_{j_1j_2}1$ *has the following properties.*

- *For* $a_{j_1j_1}0 > 0$ *and* $a_{j_2j_2}0(a_{j_2j_2s_1} - a_{j_2j_2s_2}) > 0$,

$$(a_{j_1j_1}1, a_{j_2j_2s_1}) = \underbrace{(\mathrm{II}:\mathrm{II},\mathrm{SO})}_{\text{inflection source}}. \tag{1.223}$$

  The equilibrium of $(x^*_{j_1}, x^*_{j_2}) = (a_{j_1j_1}1, a_{j_2j_2s_1})$ *is an* (II:II,SO)*-increasing-inflection source.*

- *For* $a_{j_1j_1}0 < 0$ *and* $a_{j_2j_2}0(a_{j_2j_2s_1} - a_{j_2j_2s_2}) > 0$,

$$(a_{j_1j_1}1, a_{j_2j_2s_1}) = \underbrace{(\mathrm{DI}:\mathrm{DI},\mathrm{SO})}_{\text{inflection source}}. \tag{1.224}$$

  The equilibrium of $(x^*_{j_1}, x^*_{j_2}) = (a_{j_1j_1}1, a_{j_2j_2s_1})$ *is a* (DI:DI,SO)*-decreasing-inflection source.*

- *For* $a_{j_1j_1}0 > 0$ *and* $a_{j_2j_2}0(a_{j_2j_2s_1} - a_{j_2j_2s_2}) < 0$,

$$(a_{j_1j_1}1, a_{j_2j_2s_1}) = \underbrace{(\mathrm{DI}:\mathrm{DI},\mathrm{SI})}_{\text{inflection sink}}. \tag{1.225}$$

  The equilibrium of $(x^*_{j_1}, x^*_{j_2}) = (a_{j_1j_1}1, a_{j_2j_2s_1})$ *is a* (DI:DI,SI)*-decreasing-inflection sink.*

- *For* $a_{j_1j_1}0 < 0$ *and* $a_{j_2j_2}0(a_{j_2j_2s_1} - a_{j_2j_2s_2}) < 0$,

$$(a_{j_1j_1}1, a_{j_2j_2s_1}) = \underbrace{(\mathrm{II}:\mathrm{II},\mathrm{SI})}_{\text{inflection sink}}. \tag{1.226}$$

  The equilibrium of $(x^*_{j_1}, x^*_{j_2}) = (a_{j_1j_1}1, a_{j_2j_2s_1})$ *is an* (II:II,SI)*-increasing-inflection sink.*

(vi$_{2c}$) The equilibrium of $(x^*_{j_1}, x^*_{j_2}) = (a_{j_1j_1}1, a_{j_2j_2s_2})$ *with* $a_{j_2j_2s_1} = a_{j_1j_2}1$ *has the following properties.*

- *For* $a_{j_1j_1}0(a_{j_2j_2s_2} - a_{j_1j_2}1) > 0$ *and* $a_{j_2j_2}0(a_{j_2j_2s_2} - a_{j_2j_2s_1}) > 0$,

$$(a_{j_1j_1}1, a_{j_2j_2s_2}) = \underbrace{(\mathrm{US},\mathrm{SO})}_{\text{upper-saddle source}}. \tag{1.227}$$

  The equilibrium of $(x^*_{j_1}, x^*_{j_2}) = (a_{j_1j_1}1, a_{j_2j_2s_2})$ *is a* (US,SO)*-upper-saddle source.*

- *For* $a_{j_1j_1}0(a_{j_2j_2s_2} - a_{j_1j_2}1) < 0$ *and* $a_{j_2j_2}0(a_{j_2j_2s_2} - a_{j_2j_2s_1}) > 0$,

$$(a_{j_1j_11}, a_{j_2j_2s_2}) = \underbrace{(\text{LS, SO})}_{\text{lower-saddle source}}. \tag{1.228}$$

The equilibrium of $(x_{j_1}^*, x_{j_2}^*) = (a_{j_1j_11}, a_{j_2j_2s_2})$ *is an* (LS,SO)-*lower-saddle source.*

- *For* $a_{j_1j_10}(a_{j_2j_2s_2} - a_{j_1j_21}) > 0$ *and* $a_{j_2j_20}(a_{j_2j_2s_2} - a_{j_2j_2s_1}) < 0$,

$$(a_{j_1j_11}, a_{j_2j_2s_2}) = \underbrace{(\text{US, SI})}_{\text{upper-saddle sink}}. \tag{1.229}$$

The equilibrium of $(x_{j_1}^*, x_{j_2}^*) = (a_{j_1j_11}, a_{j_2j_2s_2})$ *is a* (US,SI)-*upper-saddle sink.*

- *For* $a_{j_1j_10}(a_{j_2j_2s_2} - a_{j_1j_21}) < 0$ *and* $a_{j_2j_20}(a_{j_2j_2s_2} - a_{j_2j_2s_1}) < 0$,

$$(a_{j_1j_11}, a_{j_2j_2s_2}) = \underbrace{(\text{LS, SI})}_{\text{lower-saddle sink}}. \tag{1.230}$$

The equilibrium of $(x_{j_1}^*, x_{j_2}^*) = (a_{j_1j_11}, a_{j_2j_2s_2})$ *is an* (LS,SI)-*lower-saddle sink.*

(vii) For $\Delta_{j_1j_2} > 0$ *and* $\Delta_{j_2j_2} < 0$, *the standard form is*

$$\begin{aligned} \dot{x}_{j_2} &= a_{j_2j_20}\left[(x_{j_2} - a_{j_2j_2})^2 + b_{j_2j_2}\right], \\ \dot{x}_{j_1} &= a_{j_1j_10}(x_{j_1} - a_{j_1j_11})(x_{j_1} - a_{j_1j_12})(x_{j_2} - a_{j_1j_21}) \end{aligned} \tag{1.231}$$

where

$$\begin{aligned} b_{j_1j_11}, b_{j_1j_12} &= -\frac{1}{2}(B_{j_1j_1} \pm \sqrt{\Delta_{j_1j_1}}), \\ \{a_{j_1j_11}, a_{j_1j_12}\} &= \text{sort}\{b_{j_1j_11}, b_{j_1j_12}\}, a_{j_1j_11} < a_{j_1j_12}; \\ a_{j_2j_2} &= -\frac{1}{2}B_{j_2j_2}, b_{j_2j_2} = \frac{1}{4}(-\Delta_{j_2j_2}). \end{aligned} \tag{1.232}$$

The first integral manifold is

$$\begin{aligned} &\frac{1}{a_{j_1j_12} - a_{j_1j_11}}(\ln \frac{|x_{j_1} - a_{j_1j_12}|}{|x_{j_10} - a_{j_1j_12}|} - \ln \frac{|x_{j_1} - a_{j_1j_11}|}{|x_{j_10} - a_{j_1j_11}|}) \\ &= \frac{a_{j_1j_10}}{a_{j_2j_20}}\Big\{\frac{1}{2}\ln \frac{|(x_{j_2} - a_{j_2j_2})^2 + b_{j_2j_2}|}{|(x_{j_20} - a_{j_2j_2})^2 + b_{j_2j_2}|} \\ &+ \frac{a_{j_2j_2} - a_{j_1j_21}}{\sqrt{b_{j_2j_2}}}(\arctan \frac{x_{j_2} - a_{j_2j_2}}{\sqrt{b_{j_2j_2}}} - \arctan \frac{x_{j_20} - a_{j_2j_2}}{\sqrt{b_{j_2j_2}}})\Big\}. \end{aligned} \tag{1.233}$$

The equilibrium of $(x_{j_1}^*, x_{j_2}^*) = (a_{j_1j_1s_1}, a_{j_1j_21})$ $(s_1, s_2 \in \{1, 2\}, s_1 \neq s_2)$ *has the following properties.*

- *For* $a_{j_1j_10}(a_{j_1j_1s_1} - a_{j_1j_1s_2}) > 0$ *and* $a_{j_2j_20} > 0$,

$$(a_{j_1j_1s_1}, a_{j_1j_21}) = \underbrace{(\text{DP} : \text{UP}, \text{pF})}_{\text{hyperbolic flow } (+)}. \tag{1.234}$$

The equilibrium of $(x_{j_1}^*, x_{j_2}^*) = (a_{j_1j_1s_1}, a_{j_1j_21})$ *is a* (DP:UP,pF)-*positive hyperbolic flow.*

- *For* $a_{j_1j_10}(a_{j_1j_1s_1} - a_{j_1j_1s_2}) < 0$ *and* $a_{j_2j_20} > 0$,

$$(a_{j_1j_1s_1}, a_{j_1j_21}) = \underbrace{(\text{UP} : \text{DP}, \text{pF})}_{\text{hyperbolic-secant flow } (+)}. \tag{1.235}$$

The equilibrium of $(x_{j_1}^*, x_{j_2}^*) = (a_{j_1j_1s_1}, a_{j_1j_21})$ *is a* (UP:DP,pF)-*positive hyperbolic-secant flow.*

- *For* $a_{j_1j_10}(a_{j_1j_1s_1} - a_{j_1j_1s_2}) > 0$ *and* $a_{j_2j_20} < 0$,

$$(a_{j_1j_1s_1}, a_{j_1j_21}) = \underbrace{(\text{UP} : \text{DP}, \text{nF})}_{\text{hyperbolic-secant flow } (-)}. \tag{1.236}$$

The equilibrium of $(x_{j_1}^*, x_{j_2}^*) = (a_{j_1j_1s_1}, a_{j_1j_21})$ *is a* (UP:DP,nF)-*negative hyperbolic-secant flow.*

- *For* $a_{j_1j_10}(a_{j_1j_1s_1} - a_{j_1j_1s_2}) < 0$ *and* $a_{j_2j_20} < 0$,

$$(a_{j_1j_1s_1}, a_{j_1j_21}) = \underbrace{(\text{DP} : \text{UP}, \text{nF})}_{\text{hyperbolic flow } (-)}. \tag{1.237}$$

The equilibrium of $(x_{j_1}^*, x_{j_2}^*) = (a_{j_1j_1s_1}, a_{j_1j_21})$ *is a* (DP:UP,nF)-*negative hyperbolic flow.*

(viii) For $\Delta_{j_1j_1} > 0$ and $\Delta_{j_2j_2} = 0$, the standard form is

$$\begin{aligned} \dot{x}_{j_2} &= a_{j_2j_20}(x_{j_2} - a_{j_2j_21})^2, \\ \dot{x}_{j_1} &= a_{j_1j_10}(x_{j_1} - a_{j_1j_11})(x_{j_1} - a_{j_1j_12})(x_{j_2} - a_{j_1j_21}) \end{aligned} \tag{1.238}$$

where

$$\begin{aligned} & b_{j_1j_11}, b_{j_1j_12} = -\frac{1}{2}(B_{j_1j_1} \pm \sqrt{\Delta_{j_1j_1}}), \\ & \{a_{j_1j_11}, a_{j_1j_12}\} = \text{sort}\{b_{j_1j_11}, b_{j_1j_12}\}, a_{j_1j_11} < a_{j_1j_12}; \\ & a_{j_2j_21} = -\frac{1}{2}B_{j_2j_2}. \end{aligned} \tag{1.239}$$

(viii₁) The first integral manifold for $a_{j_2j_21} \neq a_{j_1j_21}, a_{j_1j_22}$ *is*

$$\frac{1}{a_{j_1j_12}-a_{j_1j_11}}(\ln\frac{|x_{j_1}-a_{j_1j_12}|}{|x_{j_10}-a_{j_1j_12}|}-\ln\frac{|x_{j_1}-a_{j_1j_11}|}{|x_{j_10}-a_{j_1j_11}|})$$
$$=\frac{a_{j_1j_10}}{a_{j_2j_20}}\left[\ln\frac{|x_{j_2}-a_{j_2j_21}|}{|x_{j_20}-a_{j_2j_21}|}-(a_{j_2j_21}-a_{j_1j_21})(\frac{1}{x_{j_2}-a_{j_2j_21}}-\frac{1}{x_{j_20}-a_{j_2j_21}})\right]. \tag{1.240}$$

(viii$_{1a}$) The equilibrium of $(x^*_{j_1},x^*_{j_2})=(a_{j_1j_1s_1},a_{j_2j_21})$ *has the following properties.*

- *For* $a_{j_1j_10}(a_{j_1j_1s_1}-a_{j_1j_1s_2})(a_{j_2j_21}-a_{j_1j_21})>0$ *and* $a_{j_2j_20}>0$,

$$(a_{j_1j_1s_1},a_{j_2j_21})=\underbrace{(\text{SO},\text{US})}_{\text{upper-saddle source}}. \tag{1.241}$$

The equilibrium of $(x^*_{j_1},x^*_{j_2})=(a_{j_1j_1s_1},a_{j_2j_21})$ *is an* (SO,US)-*upper-saddle source.*

- *For* $a_{j_1j_10}(a_{j_1j_1s_1}-a_{j_1j_1s_2})(a_{j_2j_21}-a_{j_1j_21})<0$ *and* $a_{j_2j_20}>0$,

$$(a_{j_1j_1s_1},a_{j_2j_21})=\underbrace{(\text{SI},\text{US})}_{\text{upper-saddle-sink}}. \tag{1.242}$$

The equilibrium of $(x^*_{j_1},x^*_{j_2})=(a_{j_1j_1s_1},a_{j_2j_21})$ *is an* (SI,US)-*upper-saddle sink.*

- *For* $a_{j_1j_10}(a_{j_1j_1s_1}-a_{j_1j_1s_2})(a_{j_2j_21}-a_{j_1j_21})>0$ *and* $a_{j_2j_20}<0$,

$$(a_{j_1j_1s_1},a_{j_2j_21})=\underbrace{(\text{SO},\text{LS})}_{\text{lower-saddle source}}. \tag{1.243}$$

The equilibrium of $(x^*_{j_1},x^*_{j_2})=(a_{j_1j_1s_1},a_{j_2j_21})$ *is an* (SO,LS)-*lower-saddle source.*

- *For* $a_{j_1j_10}(a_{j_1j_1s_1}-a_{j_1j_1s_2})(a_{j_2j_21}-a_{j_1j_21})<0$ *and* $a_{j_2j_20}<0$,

$$(a_{j_1j_1s_1},a_{j_2j_21})=\underbrace{(\text{SI},\text{LS})}_{\text{lower-saddle source}}. \tag{1.244}$$

The equilibrium of $(x^*_{j_1},x^*_{j_2})=(a_{j_1j_1s_1},a_{j_2j_21})$ *is an* (SI,LS)-*lower-saddle sink.*

– *The saddle-source and saddle-sink are the appearing bifurcations of saddle-source and saddle-sink.*

(viii$_{1b}$) The equilibrium of $(x^*_{j_1},x^*_{j_2})=(a_{j_1j_1s_1},a_{j_1j_21})$*with* $a_{j_1j_21}\neq a_{j_2j_21}$ $(s_1,\ s_2\in\{1,2\},$ $s_1\neq s_2)$ *has the following properties.*

- *For* $a_{j_1j_10}(a_{j_1j_1s_1}-a_{j_1j_1s_2})>0$ *and* $a_{j_2j_20}(a_{j_1j_21}-a_{j_2j_21})^2>0$,

$$(a_{j_1j_1s_1},a_{j_1j_21})=\underbrace{(\text{DP}:\text{UP},\text{pF})}_{\text{hyperbolic flow }(+)}. \tag{1.245}$$

The equilibrium of $(x^*_{j_1},x^*_{j_2})=(a_{j_1j_1s_1},a_{j_1j_21})$ *is a* (DP:UP, pF)-*positive hyperbolic flow.*

- *For* $a_{j_1j_10}(a_{j_1j_1s_1} - a_{j_1j_1s_2}) < 0$ *and* $a_{j_2j_20}(a_{j_1j_21} - a_{j_2j_21})^2 > 0$,

$$(a_{j_1j_1s_1}, a_{j_1j_21}) = \underbrace{(\text{UP : DP, pF})}_{\text{hyperbolic-secant flow } (+)}. \tag{1.246}$$

The equilibrium of $(x^*_{j_1}, x^*_{j_2}) = (a_{j_1j_1s_1}, a_{j_1j_21})$ *is a* (UP:DP, pF)-*positive hyperbolic-secant flow.*

- *For* $a_{j_1j_10}(a_{j_1j_2s_1} - a_{j_1j_2s_2}) > 0$ *and* $a_{j_2j_20}(a_{j_1j_2s_1} - a_{j_2j_21})^2 < 0$,

$$(a_{j_1j_1s_1}, a_{j_1j_21}) = \underbrace{(\text{UP : DP, nF})}_{\text{hyperbolic-secant flow } (-)}. \tag{1.247}$$

The equilibrium of $(x^*_{j_1}, x^*_{j_2}) = (a_{j_1j_1s_1}, a_{j_1j_21})$ *is a* (UP:DP,nF)-*negative hyperbolic-secant flow.*

- *For* $a_{j_1j_10}(a_{j_1j_2s_1} - a_{j_1j_2s_2}) < 0$ *and* $a_{j_2j_20}(a_{j_1j_2s_1} - a_{j_2j_21})^2 < 0$,

$$(a_{j_1j_1s_1}, a_{j_1j_21}) = \underbrace{(\text{DP : UP, nF})}_{\text{hyperbolic flow } (-)}. \tag{1.248}$$

The equilibrium of $(x^*_{j_1}, x^*_{j_2}) = (a_{j_1j_1s_1}, a_{j_1j_21})$ *is a* (DP:UP,nF)-*negative hyperbolic flow.*

(viii$_2$) The first integral manifold for $a_{j_2j_21} = a_{j_1j_21}$ *is*

$$\begin{aligned} &\frac{1}{a_{j_1j_12} - a_{j_1j_11}}\left(\ln\frac{|x_{j_1} - a_{j_1j_12}|}{|x_{j_10} - a_{j_1j_12}|} - \ln\frac{|x_{j_1} - a_{j_1j_11}|}{|x_{j_10} - a_{j_1j_11}|}\right) \\ &= \frac{a_{j_1j_10}}{a_{j_2j_20}} \ln\frac{|x_{j_2} - a_{j_2j_21}|}{|x_{j_20} - a_{j_2j_21}|}. \end{aligned} \tag{1.249}$$

(viii$_{2a}$) The infinite-equilibrium of $x^*_{j_2} = a_{j_1j_21} = a_{j_2j_21}$ *with* $\bar{x}_{j_1} \neq a_{j_1j_1s_1}$ $(s_1 = 1, 2)$ *has the following properties.*

- *For* $a_{j_1j_10}\prod_{s_1=1}^{2}(\bar{x}_{j_1} - a_{j_1j_1s_1}) > 0$ *and* $a_{j_2j_20} > 0$,

$$(\bar{x}_{j_1}, a_{j_2j_21}) = \underbrace{(\text{DU, US})}_{\text{down-up upper-saddle}}. \tag{1.250}$$

The infinite-equilibrium of $x^*_{j_2} = a_{j_1j_21} = a_{j_2j_21}$ *is an* (DU,US)-*down-up upper-saddle.*

- *For* $a_{j_1j_10}\prod_{s_1=1}^{2}(\bar{x}_{j_1} - a_{j_1j_1s_1}) < 0$ *and* $a_{j_2j_20} > 0$,

$$(\bar{x}_{j_1}, a_{j_2 j_2 1}) = \underbrace{(\text{UD}, \text{US})}_{\text{up-down upper-saddle}}. \tag{1.251}$$

The infinite-equilibrium of $x^*_{j_2} = a_{j_1 j_2 1} = a_{j_2 j_2 1}$ *is a* (UD,US)*-up-down upper-saddle.*

- *For* $a_{j_1 j_1 0} \prod_{s_1=1}^{2} (\bar{x}_{j_1} - a_{j_1 j_1 s_1}) > 0$ *and* $a_{j_2 j_2 0} < 0$,

$$(\bar{x}_{j_1}, a_{j_2 j_2 1}) = \underbrace{(\text{UD}, \text{LS})}_{\text{up-down lower-saddle}}. \tag{1.252}$$

The infinite-equilibrium of $x^*_{j_2} = a_{j_1 j_2 1} = a_{j_2 j_2 1}$ *is a* (UD,LS)*-up-down lower-saddle.*

- *For* $a_{j_1 j_1 0} \prod_{s_1=1}^{2} (\bar{x}_{j_1} - a_{j_1 j_1 s_1}) < 0$ *and* $a_{j_2 j_2 0} < 0$,

$$(\bar{x}_{j_1}, a_{j_2 j_2 1}) = \underbrace{(\text{DU}, \text{LS})}_{\text{down-up lower-saddle}}. \tag{1.253}$$

The infinite-equilibrium of $x^*_{j_2} = a_{j_1 j_2 1} = a_{j_2 j_2 1}$ *is a* (DU,LS)*-down-up lower-saddle.*

(viii$_{2b}$) The equilibrium of $(x^*_{j_1}, x^*_{j_2}) = (a_{j_1 j_1 s_1}, a_{j_2 j_2 1})$ *with* $a_{j_2 j_2 1} = a_{j_1 j_2 1}$ *has the following properties.*

- *For* $a_{j_1 j_1 0}(a_{j_1 j_1 s_1} - a_{j_1 j_1 s_2}) > 0$ *and* $a_{j_2 j_2 0} > 0$,

$$(a_{j_1 j_1 s_1}, a_{j_2 j_2 1}) = \underbrace{(\text{DU} : \text{UD}, \text{US})}_{\text{down-up-to-up-down upper-saddle}}. \tag{1.254}$$

The equilibrium of $(x^*_{j_1}, x^*_{j_2}) = (a_{j_1 j_1 s_1}, a_{j_2 j_2 1})$ *is a* (DU:UD,US)*-down-up-to-up-down upper-saddle.*

- *For* $a_{j_1 j_1 0}(a_{j_1 j_1 s_1} - a_{j_1 j_1 s_2}) < 0$ *and* $a_{j_2 j_2 0} > 0$,

$$(a_{j_1 j_1 s_1}, a_{j_2 j_2 1}) = \underbrace{(\text{UD} : \text{DU}, \text{US})}_{\text{up-down-to-down-up upper-saddle}}. \tag{1.255}$$

The equilibrium of $(x^*_{j_1}, x^*_{j_2}) = (a_{j_1 j_1 s_1}, a_{j_2 j_2 1})$ *is a* (UD:DU,US)*-up-down-to-down-up upper-saddle.*

- *For* $a_{j_1 j_1 0}(a_{j_1 j_1 s_1} - a_{j_1 j_1 s_2}) > 0$ *and* $a_{j_2 j_2 0} < 0$,

$$(a_{j_1j_1s_1}, a_{j_2j_21}) = \underbrace{(\text{UD} : \text{DU}, \text{LS})}_{\text{up-down-to-down-up lower-saddle}}. \tag{1.256}$$

The equilibrium of $(x^*_{j_1}, x^*_{j_2}) = (a_{j_1j_1s_1}, a_{j_2j_21})$ *is a* (UD:DU,LS)-*up-down-to-down-up lower-saddle.*

- *For* $a_{j_1j_10}(a_{j_1j_1s_1} - a_{j_1j_1s_2}) < 0$ *and* $a_{j_2j_20} < 0$,

$$(a_{j_1j_1s_1}, a_{j_2j_21}) = \underbrace{(\text{DU} : \text{UD}, \text{LS})}_{\text{down-up-to-up-down lower-saddle}}. \tag{1.257}$$

The equilibrium of $(x^*_{j_1}, x^*_{j_2}) = (a_{j_1j_1s_1}, a_{j_2j_21})$ *is a* (DU:UD,LS)-*down-up-to-up-down lower-saddle.*

(ix) For $\Delta_{j_1j_1} > 0$ and $\Delta_{j_2j_2} > 0$, the standard form is

$$\begin{aligned} \dot{x}_{j_2} &= a_{j_2j_20}(x_{j_2} - a_{j_2j_21})(x_{j_2} - a_{j_2j_22}), \\ \dot{x}_{j_1} &= a_{j_1j_10}(x_{j_1} - a_{j_1j_11})(x_{j_1} - a_{j_1j_12})(x_{j_2} - a_{j_1j_21}) \end{aligned} \tag{1.258}$$

where

$$\begin{aligned} & b_{j_1j_11}, b_{j_1j_12} = -\frac{1}{2}(B_{j_1j_1} \pm \sqrt{\Delta_{j_1j_1}}), \\ & \{a_{j_1j_11}, a_{j_1j_12}\} = \text{sort}\{b_{j_1j_11}, b_{j_1j_12}\}, a_{j_1j_11} < a_{j_1j_12}; \\ & b_{j_2j_21}, b_{j_2j_22} = -\frac{1}{2}(B_{j_2j_2} \pm \sqrt{\Delta_{j_2j_2}}), \\ & \{a_{j_2j_21}, a_{j_2j_22}\} = \text{sort}\{b_{j_2j_21}, b_{j_2j_22}\}, a_{j_2j_21} < a_{j_2j_22}. \end{aligned} \tag{1.259}$$

(ix$_1$) The first integral manifold $(a_{j_2j_2s_1} \neq a_{j_1j_21})$ *is*

$$\begin{aligned} & \frac{1}{a_{j_1j_12} - a_{j_1j_11}}\left(\ln \frac{|x_{j_1} - a_{j_1j_12}|}{|x_{j_10} - a_{j_1j_12}|} - \ln \frac{|x_{j_1} - a_{j_1j_11}|}{|x_{j_10} - a_{j_1j_11}|}\right) \\ & = \frac{a_{j_1j_10}}{a_{j_2j_20}} \sum_{l_1=1}^{2} \frac{a_{j_2j_2l_1} - a_{j_1j_21}}{a_{j_2j_2l_1} - a_{j_2j_2l_2}} \ln \frac{|x_{j_2} - a_{j_2j_2l_1}|}{|x_{j_20} - a_{j_2j_2l_1}|}. \end{aligned} \tag{1.260}$$

(ix$_{1a}$) The equilibrium of $(x^*_{j_1}, x^*_{j_2}) = (a_{j_1j_1s_1}, a_{j_2j_2l_1})$ $(s_1, s_2, l_1, l_2 \in \{1, 2\}, s_1 \neq s_2, l_1 \neq l_2)$ *has the following properties.*

- *For* $a_{j_1j_10}(a_{j_2j_2l_1} - a_{j_1j_21})(a_{j_1j_1s_1} - a_{j_1j_1s_2}) > 0$ *and* $a_{j_2j_20}(a_{j_2j_2l_1} - a_{j_2j_2l_2}) > 0$,

$$(a_{j_1j_1s_1}, a_{j_2j_2l_1}) = \underbrace{(\text{SO}, \text{SO})}_{\text{source}}. \tag{1.261}$$

The equilibrium of $(x^*_{j_1}, x^*_{j_2}) = (a_{j_1j_1s_1}, a_{j_2j_2l_1})$ *is an* (SO,SO)-*source.*

- *For* $a_{j_1j_10}(a_{j_2j_2l_1}-a_{j_1j_21})(a_{j_1j_1s_1}-a_{j_1j_1s_2})<0$ *and* $a_{j_2j_20}(a_{j_2j_2l_1}-a_{j_2j_2l_2})>0$,

$$(a_{j_1j_1s_1},a_{j_2j_2l_1})=\underbrace{(\text{SI},\text{SO})}_{\text{saddle}}. \tag{1.262}$$

The equilibrium of $(x^*_{j_1},x^*_{j_2})=(a_{j_1j_1s_1},a_{j_2j_2l_1})$ *is an* (SI,SO)*-saddle.*

- *For* $a_{j_1j_10}(a_{j_2j_2l_1}-a_{j_1j_21})(a_{j_1j_1s_1}-a_{j_1j_1s_2})>0$ *and* $a_{j_2j_20}(a_{j_2j_2l_1}-a_{j_2j_2l_2})<0$,

$$(a_{j_1j_1s_1},a_{j_2j_2l_1})=\underbrace{(\text{SO},\text{SI})}_{\text{saddle}}. \tag{1.263}$$

The equilibrium of $(x^*_{j_1},x^*_{j_2})=(a_{j_1j_1s_1},a_{j_2j_2l_1})$ *is an* (SO,SI)*-saddle.*

- *For* $a_{j_1j_10}(a_{j_2j_2l_1}-a_{j_1j_21})(a_{j_1j_1s_1}-a_{j_1j_1s_2})<0$ *and* $a_{j_2j_20}(a_{j_2j_2l_1}-a_{j_2j_2l_2})<0$,

$$(a_{j_1j_1s_1},a_{j_2j_2l_1})=\underbrace{(\text{SI},\text{SI})}_{\text{sink}}. \tag{1.264}$$

The equilibrium of $(x^*_{j_1},x^*_{j_2})=(a_{j_1j_1s_1},a_{j_2j_2l_1})$ *is an* (SI,SI)*-sink.*

(ix_{1b}*) The equilibrium of* $(x^*_{j_1},x^*_{j_2})=(a_{j_1j_1s_1},a_{j_1j_21})$ $(s_1,\ s_2\in\{1,2\},\ s_1\neq s_2)$ *with* $a_{j_1j_21}\neq a_{j_2j_21},a_{j_2j_22}$ *has the following properties.*

- *For* $a_{j_1j_10}(a_{j_1j_1s_1}-a_{j_1j_1s_2})>0$ *and* $a_{j_2j_20}\prod_{l_1=1}^{2}(a_{j_1j_21}-a_{j_2j_2l_1})>0$,

$$(a_{j_1j_1s_1},a_{j_1j_21})=\underbrace{(\text{DP}:\text{UP},\text{pF})}_{\text{hyperbolic flow }(+)}. \tag{1.265}$$

The equilibrium of $(x^*_{j_1},x^*_{j_2})=(a_{j_1j_1s_1},a_{j_1j_21})$ *is a* (DP:UP,pF)*-positive hyperbolic flow.*

- *For* $a_{j_1j_10}(a_{j_1j_1s_1}-a_{j_1j_1s_2})<0$ *and* $a_{j_2j_20}\prod_{l_1=1}^{2}(a_{j_1j_21}-a_{j_2j_2l_1})>0$,

$$(a_{j_1j_1s_1},a_{j_1j_21})=\underbrace{(\text{UP}:\text{DP},\text{pF})}_{\text{hyperbolic-secant flow }(+)}. \tag{1.266}$$

The equilibrium of $(x^*_{j_1},x^*_{j_2})=(a_{j_1j_1s_1},a_{j_1j_21})$ *is a* (UP:DP,pF)*-positive hyperbolic-secant flow.*

- *For* $a_{j_1j_10}(a_{j_1j_1s_1}-a_{j_1j_1s_2})>0$ *and* $a_{j_2j_20}\prod_{l_1=1}^{2}(a_{j_1j_21}-a_{j_2j_2l_1})<0$,

$$(a_{j_1j_1s_1},a_{j_1j_21})=\underbrace{(\text{UP}:\text{DP},\text{nF})}_{\text{hyperbolic-secant flow }(-)}. \tag{1.267}$$

The equilibrium of $(x^*_{j_1}, x^*_{j_2}) = (a_{j_1j_1s_1}, a_{j_1j_21})$ *is a* (UP:DP,nF)-*negative hyperbolic-secant flow.*

- *For* $a_{j_1j_10}(a_{j_1j_1s_1} - a_{j_1j_1s_2}) < 0$ *and* $a_{j_2j_20}\prod_{l_1=1}^{2}(a_{j_1j_21} - a_{j_2j_2l_1}) < 0$,

$$(a_{j_1j_1s_1}, a_{j_1j_21}) = \underbrace{(\mathrm{DP} : \mathrm{UP}, \mathrm{nF})}_{\text{hyperbolic flow }(-)}. \tag{1.268}$$

The equilibrium of $(x^*_{j_1}, x^*_{j_2}) = (a_{j_1j_1s_1}, a_{j_1j_21})$ *is a* (DP:UP,nF)-*negative hyperbolic flow.*

(ix$_2$) If $a_{j_1j_21} = a_{j_2j_2l_1}$ $(l_1, l_2 \in \{1, 2\}, l_1 \neq l_2)$*, the first integral manifold is*

$$\frac{1}{a_{j_1j_12} - a_{j_1j_11}}\left(\ln\frac{|x_{j_1} - a_{j_1j_12}|}{|x_{j_10} - a_{j_1j_12}|} - \ln\frac{|x_{j_1} - a_{j_1j_11}|}{|x_{j_10} - a_{j_1j_11}|}\right) = \frac{a_{j_1j_10}}{a_{j_2j_20}}\ln\frac{|x_{j_2} - a_{j_2j_2s_2}|}{|x_{j_2} - a_{j_2j_2s_2}|}. \tag{1.269}$$

(ix$_{2a}$) The infinite-equilibrium of $x^*_{j_2} = a_{j_1j_21} = a_{j_2j_2l_1}$ *with* $\bar{x}_{j_1} \neq a_{j_1j_1s_1}, a_{j_1j_1s_2}$ *has the following properties.*

- *For* $a_{j_1j_10}\prod_{s_1=1}^{2}(\bar{x}_{j_1} - a_{j_1j_1s_1}) > 0$ *and* $a_{j_2j_20}(a_{j_2j_2l_1} - a_{j_2j_2l_2}) > 0$,

$$(\bar{x}_{j_1}, a_{j_2j_2l_1}) = \underbrace{(\mathrm{II}, \mathrm{SO})}_{\text{increasing-inflection source}}. \tag{1.270}$$

The infinite-equilibrium of $x^*_{j_2} = a_{j_1j_21} = a_{j_2j_2l_1}$ *is an* (II,SO)-*increasing-inflection source.*

- *For* $a_{j_1j_10}\prod_{s_1=1}^{2}(\bar{x}_{j_1} - a_{j_1j_1s_1}) < 0$ *and* $a_{j_2j_20}(a_{j_2j_2l_1} - a_{j_2j_2l_2}) > 0$,

$$(\bar{x}_{j_1}, a_{j_2j_2l_1}) = \underbrace{(\mathrm{DI}, \mathrm{SO})}_{\text{decreasing-inflection source}}. \tag{1.271}$$

The infinite-equilibrium of $x^*_{j_2} = a_{j_1j_21} = a_{j_2j_2l_1}$ *is a* (DI,SO)-*decreasing-inflection source.*

- *For* $a_{j_1j_10}\prod_{s_1=1}^{2}(\bar{x}_{j_1} - a_{j_1j_1s_1}) > 0$ *and* $a_{j_2j_20}(a_{j_2j_2l_1} - a_{j_2j_2l_2}) < 0$,

$$(\bar{x}_{j_1}, a_{j_2j_2l_1}) = \underbrace{(\mathrm{DI}, \mathrm{SI})}_{\text{decreasing-inflection sink}}. \tag{1.272}$$

The infinite-equilibrium of $x^*_{j_2} = a_{j_1j_21} = a_{j_2j_2l_1}$ *is a* (DI,SI)-*decreasing-inflection sink.*

- *For* $a_{j_1j_10}\prod_{s_1=1}^{2}(\bar{x}_{j_1} - a_{j_1j_1s_1}) < 0$ *and* $a_{j_2j_20}(a_{j_2j_2l_1} - a_{j_2j_2l_2}) < 0$,

$$(\bar{x}_{j_1}, a_{j_2 j_2 l_1}) = \underbrace{(\text{II}, \text{SI})}_{\text{increasing-inflection sink}}. \quad (1.273)$$

The infinite-equilibrium of $x^*_{j_2} = a_{j_1 j_2 1} = a_{j_2 j_2 l_1}$ *is an* (II,SI)*-increasing-inflection sink.*

(ix$_{2b}$) The equilibrium of $(x^*_{j_1}, x^*_{j_2}) = (a_{j_1 j_1 s_1}, a_{j_2 j_2 l_1})$ *with* $a_{j_2 j_2 s_1} = a_{j_1 j_2 l_1}$ *has the following properties.*

- *For* $a_{j_1 j_1 0}(a_{j_1 j_1 s_1} - a_{j_1 j_1 s_2}) > 0$ *and* $a_{j_2 j_2 0}(a_{j_2 j_2 l_1} - a_{j_2 j_2 l_2}) > 0$,

$$(a_{j_1 j_1 s_1}, a_{j_2 j_2 l_1}) = \underbrace{(\text{DI} : \text{II}, \text{SO})}_{\text{inflection source}}. \quad (1.274)$$

The equilibrium of $(x^*_{j_1}, x^*_{j_2}) = (a_{j_1 j_1 s_1}, a_{j_2 j_2 l_1})$ *is a* (DI:II,SO)*-inflection source.*

- *For* $a_{j_1 j_1 0}(a_{j_1 j_1 s_1} - a_{j_1 j_1 s_2}) < 0$ *and* $a_{j_2 j_2 0}(a_{j_2 j_2 l_1} - a_{j_2 j_2 l_2}) > 0$,

$$(a_{j_1 j_1 s_1}, a_{j_2 j_2 l_1}) = \underbrace{(\text{II} : \text{DI}, \text{SO})}_{\text{inflection source}}. \quad (1.275)$$

The equilibrium of $(x^*_{j_1}, x^*_{j_2}) = (a_{j_1 j_1 s_1}, a_{j_2 j_2 l_1})$ *is an* (II:DI,SO)*-inflection source.*

- *For* $a_{j_1 j_1 0}(a_{j_1 j_1 s_1} - a_{j_1 j_1 s_2}) > 0$ *and* $a_{j_2 j_2 0}(a_{j_2 j_2 l_1} - a_{j_2 j_2 l_2}) < 0$,

$$(a_{j_1 j_1 s_1}, a_{j_2 j_2 l_1}) = \underbrace{(\text{II} : \text{DI}, \text{SI})}_{\text{inflection sink}}. \quad (1.276)$$

The equilibrium of $(x^*_{j_1}, x^*_{j_2}) = (a_{j_1 j_1 s_1}, a_{j_2 j_2 l_1})$ *is an* (II:DI,SI)*-inflection sink.*

- *For* $a_{j_1 j_1 0}(a_{j_1 j_1 s_1} - a_{j_1 j_1 s_2}) < 0$ *and* $a_{j_2 j_2 0}(a_{j_2 j_2 l_1} - a_{j_2 j_2 l_2}) < 0$,

$$(a_{j_1 j_1 s_1}, a_{j_2 j_2 l_1}) = \underbrace{(\text{DI} : \text{II}, \text{SI})}_{\text{inflection sink}}. \quad (1.277)$$

The equilibrium of $(x^*_{j_1}, x^*_{j_2}) = (a_{j_1 j_1 s_1}, a_{j_2 j_2 l_1})$ *is a* (DI:II,SI)*-inflection sink.*

(ix$_{2c}$) The equilibrium of $(x^*_{j_1}, x^*_{j_2}) = (a_{j_1 j_1 s_1}, a_{j_2 j_2 l_2})$ *with* $a_{j_1 j_2 1} = a_{j_2 j_2 l_1}$ *has the following properties.*

- *For* $a_{j_1 j_1 0}(a_{j_1 j_1 s_1} - a_{j_1 j_1 s_2})(a_{j_2 j_2 l_2} - a_{j_1 j_2 1}) > 0$ *and* $a_{j_2 j_2 0}(a_{j_2 j_2 l_2} - a_{j_2 j_2 l_1}) > 0$,

$$(a_{j_1 j_1 s_1}, a_{j_2 j_2 l_2}) = \underbrace{(\text{SO}, \text{SO})}_{\text{source}}. \quad (1.278)$$

The equilibrium of $(x^*_{j_1}, x^*_{j_2}) = (a_{j_1 j_1 s_1}, a_{j_2 j_2 l_2})$ *is an* (SO,SO)*-source.*

- *For* $a_{j_1 j_1 0}(a_{j_1 j_1 s_1} - a_{j_1 j_1 s_2})(a_{j_2 j_2 l_2} - a_{j_1 j_2 1}) < 0$ *and* $a_{j_2 j_2 0}(a_{j_2 j_2 l_2} - a_{j_2 j_2 l_1}) > 0$,

$$(a_{j_1j_1s_1}, a_{j_2j_2l_2}) = \underbrace{(\text{SI}, \text{SO})}_{\text{saddle}}. \tag{1.279}$$

The equilibrium of $(x^*_{j_1}, x^*_{j_2}) = (a_{j_1j_1s_1}, a_{j_2j_2l_2})$ *is an* (SI,SO)*-saddle.*

- *For* $a_{j_1j_10}(a_{j_1j_1s_1} - a_{j_1j_1s_2})(a_{j_2j_2l_2} - a_{j_1j_21}) > 0$ *and* $a_{j_2j_20}(a_{j_2j_2l_2} - a_{j_2j_2l_1}) < 0$,

$$(a_{j_1j_1s_1}, a_{j_2j_2l_2}) = \underbrace{(\text{SO}, \text{SI})}_{\text{saddle}}. \tag{1.280}$$

The equilibrium of $(x^*_{j_1}, x^*_{j_2}) = (a_{j_1j_1s_1}, a_{j_2j_2l_2})$ *is an* (SO,SI)*-saddle.*

- *For* $a_{j_1j_10}(a_{j_1j_1s_1} - a_{j_1j_1s_2})(a_{j_2j_2l_2} - a_{j_1j_21}) < 0$ *and* $a_{j_2j_20}(a_{j_2j_2l_2} - a_{j_2j_2l_1}) < 0$,

$$(a_{j_1j_1s_1}, a_{j_2j_2l_2}) = \underbrace{(\text{SI}, \text{SI})}_{\text{sink}}. \tag{1.281}$$

The equilibrium of $(x^*_{j_1}, x^*_{j_2}) = (a_{j_1j_1s_1}, a_{j_2j_2l_2})$ *is an* (SI,SI)*-sink.*

1.4 Proof of Theorem 1.2

Consider a two-dimensional, nonlinear dynamical system as

$$\begin{aligned}\dot{x}_{j_2} &= a_{j_2j_20}(x^2_{j_2} + B_{j_2j_2}x_{j_2} + C_{j_2j_2}),\\ \dot{x}_{j_1} &= a_{j_1j_10}(x_{j_2} - a_{j_1j_21})(x^2_{j_1} + B_{j_1j_1}x_{j_1} + C_{j_1j_1})\end{aligned}$$

with

$$\begin{aligned}\Delta_{j_2j_2} &= B^2_{j_2j_2} - 4C_{j_2j_2},\\ \Delta_{j_1j_1} &= B^2_{j_1j_1} - 4C_{j_1j_1}.\end{aligned}$$

(i) For $\Delta_{j_1j_1} < 0$ and $\Delta_{j_2j_2} < 0$, the standard form is

$$\begin{aligned}\dot{x}_{j_2} &= a_{j_2j_20}\left[(x_{j_2} - a_{j_2j_2})^2 + b_{j_2j_2}\right],\\ \dot{x}_{j_1} &= a_{j_1j_10}\left[(x_{j_1} - a_{j_1j_1})^2 + b_{j_1j_1}\right](x_{j_2} - a_{j_1j_21})\end{aligned}$$

where

$$\begin{aligned}a_{j_2j_2} &= -\frac{1}{2}B_{j_2j_2}, b_{j_2j_2} = \frac{1}{4}(-\Delta_{j_1j_2});\\ a_{j_1j_1} &= -\frac{1}{2}B_{j_1j_1}, b_{j_1j_1} = \frac{1}{4}(-\Delta_{j_1j_1}).\end{aligned}$$

In phase space,

$$\frac{dx_{j_1}}{dx_{j_2}} = \frac{a_{j_1j_10}}{a_{j_2j_20}} \frac{(x_{j_2} - a_{j_1j_21})\left[(x_{j_1} - a_{j_1j_1})^2 + b_{j_1j_1}\right]}{(x_{j_2} - a_{j_2j_2})^2 + b_{j_2j_2}},$$

and the deformation of the foregoing equation is

$$\frac{1}{(x_{j_1} - a_{j_1j_1})^2 + b_{j_1j_1}} dx_{j_1} = \frac{a_{j_1j_10}}{a_{j_2j_20}} \left[\frac{(x_{j_2} - a_{j_2j_2})}{(x_{j_2} - a_{j_2j_2})^2 + b_{j_2j_2}} + \frac{(a_{j_2j_2} - a_{j_1j_21})}{(x_{j_2} - a_{j_2j_2})^2 + b_{j_2j_2}}\right] dx_{j_2}.$$

With an initial condition of (x_{j_10}, x_{j_20}) at $t = t_0$, the integration of the above equation gives

$$\begin{aligned}
&\frac{1}{\sqrt{b_{j_1j_1}}}\left(\arctan\frac{x_{j_1} - a_{j_1j_1}}{\sqrt{b_{j_1j_1}}} - \arctan\frac{x_{j_10} - a_{j_1j_1}}{\sqrt{b_{j_1j_1}}}\right)\\
&= \frac{a_{j_1j_10}}{a_{j_2j_20}}\left[\frac{1}{2}\ln\frac{\left|(x_{j_2} - a_{j_2j_2})^2 + b_{j_2j_2}\right|}{\left|(x_{j_20} - a_{j_2j_2})^2 + b_{j_2j_2}\right|}\right.\\
&\quad \left. + \frac{a_{j_2j_2} - a_{j_1j_21}}{\sqrt{b_{j_2j_2}}}\left(\arctan\frac{x_{j_2} - a_{j_2j_2}}{\sqrt{b_{j_2j_2}}} - \arctan\frac{x_{j_20} - a_{j_2j_2}}{\sqrt{b_{j_2j_2}}}\right)\right].
\end{aligned}$$

At $x_{j_1}^* = a_{j_1j_21}$, in phase space,

$$\left.\frac{dx_{j_1}}{dx_{j_2}}\right|_{x_{j_2}^* = a_{j_1j_21}} = \frac{a_{j_1j_10}}{a_{j_2j_20}} \left.\frac{(x_{j_2} - a_{j_1j_21})\left[(x_{j_1} - a_{j_1j_1})^2 + b_{j_1j_1}\right]}{(x_{j_2} - a_{j_2j_2})^2 + b_{j_2j_2}}\right|_{x_{j_2}^* = a_{j_1j_21}} = 0.$$

If

$$\left.\frac{d^2x_{j_1}}{dx_{j_2}^2}\right|_{x_{j_2}^* = a_{j_1j_21}} = \frac{a_{j_1j_10}}{a_{j_2j_20}} \frac{\left[(\bar{x}_{j_1} - a_{j_1j_1})^2 + b_{j_1j_1}\right]}{(a_{j_1j_21} - a_{j_2j_2})^2 + b_{j_2j_2}} > 0,$$

there is an up-parabola flow. If

$$\left.\frac{d^2x_{j_1}}{dx_{j_2}^2}\right|_{x_{j_2}^* = a_{j_1j_21}} = \frac{a_{j_1j_10}}{a_{j_2j_20}} \frac{\left[(\bar{x}_{j_1} - a_{j_1j_1})^2 + b_{j_1j_1}\right]}{(a_{j_1j_21} - a_{j_2j_2})^2 + b_{j_2j_2}} < 0,$$

there is a down-parabola flow. Let

$$\dot{x}_{j_2} = a_{j_2j_20}\left[(a_{j_1j_21} - a_{j_2j_2})^2 + b_{j_2j_2}\right].$$

Because

$$(a_{j_1j_21} - a_{j_2j_2})^2 + b_{j_2j_2} > 0,$$

the parabola flows at $x_{j_1}^* = a_{j_1j_1}1$ are positive and negative for $a_{j_2j_20} > 0$ and $a_{j_2j_20} < 0$, respectively. In addition,

$$(\bar{x}_{j_1} - a_{j_1j_1})^2 + b_{j_1j_1} > 0.$$

Therefore, the equilibrium of $x_{j_1}^* = a_{j_1j_2}1$ has the following properties as in Eqs. (1.144)–(1.147).

- For $a_{j_1j_10} > 0$ and $a_{j_2j_20} > 0$,

$$(a_{j_1j_2}1, \dot{x}_{j_2}) = \underbrace{(\mathrm{UP}, \mathrm{pF})}_{\text{up-parabola flow } (+)}.$$

  The equilibrium of $x_{j_1}^* = a_{j_1j_2}1$ is a (UP,pF)-positive up-parabola flow.
- For $a_{j_1j_10} < 0$ and $a_{j_2j_20} > 0$,

$$(a_{j_1j_2}1, \dot{x}_{j_2}) = \underbrace{(\mathrm{DP}, \mathrm{pF})}_{\text{down-parabola flow } (+)}.$$

  The equilibrium of $x_{j_1}^* = a_{j_1j_2}1$ is a (DP,pF)-positive down-parabola flow.
- For $a_{j_1j_10} > 0$ and $a_{j_2j_20} < 0$,

$$(a_{j_1j_2}1, \dot{x}_{j_2}) = \underbrace{(\mathrm{DP}, \mathrm{nF})}_{\text{down-parabola flow } (-)}.$$

  The equilibrium of $x_{j_1}^* = a_{j_1j_2}1$ is a (DP,nF)-negative down-parabola flow.
- For $a_{j_1j_10} < 0$ and $a_{j_2j_20} < 0$,

$$(a_{j_1j_2}1, \dot{x}_{j_2}) = \underbrace{(\mathrm{UP}, \mathrm{nF})}_{\text{up-parabola flow } (-)}.$$

  The equilibrium of $x_{j_1}^* = a_{j_1j_2}1$ is a (UP,nF)-negative up-parabola flow.

(ii) For $\Delta_{j_1j_2} < 0$ and $\Delta_{j_2j_2} = 0$, the standard form is

$$\begin{aligned}
\dot{x}_{j_2} &= a_{j_2j_20}(x_{j_2} - a_{j_2j_2}1)^2, \\
\dot{x}_{j_1} &= a_{j_1j_10}(x_{j_2} - a_{j_1j_2}1)\left[(x_{j_1} - a_{j_1j_1})^2 + b_{j_1j_1}\right]
\end{aligned}$$

where

$$\begin{aligned} a_{j_1j_1} &= -\frac{1}{2}B_{j_1j_1}, b_{j_1j_1} = \frac{1}{4}(-\Delta_{j_1j_1}), \\ a_{j_2j_2 1} &= -\frac{1}{2}B_{j_2j_2}. \end{aligned}$$

In phase space,

$$\frac{dx_{j_1}}{dx_{j_2}} = \frac{a_{j_1j_1 0}}{a_{j_2j_2 0}} \frac{(x_{j_2} - a_{j_1j_2 1})\left[(x_{j_1} - a_{j_1j_1})^2 + b_{j_1j_1}\right]}{(x_{j_2} - a_{j_2j_2 1})^2},$$

and the deformation of the foregoing equation is

$$\frac{1}{(x_{j_1} - a_{j_1j_1})^2 + b_{j_1j_1}} dx_{j_1} = \frac{a_{j_1j_1 0}}{a_{j_2j_2 0}} \left[\frac{1}{(x_{j_2} - a_{j_2j_2 1})} + \frac{(a_{j_2j_2 1} - a_{j_1j_2 1})}{(x_{j_2} - a_{j_2j_2 1})^2}\right] dx_{j_2}.$$

With an initial condition of $(x_{j_1 0}, x_{j_2 0})$ at $t = t_0$, the integration of the above equation gives

$$\begin{aligned} &\frac{1}{\sqrt{b_{j_1j_1}}}\left(\arctan\frac{x_{j_1} - a_{j_1j_1}}{\sqrt{b_{j_1j_1}}} - \arctan\frac{x_{j_1 0} - a_{j_1j_1}}{\sqrt{b_{j_1j_1}}}\right) \\ &= \frac{a_{j_1j_1 0}}{a_{j_2j_2 0}}\left[\ln\frac{|x_{j_2} - a_{j_2j_2 1}|}{|x_{j_2 0} - a_{j_2j_2 1}|} - (a_{j_2j_2 1} - a_{j_1j_2 1})\left(\frac{1}{x_{j_2} - a_{j_2j_2 1}} - \frac{1}{x_{j_2 0} - a_{j_2j_2 1}}\right)\right]. \end{aligned}$$

($\text{ii}_{1\text{a}}$) At $x_{j_2}^* = a_{j_1j_2 1} \neq a_{j_2j_2 1}$, similar to case (i), in phase space,

$$\left.\frac{dx_{j_1}}{dx_{j_2}}\right|_{x_{j_2}^* = a_{j_1j_2 1}} = \frac{a_{j_1j_1 0}}{a_{j_2j_2 0}} \left.\frac{(x_{j_2} - a_{j_1j_2 1})\left[(x_{j_1} - a_{j_1j_1})^2 + b_{j_1j_1}\right]}{(x_{j_2} - a_{j_2j_2 1})^2}\right|_{x_{j_2}^* = a_{j_1j_2 1}} = 0.$$

If

$$\left.\frac{d^2x_{j_1}}{dx_{j_2}^2}\right|_{x_{j_2}^* = a_{j_1j_2 1}} = \frac{a_{j_1j_1 0}}{a_{j_2j_2 0}} \frac{\left[(\bar{x}_{j_1} - a_{j_1j_1})^2 + b_{j_1j_1}\right]}{(a_{j_1j_2 1} - a_{j_2j_2 1})^2} > 0,$$

there is an up-parabola flow. If

$$\left.\frac{d^2x_{j_1}}{dx_{j_2}^2}\right|_{x_{j_2}^* = a_{j_1j_2 1}} = \frac{a_{j_1j_1 0}}{a_{j_2j_2 0}} \frac{\left[(\bar{x}_{j_1} - a_{j_1j_1})^2 + b_{j_1j_1}\right]}{(a_{j_1j_2 1} - a_{j_2j_2 1})^2} < 0,$$

there is a down-parabola flow. Let

$$\dot{x}_{j_2} = a_{j_2j_20}(a_{j_1j_21} - a_{j_2j_21})^2 \neq 0.$$

The parabola flows at $x_{j_1}^* = a_{j_1j_11}$ in the x_{j_2}-direction are positive and negative for $a_{j_2j_20}(a_{j_1j_21} - a_{j_2j_21})^2 > 0$ and $a_{j_2j_20}(a_{j_1j_21} - a_{j_2j_21})^2 < 0$, respectively. In addition,

$$(\overline{x}_{j_1} - a_{j_1j_1})^2 + b_{j_1j_1} > 0.$$

Therefore, the equilibrium of $x_{j_2}^* = a_{j_1j_21} \neq a_{j_2j_21}$ has the following properties as in Eqs. (1.151)–(1.154).

- For $a_{i_1i_10} > 0$ and $a_{j_2j_20}(a_{j_1j_21} - a_{j_2j_21})^2 > 0$,

$$(a_{j_1j_21}, \dot{x}_{j_2}) = \underbrace{(\text{UP}, \text{pF})}_{\text{up-parabola flow } (+)}.$$

  The equilibrium of $x_{j_1}^* = a_{j_1j_11}$ is a (UP,pF)-positive up-parabola flow.
- For $a_{j_1j_10} < 0$ and $a_{j_2j_20}(a_{j_1j_21} - a_{j_2j_21})^2 > 0$,

$$(a_{j_1j_21}, \dot{x}_{j_2}) = \underbrace{(\text{DP}, \text{pF})}_{\text{down-parabola flow } (+)}.$$

  The equilibrium of $x_{j_1}^* = a_{j_1j_11}$ is a (DP,pF)-positive down-parabola flow.
- For $a_{j_1j_10} > 0$ and $a_{j_2j_20}(a_{j_1j_21} - a_{j_2j_21})^2 < 0$,

$$(a_{j_1j_21}, \dot{x}_{j_2}) = \underbrace{(\text{DP}, \text{nF})}_{\text{down-parabola flow } (-)}.$$

  The equilibrium of $x_{j_1}^* = a_{j_1j_11}$ is a (DP,nF)-negative down-parabola flow.
- For $a_{j_1j_10} < 0$ and $a_{j_2j_20}(a_{j_1j_21} - a_{j_2j_21})^2 < 0$,

$$(a_{j_1j_21}, \dot{x}_{j_2}) = \underbrace{(\text{UP}, \text{nF})}_{\text{up-parabola flow } (-)}.$$

  The equilibrium of $x_{j_1}^* = a_{j_1j_11}$ is a (UP,nF)-negative up-parabola flow.

($\text{ii}_{1\text{b}}$) At $x_{j_2}^* = a_{j_2j_21} \neq a_{j_1j_21}$, the variational equation is

$$\Delta\dot{x}_{j_2} = a_{j_2j_20}(\Delta x_{j_2})^2.$$

The flows at $x_{j_1}^* = a_{j_1j_11}$ in the x_{j_2}-direction are upper-saddle and lower-saddle for $a_{j_2j_20} > 0$ and $a_{j_2j_20} < 0$, respectively. Let

$$\dot{x}_{j_1} = a_{j_1 j_1 0}(a_{j_2 j_2 1} - a_{j_1 j_1 1})\left[(\overline{x}_{j_1} - a_{j_1 j_1})^2 + b_{j_1 j_1}\right],$$

with

$$(\overline{x}_{j_1} - a_{j_1 j_1})^2 + b_{j_1 j_1} > 0.$$

Therefore, the equilibrium of $x^*_{j_2} = a_{j_2 j_2 1} \neq a_{j_1 j_2 1}$ has the following properties as in (1.155)–(1.159).

- For $a_{j_1 j_1 0}(a_{j_2 j_2 1} - a_{j_1 j_2 1}) > 0$ and $a_{j_2 j_2 0} > 0$,

$$(\dot{x}_{j_1}, a_{j_2 j_2 1}) = \underbrace{(\text{pF}, \text{US})}_{\text{upper-saddle flow }(+)}.$$

  The equilibrium of $x^*_{j_2} = a_{j_2 j_2 1} \neq a_{j_1 j_2 1}$ is a (pF,US)-positive upper-saddle flow.
- For $a_{j_1 j_1 0}(a_{j_2 j_2 1} - a_{j_1 j_2 1}) < 0$ and $a_{j_2 j_2 0} > 0$,

$$(\dot{x}_{j_1}, a_{j_2 j_2 1}) = \underbrace{(\text{nF}, \text{US})}_{\text{upper-saddle flow }(-)}.$$

  The equilibrium of $x^*_{j_2} = a_{j_2 j_2 1} \neq a_{j_1 j_2 1}$ is an (nF,US)-negative upper-saddle flow.
- For $a_{j_1 j_1 0}(a_{j_2 j_2 1} - a_{j_1 j_2 1}) > 0$ and $a_{j_2 j_2 0} < 0$,

$$(\dot{x}_{j_1}, a_{j_2 j_2 1}) = \underbrace{(\text{pF}, \text{LS})}_{\text{lower-saddle flow }(+)}.$$

  The equilibrium of $x^*_{j_2} = a_{j_2 j_2 1} \neq a_{j_1 j_2 1}$ is a (pF,LS)-positive lower-saddle flow.
- For $a_{j_1 j_1 0}(a_{j_2 j_2 1} - a_{j_1 j_2 1}) < 0$ and $a_{j_2 j_2 0} < 0$,

$$(\dot{x}_{j_1}, a_{j_2 j_2 1}) = \underbrace{(\text{nF}, \text{LS})}_{\text{lower-saddle flow }(-)}.$$

  The equilibrium of $x^*_{j_2} = a_{j_2 j_2 1} \neq a_{j_1 j_2 1}$ is an (nF,LS)-negative lower-saddle flow.

– The upper-saddle and lower-saddle flows are the appearing bifurcations of sink and source flows.

(ii$_2$) For $a_{j_1 j_2 1} = a_{j_2 j_2 1}$, in phase space,

$$\frac{dx_{j_1}}{dx_{j_2}} = \frac{a_{j_1 j_1 0}}{a_{j_2 j_2 0}} \frac{\left[(x_{j_1} - a_{j_1 j_1})^2 + b_{j_1 j_1}\right]}{(x_{j_2} - a_{j_2 j_2 1})},$$

and the deformation of the foregoing equation is

$$\frac{dx_{j_1}}{(x_{j_1}-a_{j_1j_1})^2+b_{j_1j_1}}=\frac{a_{j_1j_10}}{a_{j_2j_20}}\frac{dx_{j_2}}{x_{j_2}-a_{j_2j_21}}.$$

With an initial condition of (x_{j_10}, x_{j_20}) at $t=t_0$, the integration of the above equation gives the first integral manifold for $a_{j_1j_21}=a_{j_2j_21}$ as

$$\frac{1}{\sqrt{b_{j_1j_1}}}(\arctan\frac{x_{j_1}-a_{j_1j_1}}{\sqrt{b_{j_1j_1}}}-\arctan\frac{x_{j_10}-a_{j_1j_1}}{\sqrt{b_{j_1j_1}}})=\frac{a_{j_1j_10}}{a_{j_2j_20}}\ln\frac{|x_{j_2}-a_{j_2j_21}|}{|x_{j_20}-a_{j_2j_21}|}.$$

In phase space, at $x_{j_2}^*=a_{j_1j_21}=a_{j_2j_21}$ with $\bar{x}_{j_1}\in\{-\infty,\infty\}$,

$$\frac{dx_{j_1}}{dx_{j_2}}\Big|_{x_{j_2}^*=a_{j_2j_21}\pm\varepsilon}=\pm\frac{a_{j_1j_10}}{a_{j_2j_20}}\frac{\left[(\bar{x}_{j_1}-a_{j_1j_1})^2+b_{j_1j_1}\right]}{\varepsilon}.$$

If

$$\frac{a_{j_1j_10}}{a_{j_2j_20}}\left[(\bar{x}_{j_1}-a_{j_1j_1})^2+b_{j_1j_1}\right]>0,$$

there is a down-up flow with a separatrix of $x_{j_2}^*=a_{j_2j_21}$. If

$$\frac{a_{j_1j_10}}{a_{j_2j_20}}\left[(\bar{x}_{j_1}-a_{j_1j_1})^2+b_{j_1j_1}\right]<0,$$

there is an up-down flow with a separatrix of $x_{j_2}^*=a_{j_2j_21}$. The variational equation is

$$\Delta\dot{x}_{j_2}=a_{j_2j_20}(\Delta x_{j_2})^2.$$

The up-down and down-up flows at $x_{j_1}^*=a_{j_1j_11}$ in the x_{j_2}-direction are upper-saddle and lower-saddle for $a_{j_2j_20}>0$ and $a_{j_2j_20}<0$, respectively. In addition,

$$(\bar{x}_{j_1}-a_{j_1j_1})^2+b_{j_1j_1}>0.$$

Therefore, the infinite-equilibrium of $x_{j_2}^*=a_{j_1j_21}=a_{j_2j_21}$ with $\bar{x}_{j_1}\in(-\infty,\infty)$ has the following properties as in Eqs. (1.160)–(1.163).

- For $a_{j_1j_10}>0$ and $a_{j_2j_20}>0$,

$$(\bar{x}_{j_1},a_{j_2j_21})=\underbrace{(\text{DU},\text{US})}_{\text{down-up upper-saddle}}.$$

  The infinite-equilibrium of $x_{j_2}^*=a_{j_1j_21}=a_{j_2j_21}$ is a (DU,US)-down-up upper-saddle.

- For $a_{j_1j_10} < 0$ and $a_{j_2j_20} > 0$,

$$(\bar{x}_{j_1}, a_{j_2j_21}) = \underbrace{(\text{UD}, \text{US})}_{\text{up-down upper-saddle}}.$$

The infinite-equilibrium of $x^*_{j_2} = a_{j_1j_21} = a_{j_2j_21}$ is a (UD,US)-up-down upper-saddle.

- For $a_{j_1j_10} > 0$ and $a_{j_2j_20} < 0$,

$$(\bar{x}_{j_1}, a_{j_2j_21}) = \underbrace{(\text{UD}, \text{LS})}_{\text{up-down lower-saddle}}.$$

The infinite-equilibrium of $x^*_{j_2} = a_{j_1j_21} = a_{j_2j_21}$ is a (UD,LS)-up-down-lower-saddle.

- For $a_{j_1j_10} < 0$ and $a_{j_2j_20} < 0$,

$$(\bar{x}_{j_1}, a_{j_2j_21}) = \underbrace{(\text{DU}, \text{LS})}_{\text{down-up lower-saddle}}.$$

The infinite-equilibrium of $x^*_{j_2} = a_{j_1j_21} = a_{j_2j_21}$ is a (DU,LS)-down-up lower-saddle.

(iii) For $\Delta_{j_1j_2} < 0$ and $\Delta_{j_2j_2} > 0$, the standard form is

$$\begin{aligned}\dot{x}_{j_2} &= a_{j_2j_20}(x_{j_2} - a_{j_2j_21})(x_{j_2} - a_{j_2j_22}),\\ \dot{x}_{j_1} &= a_{j_1j_10}\left[(x_{j_1} - a_{j_1j_1})^2 + b_{j_1j_1}\right](x_{j_2} - a_{j_1j_21})\end{aligned}$$

where

$$\begin{aligned}a_{j_1j_1} &= -\frac{1}{2}B_{j_1j_1}, b_{j_1j_1} = \frac{1}{4}(-\Delta_{j_1j_1}),\\ b_{j_2j_21}, b_{j_2j_22} &= -\frac{1}{2}(B_{j_2j_2} \pm \sqrt{\Delta_{j_2j_2}}),\\ \{a_{j_2j_21}, a_{j_2j_22}\} &= \text{sort}\{b_{j_2j_21}, b_{j_2j_22}\}, a_{j_2j_21} < a_{j_2j_22}.\end{aligned}$$

(iii_1) In phase space, for $a_{j_1j_21} \neq a_{j_2j_2s_1}$ $(s_1 = 1, 2)$,

$$\frac{dx_{j_1}}{dx_{j_2}} = \frac{a_{j_1j_10}}{a_{j_2j_20}} \frac{(x_{j_2} - a_{j_1j_21})\left[(x_{j_1} - a_{j_1j_1})^2 + b_{j_1j_1}\right]}{(x_{j_2} - a_{j_2j_21})(x_{j_2} - a_{j_2j_22})},$$

and the deformation of the foregoing equation is

$$\frac{1}{(x_{j_1}-a_{j_1j_1})^2+b_{j_1j_1}}dx_{j_1}=\frac{a_{j_1j_10}}{a_{j_2j_20}}\left[\frac{(a_{j_2j_21}-a_{j_1j_21})}{(a_{j_2j_21}-a_{j_2j_22})}\frac{1}{(x_{j_2}-a_{j_2j_21})}\right.$$
$$\left.+\frac{(a_{j_2j_22}-a_{j_1j_21})}{(a_{j_2j_22}-a_{j_2j_21})}\frac{1}{(x_{j_2}-a_{j_2j_22})}\right]dx_{j_2}.$$

With an initial condition of (x_{j_10}, x_{j_20}) at $t=t_0$, the integration of the above equation gives

$$\frac{1}{\sqrt{b_{j_1j_1}}}\left(\arctan\frac{x_{j_1}-a_{j_1j_1}}{\sqrt{b_{j_1j_1}}}-\arctan\frac{x_{j_10}-a_{j_1j_1}}{\sqrt{b_{j_1j_1}}}\right)$$
$$=\frac{a_{j_1j_10}}{a_{j_2j_20}}\sum\nolimits_{s_1=1}^{2}\frac{(a_{j_2j_2s_1}-a_{j_1j_21})}{(a_{j_2j_2s_1}-a_{j_2j_2s_2})}\ln\frac{|x_{j_2}-a_{j_2j_2s_1}|}{|x_{j_2}-a_{j_2j_2s_1}|}.$$

(iii$_{1a}$) At $x_{j_2}^*=a_{j_1j_21}\neq a_{j_2j_2s_1}, a_{j_2j_2s_2}$ $(s_1, s_2\in\{1,2\}, s_1\neq s_2)$, in phase space,

$$\left.\frac{dx_{j_1}}{dx_{j_2}}\right|_{x_{j_2}^*=a_{j_1j_21}}=\frac{a_{j_1j_10}}{a_{j_2j_20}}\left.\frac{(x_{j_2}-a_{j_1j_21})[(x_{j_1}-a_{j_1j_1})^2+b_{j_1j_1}]}{\prod_{s_1=1}^{2}(x_{j_2}-a_{j_2j_2s_1})}\right|_{x_{j_2}^*=a_{j_1j_21}}=0.$$

If

$$\left.\frac{d^2x_{j_1}}{dx_{j_2}^2}\right|_{x_{j_2}^*=a_{j_1j_21}}=\frac{a_{j_1j_10}}{a_{j_2j_20}}\frac{[(\bar{x}_{j_1}-a_{j_1j_1})^2+b_{j_1j_1}]}{\prod_{s_1=1}^{2}(a_{j_1j_21}-a_{j_2j_2s_1})}>0,$$

there is an up-parabola flow. If

$$\left.\frac{d^2x_{j_1}}{dx_{j_2}^2}\right|_{x_{j_2}^*=a_{j_1j_21}}=\frac{a_{j_1j_10}}{a_{j_2j_20}}\frac{[(\bar{x}_{j_1}-a_{j_1j_1})^2+b_{j_1j_1}]}{\prod_{s_1=1}^{2}(a_{j_1j_21}-a_{j_2j_2s_1})}>0,$$

there is a down-parabola flow. Let

$$\dot{x}_{j_2}=a_{j_2j_20}\prod\nolimits_{s_1=1}^{2}(a_{j_1j_21}-a_{j_2j_2s_1})\neq 0.$$

The parabola flows at $x_{j_1}^*=a_{j_1j_11}$ in the x_{j_2}-direction are positive and negative for $a_{j_2j_20}\prod_{s_1=1}^{2}(a_{j_1j_21}-a_{j_2j_2s_1})>0$ and $a_{j_2j_20}\prod_{s_1=1}^{2}(a_{j_1j_21}-a_{j_2j_2s_1})<0$, respectively. We have,

$$(\bar{x}_{j_1}-a_{j_1j_1})^2+b_{j_1j_1}>0.$$

Therefore, the equilibrium of $x_{j_2}^*=a_{j_1j_21}\neq a_{j_2j_2s_1}, a_{j_2j_2s_2}$ $(s_1, s_2\in\{1,2\}, s_1\neq s_2)$ has the following properties as in Eqs. (1.167)–(1.170).

- For $a_{j_1j_10}>0$ and $a_{j_2j_20}\prod_{s_1=1}^{2}(a_{j_1j_21}-a_{j_2j_2s_1})>0$,

$$(a_{j_1j_21},\dot{x}_{j_2})=\underbrace{(\text{UP},\text{pF})}_{\text{up-parabola flow }(+)}.$$

The equilibrium of $x_{j_1}^*=a_{j_1j_21}$ is a (UP,pF)-positive up-parabola flow.

- For $a_{j_1j_10}<0$ and $a_{j_2j_20}\prod_{s_1=1}^{2}(a_{j_1j_21}-a_{j_2j_2s_1})>0$,

$$(a_{j_1j_21},\dot{x}_{j_2})=\underbrace{(\text{DP},\text{pF})}_{\text{down-parabola flow }(+)}.$$

The equilibrium of $x_{j_1}^*=a_{j_1j_21}$ is a (DP,pF)-positive down-parabola flow.

- For $a_{j_1j_10}>0$ and $a_{j_2j_20}\prod_{s_1=1}^{2}(a_{j_1j_21}-a_{j_2j_2s_1})<0$,

$$(a_{j_1j_21},\dot{x}_{j_2})=\underbrace{(\text{DP},\text{nF})}_{\text{down-parabola flow }(-)}.$$

The equilibrium of $x_{j_1}^*=a_{j_1j_21}$ is a (DP,nF)-negative down-parabola flow.

- For $a_{j_1j_10}<0$ and $a_{j_2j_20}\prod_{s_1=1}^{2}(a_{j_1j_21}-a_{j_2j_2s_1})<0$,

$$(a_{j_1j_21},\dot{x}_{j_2})=\underbrace{(\text{UP},\text{nF})}_{\text{up-parabola flow }(-)}.$$

The equilibrium of $x_{j_1}^*=a_{j_1j_21}$ is a (UP,nF)-negative up-parabola flow.

(iii$_{1b}$) At $x_{j_2}^*=a_{j_2j_2s_1}\neq a_{j_1j_21}$ ($s_1,\ s_2\in\{1,2\},\ s_1\neq s_2$) with $\bar{x}_{j_1}\in(-\infty,\infty)$, the variational equation is

$$\begin{aligned}\Delta\dot{x}_{j_2}&=a_{j_2j_20}(a_{j_2j_2s_1}-a_{j_2j_2s_2})\Delta x_{j_2},\\ \dot{x}_{j_1}&=a_{j_1j_10}(a_{j_2j_2s_1}-a_{j_1j_21})\left[(\bar{x}_{j_1}-a_{j_1j_1})^2+b_{j_1j_1}\right]\end{aligned}$$

with

$$(\bar{x}_{j_1}-a_{j_1j_1})^2+b_{j_1j_1}>0.$$

The flows at $x_{j_2}^*=a_{j_2j_2s_1}$ for $a_{j_2j_20}(a_{j_2j_2s_1}-a_{j_2j_2s_2})>0$ and $a_{j_2j_20}(a_{j_2j_2s_1}-a_{j_2j_2s_2})<0$ are source and sink in the x_{j_2}-direction, respectively. The flows at $x_{j_2}^*=a_{j_2j_2s_1}$ in the x_{j_1}-direction are positive and negative for $a_{j_1j_10}(a_{j_2j_2s_1}-a_{j_1j_21})>0$ and $a_{j_1j_10}(a_{j_2j_2s_1}-a_{j_1j_21})<0$, respectively.

Therefore, the equilibrium of $x_{j_2}^* = a_{j_2j_2s_1} \neq a_{j_1j_2 1}$ (s_1, $s_2 \in \{1,2\}$, $s_1 \neq s_2$) has the following properties as in Eqs. (1.171)–(1.174).

- For $a_{j_1j_1 0}(a_{j_2j_2s_1} - a_{j_1j_2 1}) > 0$ and $a_{j_2j_2 0}(a_{j_2j_2s_1} - a_{j_2j_2s_2}) > 0$,

$$(\dot{x}_{j_1}, a_{j_2j_2s_1}) = \underbrace{(\text{pF}, \text{SO})}_{\text{source flow }(+)}.$$

The equilibrium of $x_{j_2}^* = a_{j_2j_2s_1} \neq a_{j_1j_2 1}$ is a (pF,SO)-positive source flow.

- For $a_{j_1j_1 0}(a_{j_2j_2s_1} - a_{j_1j_2 1}) < 0$ and $a_{j_2j_2 0}(a_{j_2j_2s_1} - a_{j_2j_2s_2}) > 0$,

$$(\dot{x}_{j_1}, a_{j_2j_2s_1}) = \underbrace{(\text{nF}, \text{SO})}_{\text{source flow }(-)}.$$

The equilibrium of $x_{j_2}^* = a_{j_2j_2s_1} \neq a_{j_1j_2 1}$ is an (nF,SO)-negative source flow.

- For $a_{j_1j_1 0}(a_{j_2j_2s_1} - a_{j_1j_2 1}) > 0$ and $a_{j_2j_2 0}(a_{j_2j_2s_1} - a_{j_2j_2s_2}) < 0$,

$$(\dot{x}_{j_1}, a_{j_2j_2s_1}) = \underbrace{(\text{pF}, \text{SI})}_{\text{sink flow }(+)}.$$

The equilibrium of $x_{j_2}^* = a_{j_2j_2s_1} \neq a_{j_1j_2 1}$ is a (pF,SI)-positive sink flow.

- For $a_{j_1j_1 0}(a_{j_2j_2s_1} - a_{j_1j_2 1}) < 0$ and $a_{j_2j_2 0}(a_{j_2j_2s_1} - a_{j_2j_2s_2}) < 0$,

$$(\dot{x}_{j_1}, a_{j_2j_2s_1}) = \underbrace{(\text{nF}, \text{SI})}_{\text{sink flow }(-)}.$$

The equilibrium of $x_{j_2}^* = a_{j_2j_2s_1} \neq a_{j_1j_2 1}$ is an (nF,SI)-negative sink flow.

(iii$_2$) At $a_{j_1j_2 1} = a_{j_2j_2s_1}$ (s_1, $s_2 \in \{1,2\}$, $s_1 \neq s_2$), In phase space,

$$\frac{dx_{j_1}}{dx_{j_2}} = \frac{a_{j_1j_1 0}}{a_{j_2j_2 0}} \frac{\left[(x_{j_1} - a_{j_1j_1})^2 + b_{j_1j_1}\right]}{(x_{j_2} - a_{j_2j_2s_2})},$$

and the deformation of the foregoing equation is

$$\frac{dx_{j_1}}{(x_{j_1} - a_{j_1j_1})^2 + b_{j_1j_1}} = \frac{a_{j_1j_1 0}}{a_{j_2j_2 0}} \frac{dx_{j_2}}{(x_{j_2} - a_{j_2j_2s_2})}.$$

With an initial condition of $(x_{j_1 0}, x_{j_2 0})$ at $t = t_0$, the integration of the above equation gives the first integral manifold for $a_{j_1j_2 1} = a_{j_2j_2s_1}$ (s_1, $s_2 \in \{1,2\}$, $s_1 \neq s_2$) as

$$\frac{1}{\sqrt{b_{j_1j_1}}}(\arctan\frac{x_{j_1}-a_{j_1j_1}}{\sqrt{b_{j_1j_1}}}-\arctan\frac{x_{j_10}-a_{j_1j_1}}{\sqrt{b_{j_1j_1}}})=\frac{a_{j_1j_10}}{a_{j_2j_20}}\ln\frac{|x_{j_2}-a_{j_2j_2s_2}|}{|x_{j_2}-a_{j_2j_2s_2}|}.$$

(iii$_{2a}$) At $x_{j_2}^{*}=a_{j_1j_21}=a_{j_2j_2s_1}$ ($s_1, s_2\in\{1,2\}$, $s_1\neq s_2$) with $\bar{x}_{j_1}\in(-\infty,\infty)$, in phase space, if

$$\frac{dx_{j_1}}{dx_{j_2}}\Big|_{x_{j_2}^{*}=a_{j_1j_21}}=\frac{a_{j_1j_10}}{a_{j_2j_20}}\frac{\left[(\bar{x}_{j_1}-a_{j_1j_1})^2+b_{j_1j_1}\right]}{(a_{j_1j_2s_1}-a_{j_2j_2s_2})}>0,$$

there is an increasing-inflection flow. If

$$\frac{dx_{j_1}}{dx_{j_2}}\Big|_{x_{j_2}^{*}=a_{j_1j_21}}=\frac{a_{j_1j_10}}{a_{j_2j_20}}\frac{\left[(\bar{x}_{j_1}-a_{j_1j_1})^2+b_{j_1j_1}\right]}{(a_{j_1j_2s_1}-a_{j_2j_2s_2})}<0,$$

there is a decreasing-inflection flow. The variational equation is

$$\Delta\dot{x}_{j_2}=a_{j_2j_20}(a_{j_2j_2s_1}-a_{j_2j_2s_2})\Delta x_{j_2}.$$

The inflection flows at $x_{j_2}^{*}=a_{j_1j_21}=a_{j_2j_2s_1}$ in the x_{j_2}-direction are positive and negative for $a_{j_2j_20}(a_{j_2j_2s_1}-a_{j_2j_2s_2})>0$ and $a_{j_2j_20}(a_{j_2j_2s_1}-a_{j_2j_2s_2})<0$, respectively.

Therefore, the infinite-equilibrium of $x_{j_2}^{*}=a_{j_1j_21}=a_{j_2j_2s_1}$ with $\bar{x}_{j_1}\in(-\infty,\infty)$ has the following properties as in Eqs. (1.176)–(1.179).

- For $a_{j_1j_10}>0$ and $a_{j_2j_20}(a_{j_2j_2s_1}-a_{j_2j_2s_2})>0$,

$$(\bar{x}_{j_1},a_{j_2j_2s_1})=\underbrace{(\mathrm{II,SO})}_{\text{increasing-inflection source}}.$$

  The infinite-equilibrium of $x_{j_2}^{*}=a_{j_1j_21}=a_{j_2j_2s_1}$ is an (II,SO)-increasing-inflection source.
- For $a_{j_1j_10}<0$ and $a_{j_2j_20}(a_{j_2j_2s_1}-a_{j_2j_2s_2})>0$,

$$(\bar{x}_{j_1},a_{j_2j_2s_1})=\underbrace{(\mathrm{DI,SO})}_{\text{decreasing-inflection source}}.$$

  The infinite-equilibrium of $x_{j_2}^{*}=a_{j_1j_21}=a_{j_2j_2s_1}$ is a (DI,SO)-decreasing-inflection source.
- For $a_{j_1j_10}>0$ and $a_{j_2j_20}(a_{j_2j_2s_1}-a_{j_2j_2s_2})<0$,

$$(\bar{x}_{j_1},a_{j_2j_2s_1})=\underbrace{(\mathrm{DI,SI})}_{\text{decreasing-inflection sink}}.$$

  The infinite-equilibrium of $x_{j_2}^{*}=a_{j_1j_21}=a_{j_2j_2s_1}$ is a (DI,SI)-decreasing-inflection sink.

- For $a_{j_1j_10} < 0$ and $a_{j_2j_20}(a_{j_1j_21} - a_{j_2j_21})^2 < 0$,

$$(\overline{x}_{j_1}, a_{j_2j_2s_1}) = \underbrace{(\text{II, SI})}_{\text{increasing-inflection sink}}.$$

The infinite-equilibrium of $x_{j_2}^* = a_{j_1j_21} = a_{j_2j_2s_1}$ is an (II,SI)-increasing-infection sink.

(iv) For $\Delta_{j_1j_2} = 0$ and $\Delta_{j_2j_2} < 0$, the standard form is

$$\begin{aligned} \dot{x}_{j_2} &= a_{j_2j_20}\left[(x_{j_2} - a_{j_2j_2})^2 + b_{j_2j_2}\right], \\ \dot{x}_{j_1} &= a_{j_1j_10}(x_{j_1} - a_{j_1j_11})^2(x_{j_2} - a_{j_1j_21}) \end{aligned}$$

where

$$\begin{aligned} a_{j_1j_11} &= -\frac{1}{2}B_{j_1j_1}, \\ a_{j_2j_2} &= -\frac{1}{2}B_{j_2j_2}, b_{j_2j_2} = \frac{1}{4}(-\Delta_{j_2j_2}). \end{aligned}$$

In phase space,

$$\frac{dx_{j_1}}{dx_{j_2}} = \frac{a_{j_1j_10}}{a_{j_2j_20}} \frac{(x_{j_1} - a_{j_1j_11})^2(x_{j_2} - a_{j_1j_21})}{(x_{j_2} - a_{j_2j_2})^2 + b_{j_2j_2}},$$

and the deformation of the foregoing equation is

$$\frac{1}{(x_{j_1} - a_{j_1j_11})^2}dx_{j_1} = \frac{a_{j_1j_10}}{a_{j_2j_20}}\left[\frac{x_{j_2} - a_{j_2j_2}}{(x_{j_2} - a_{j_2j_2})^2 + b_{j_2j_2}} + \frac{a_{j_2j_2} - a_{j_1j_21}}{(x_{j_2} - a_{j_2j_2})^2 + b_{j_2j_2}}\right]dx_{j_2}.$$

With an initial condition of (x_{j_10}, x_{j_20}) at $t = t_0$, the integration of the above equation gives

$$\begin{aligned} &-\left(\frac{1}{x_{j_1} - a_{j_1j_11}} - \frac{1}{x_{j_10} - a_{j_1j_11}}\right) = \frac{a_{j_1j_10}}{a_{j_2j_20}}\left[\frac{1}{2}\ln\frac{|(x_{j_2} - a_{j_2j_2})^2 + b_{j_2j_2}|}{|(x_{j_20} - a_{j_2j_2})^2 + b_{j_2j_2}|}\right. \\ &\left. + \frac{(a_{j_2j_2} - a_{j_1j_21})}{\sqrt{b_{j_2j_2}}}\left(\arctan\frac{x_{j_2} - a_{j_2j_2}}{\sqrt{b_{j_2j_2}}} - \arctan\frac{x_{j_20} - a_{j_2j_2}}{\sqrt{b_{j_2j_2}}}\right)\right]. \end{aligned}$$

In phase space, at $x_{j_2}^* = a_{j_1j_21}$ with $\overline{x}_{j_1} \neq a_{j_1j_11}$,

$$\frac{dx_{j_1}}{dx_{j_2}}\bigg|_{x_{j_2}^*=a_{j_1j_2 1}} = \frac{a_{j_1j_1 0}}{a_{j_2j_2 0}} \frac{(\overline{x}_{j_1} - a_{j_1j_1 1})^2(x_{j_2} - a_{j_1j_2 1})}{(x_{j_2} - a_{j_2j_2})^2 + b_{j_2j_2}}\bigg|_{x_{j_2}^*=a_{j_1j_2 1}} = 0.$$

If

$$\frac{d^2x_{j_1}}{dx_{j_2}^2}\bigg|_{x_{j_2}^*=a_{j_1j_2 1}} = \frac{a_{j_1j_1 0}}{a_{j_2j_2 0}} \frac{(\overline{x}_{j_1} - a_{j_1j_1 1})^2}{(a_{j_1j_2 1} - a_{j_2j_2})^2 + b_{j_2j_2}} > 0,$$

there is an up-parabola flow at $x_{j_2}^* = a_{j_1j_2 1}$ with $\overline{x}_{j_1} \neq a_{j_1j_1 1}$. If

$$\frac{d^2x_{j_1}}{dx_{j_2}^2}\bigg|_{x_{j_2}^*=a_{j_1j_2 1}} = \frac{a_{j_1j_1 0}}{a_{j_2j_2 0}} \frac{(\overline{x}_{j_1} - a_{j_1j_1 1})^2}{(a_{j_1j_2 1} - a_{j_2j_2})^2 + b_{j_2j_2}} < 0,$$

there is a down-parabola flow at $x_{j_2}^* = a_{j_1j_2 1}$ with $\overline{x}_{j_1} \neq a_{j_1j_1 1}$. Let

$$\dot{x}_{j_2} = a_{j_2j_2 0}\left[(a_{j_1j_2 1} - a_{j_2j_2})^2 + b_{j_2j_2}\right],$$

with

$$(a_{j_1j_2 1} - a_{j_2j_2})^2 + b_{j_2j_2} > 0.$$

The parabola flows at $x_{j_2}^* = a_{j_1j_2 1}$ with $\overline{x}_{j_1} \neq a_{j_1j_1 1}$ are positive and negative for $a_{j_2j_2 0} > 0$ and $a_{j_2j_2 0} < 0$, respectively.

Thus, the equilibrium of $x_{j_2}^* = a_{j_1j_2 1}$ with $\overline{x}_{j_1} \neq a_{j_1j_1 1}$ has the following properties.

- For $a_{j_1j_1 0}(\overline{x}_{j_1} - a_{j_1j_1 1})^2 > 0$ and $a_{j_2j_2 0} > 0$,

$$(a_{j_1j_2 1}, \dot{x}_{j_2}) = \underbrace{(\text{UP}, \text{pF})}_{\text{up-parabola flow } (+)}.$$

- For $a_{j_1j_1 0}(\overline{x}_{j_1} - a_{j_1j_1 1})^2 < 0$ and $a_{j_2j_2 0} > 0$,

$$(a_{j_1j_2 1}, \dot{x}_{j_2}) = \underbrace{(\text{DP}, \text{pF})}_{\text{down-parabola flow } (+)}.$$

- For $a_{j_1j_1 0}(\overline{x}_{j_1} - a_{j_1j_1 1})^2 > 0$ and $a_{j_2j_2 0} < 0$,

$$(a_{j_1j_2 1}, \dot{x}_{j_2}) = \underbrace{(\text{DP}, \text{nF})}_{\text{dwon-parabola flow } (-)}.$$

- For $a_{j_1j_1 0}(\overline{x}_{j_1} - a_{j_1j_1 1})^2 < 0$ and $a_{j_2j_2 0} < 0$,

$$(a_{j_1j_2}1, \dot{x}_{j_2}) = \underbrace{(\mathrm{UP}, \mathrm{nF})}_{\text{up-parabola flow } (-)}.$$

The variational equation at $x^*_{j_1} = a_{j_1j_1}1$ with $\overline{x}_{j_2} \neq a_{j_1j_2}1$ is

$$\Delta\dot{x}_{j_1} = a_{j_1j_1}0(\overline{x}_{j_2} - a_{j_1j_2}1)(\Delta x_{j_1})^2$$

and

$$\dot{x}_{j_2} = a_{j_2j_2}0\left[(\overline{x}_{j_2} - a_{j_2j_2})^2 + b_{j_2j_2}\right].$$

Thus, the equilibrium of $x^*_{j_2} = a_{j_1j_1}1$ with $\overline{x}_{j_2} \neq a_{j_1j_2}1$ has the following properties.

- For $a_{j_1j_1}0(\overline{x}_{j_2} - a_{j_1j_2}1) > 0$ and $a_{j_2j_2}0 > 0$,

$$(a_{j_1j_1}1, \dot{x}_{j_2}) = \underbrace{(\mathrm{US}, \mathrm{pF})}_{\text{upper-saddle flow } (+)}.$$

- For $a_{j_1j_1}0(\overline{x}_{j_2} - a_{j_1j_2}1) < 0$ and $a_{j_2j_2}0 > 0$,

$$(a_{j_1j_1}1, \dot{x}_{j_2}) = \underbrace{(\mathrm{LS}, \mathrm{pF})}_{\text{lower-saddle flow } (+)}.$$

- For $a_{j_1j_1}0(\overline{x}_{j_2} - a_{j_1j_2}1) > 0$ and $a_{j_2j_2}0 < 0$,

$$(a_{j_1j_1}1, \dot{x}_{j_2}) = \underbrace{(\mathrm{US}, \mathrm{nF})}_{\text{upper-saddle flow } (-)}.$$

- For $a_{j_1j_1}0(\overline{x}_{j_2} - a_{j_1j_2}1) < 0$ and $a_{j_2j_2}0 < 0$,

$$(a_{j_1j_1}1, \dot{x}_{j_2}) = \underbrace{(\mathrm{LS}, \mathrm{nF})}_{\text{lower-saddle flow } (-)}.$$

From the above analysis, therefore, the equilibrium of $(x^*_{j_1}, x^*_{j_2}) = (a_{j_1j_1}1, a_{j_1j_2}1)$ has the following properties as in Eqs. (1.183)–(1.186).

- For $a_{j_1j_1}0 > 0$ and $a_{j_2j_2}0 > 0$,

$$(a_{j_1j_1}1, a_{j_1j_2}1) = \underbrace{(\mathrm{UP} : \mathrm{UP}, \mathrm{pF})}_{\text{hyperbolic-secant-to-hyperbolic flow } (+)}.$$

The equilibrium of $(x^*_{j_1}, x^*_{j_2}) = (a_{j_1j_11}, a_{j_1j_21})$ is a (UP:UP,pF)-positive hyperbolic-secant-to-hyperbolic flow.

- For $a_{j_1j_10} < 0$ and $a_{j_2j_20} > 0$,

$$(a_{j_1j_11}, a_{j_1j_21}) = \underbrace{(\mathrm{DP}:\mathrm{DP},\mathrm{pF})}_{\text{hyperbolic-to-hyperbolic-secant flow }(+)}.$$

The equilibrium of $(x^*_{j_1}, x^*_{j_2}) = (a_{j_1j_11}, a_{j_1j_21})$ is a (DP:DP,pF)-positive hyperbolic-to-hyperbolic-secant flow.

- For $a_{j_1j_10} > 0$ and $a_{j_2j_20} < 0$,

$$(a_{j_1j_11}, a_{j_1j_21}) = \underbrace{(\mathrm{DP}:\mathrm{DP},\mathrm{nF})}_{\text{hyperbolic-to-hyperbolic-secant flow }(-)}.$$

The equilibrium of $(x^*_{j_1}, x^*_{j_2}) = (a_{j_1j_11}, a_{j_1j_21})$ is a (DP:DP, nF)-negative hyperbolic-to-hyperbolic-secant flow.

- For $a_{j_1j_10} < 0$ and $a_{j_2j_20} < 0$,

$$(a_{j_1j_11}, a_{j_1j_21}) = \underbrace{(\mathrm{UP}:\mathrm{UP},\mathrm{nF})}_{\text{hyperbolic-secant-to-hyperbolic flow }(-)}.$$

The equilibrium of $(x^*_{j_1}, x^*_{j_2}) = (a_{j_1j_11}, a_{j_1j_21})$ is a (UP:UP,nF)-negative hyperbolic-secant-to-hyperbolic flow.

(v) For $\Delta_{j_1j_2} = 0$ and $\Delta_{j_2j_2} = 0$, the standard form is

$$\begin{aligned} \dot{x}_{j_2} &= a_{j_2j_20}(x_{j_2} - a_{j_2j_21})^2, \\ \dot{x}_{j_1} &= a_{j_1j_10}(x_{j_1} - a_{j_1j_11})^2(x_{j_2} - a_{j_1j_21}) \end{aligned}$$

where

$$a_{j_1j_11} = -\frac{1}{2}B_{j_1j_1},\ a_{j_2j_21} = -\frac{1}{2}B_{j_2j_2}.$$

(v_1) For $a_{j_2j_21} \neq a_{j_1j_21}$, in phase space,

$$\frac{dx_{j_1}}{dx_{j_2}} = \frac{a_{j_1j_10}}{a_{j_2j_20}} \frac{(x_{j_1} - a_{j_1j_11})^2(x_{j_2} - a_{j_1j_21})}{(x_{j_2} - a_{j_2j_21})^2},$$

and the deformation of the foregoing equation is

$$\frac{1}{(x_{j_1}-a_{j_1j_1 1})^2}dx_{j_1} = \frac{a_{j_1j_1 0}}{a_{j_2j_2 0}}\left[\frac{1}{(x_{j_2}-a_{j_2j_2 1})} + \frac{a_{j_2j_2 1}-a_{j_1j_2 1}}{(x_{j_2}-a_{j_2j_2 1})^2}\right]dx_{j_2}.$$

With an initial condition of $(x_{j_1 0}, x_{j_2 0})$ at $t = t_0$, the integration of the above equation gives

$$-(\frac{1}{x_{j_1}-a_{j_1j_1 1}} - \frac{1}{x_{j_1 0}-a_{j_1j_1 1}}) = \frac{a_{j_1j_1 0}}{a_{j_2j_2 0}}\left[\ln\frac{|x_{j_2}-a_{j_2j_2 1}|}{|x_{j_2 0}-a_{j_2j_2 1}|} \right.$$
$$\left. -(a_{j_2j_2 1}-a_{j_1j_2 1})(\frac{1}{x_{j_2}-a_{j_2j_2 1}} - \frac{1}{x_{j_2 0}-a_{j_2j_2 1}})\right].$$

(v_{1a}) For $(x^*_{j_1}, x^*_{j_2}) = (a_{j_1j_1 1}, a_{j_2j_2 1})$, the variational equations are

$$\Delta\dot{x}_{j_2} = a_{j_2j_2 0}(\Delta x_{j_2})^2,$$
$$\Delta\dot{x}_{j_1} = a_{j_1j_1 0}(a_{j_2j_2 1}-a_{j_1j_2 1})(\Delta x_{j_1})^2.$$

The flows at $(x^*_{j_1}, x^*_{j_2}) = (a_{j_1j_1 1}, a_{j_2j_2 1})$ in the x_{j_1}-direction are upper-saddle and lower-saddle for $a_{j_1j_1 0}(a_{j_2j_2 1}-a_{j_1j_2 1}) > 0$ and $a_{j_1j_1 0}(a_{j_2j_2 1}-a_{j_1j_2 1}) < 0$, respectively. The flows at $(x^*_{j_1}, x^*_{j_2}) = (a_{j_1j_1 1}, a_{j_2j_2 1})$ in the x_{j_2}-direction are upper-saddle and lower-saddle for $a_{j_2j_2 0} > 0$ and $a_{j_2j_2 0} < 0$, respectively.

Therefore, the equilibrium of $(x^*_{j_1}, x^*_{j_2}) = (a_{j_1j_1 1}, a_{j_2j_2 1})$ with $a_{j_2j_2 1} \neq a_{j_1j_2 1}$ has the following properties as in Eqs. (1.190)–(1.193).

- For $a_{j_1j_1 0}(a_{j_2j_2 1}-a_{j_1j_2 1}) > 0$ and $a_{j_2j_2 0} > 0$,

$$(a_{j_1j_1 1}, a_{j_2j_2 1}) = \underbrace{(\mathrm{US}, \mathrm{US})}_{\text{double-saddle}}.$$

The equilibrium of $(x^*_{j_1}, x^*_{j_2}) = (a_{j_1j_1 1}, a_{j_2j_2 1})$ is a (US,US)-double-saddle.

- For $a_{j_1j_1 0}(a_{j_2j_2 1}-a_{j_1j_2 1}) < 0$ and $a_{j_2j_2 0} > 0$,

$$(a_{j_1j_1 1}, a_{j_2j_2 1}) = \underbrace{(\mathrm{LS}, \mathrm{US})}_{\text{double-saddle}}.$$

The equilibrium of $(x^*_{j_1}, x^*_{j_2}) = (a_{j_1j_1 1}, a_{j_2j_2 1})$ is an (LS,US)-double-saddle.

- For $a_{j_1j_1 0}(a_{j_2j_2 1}-a_{j_1j_2 1}) > 0$ and $a_{j_2j_2 0} < 0$,

$$(a_{j_1j_1 1}, a_{j_2j_2 1}) = \underbrace{(\mathrm{US}, \mathrm{LS})}_{\text{double-saddle}}.$$

The equilibrium of $(x^*_{j_1}, x^*_{j_2}) = (a_{j_1j_1 1}, a_{j_2j_2 1})$ is a (US,LS)-double-saddle.

- For $a_{j_1j_1 0}(a_{j_2j_2 1}-a_{j_1j_2 1}) < 0$ and $a_{j_2j_2 0} < 0$,

$$(a_{j_1j_1 1}, a_{j_2j_2 1}) = \underbrace{(\mathrm{LS}, \mathrm{LS})}_{\text{double-saddle}}.$$

The equilibrium of $(x_{j_1}^*, x_{j_2}^*) = (a_{j_1j_1 1}, a_{j_2j_2 1})$ is an (LS,LS)-double-saddle.

– The double-saddles are the appearing bifurcations of saddle-source and saddle-sink, and such double-saddles are also the appearing bifurcations of networks of saddles, source and sink.

(v_{1b}) For $(x_{j_1}^*, x_{j_2}^*) = (a_{j_1j_1 1}, a_{j_1j_2 1})$ with $a_{j_1j_2 1} \neq a_{j_2j_2 1}$, similar to case (iv), the proof can be done as follows. In phase space, at $x_{j_2}^* = a_{j_1j_2 1}$ with $\bar{x}_{j_1} \neq a_{j_1j_1 1}$,

$$\frac{dx_{j_1}}{dx_{j_2}}\Big|_{x_{j_2}^* = a_{j_1j_2 1}} = \frac{a_{j_1j_1 0}}{a_{j_2j_2 0}} \frac{(\bar{x}_{j_1} - a_{j_1j_1 1})^2 (x_{j_2} - a_{j_1j_2 1})}{(x_{j_2} - a_{j_2j_2 1})^2}\Big|_{x_{j_2}^* = a_{j_1j_2 1}} = 0.$$

If

$$\frac{d^2 x_{j_1}}{dx_{j_2}^2}\Big|_{x_{j_2}^* = a_{j_1j_2 1}} = \frac{a_{j_1j_1 0}}{a_{j_2j_2 0}} \frac{(\bar{x}_{j_1} - a_{j_1j_1 1})^2}{(a_{j_1j_2 1} - a_{j_2j_2 1})^2} > 0,$$

there is an up-parabola flow at $x_{j_2}^* = a_{j_1j_2 1}$ with $\bar{x}_{j_1} \neq a_{j_1j_1 1}$. If

$$\frac{d^2 x_{j_1}}{dx_{j_2}^2}\Big|_{x_{j_2}^* = a_{j_1j_2 1}} = \frac{a_{j_1j_1 0}}{a_{j_2j_2 0}} \frac{(\bar{x}_{j_1} - a_{j_1j_1 1})^2}{(a_{j_1j_2 1} - a_{j_2j_2 1})^2} < 0,$$

there is a down-parabola flow at $x_{j_2}^* = a_{j_1j_2 1}$ with $\bar{x}_{j_1} \neq a_{j_1j_1 1}$. Let

$$\dot{x}_{j_2} = a_{j_2j_2 0}(a_{j_1j_2 1} - a_{j_2j_2 1})^2 \neq 0.$$

The parabola flows at $x_{j_2}^* = a_{j_1j_2 1}$ with $\bar{x}_{j_1} \neq a_{j_1j_1 1}$ are positive and negative for $a_{j_2j_2 0}(a_{j_1j_2 1} - a_{j_2j_2 1})^2 > 0$ and $a_{j_2j_2 0}(a_{j_1j_2 1} - a_{j_2j_2 1})^2 < 0$, respectively.

Thus, the equilibrium of $x_{j_2}^* = a_{j_1j_2 1}$ with $\bar{x}_{j_1} \neq a_{j_1j_1 1}$ has the following properties.

- For $a_{j_1j_1 0}(\bar{x}_{j_1} - a_{j_1j_1 1})^2 > 0$ and $a_{j_2j_2 0}(a_{j_1j_2 1} - a_{j_2j_2 1})^2 > 0$,

$$(a_{j_1j_2 1}, \dot{x}_{j_2}) = \underbrace{(\mathrm{UP}, \mathrm{pF})}_{\text{up-parabola flow } (+)}.$$

- For $a_{j_1j_1 0}(\bar{x}_{j_1} - a_{j_1j_1 1})^2 < 0$ and $a_{j_2j_2 0}(a_{j_1j_2 1} - a_{j_2j_2 1})^2 > 0$,

$$(a_{j_1j_2 1}, \dot{x}_{j_2}) = \underbrace{(\mathrm{DP}, \mathrm{pF})}_{\text{down-parabola flow } (+)}.$$

- For $a_{j_1j_10}(\overline{x}_{j_1} - a_{j_1j_11})^2 > 0$ and $a_{j_2j_20}(a_{j_1j_21} - a_{j_2j_21})^2 < 0$,

$$(a_{j_1j_21}, \dot{x}_{j_2}) = \underbrace{(\mathrm{DP}, \mathrm{nF})}_{\text{dwon-parabola flow }(-)}.$$

- For $a_{j_1j_10}(\overline{x}_{j_1} - a_{j_1j_11})^2 < 0$ and $a_{j_2j_20}(a_{j_1j_21} - a_{j_2j_21})^2 < 0$,

$$(a_{j_1j_21}, \dot{x}_{j_2}) = \underbrace{(\mathrm{UP}, \mathrm{nF})}_{\text{up-parabola flow }(-)}.$$

The variational equation at $x_{j_1}^* = a_{j_1j_11}$ with $\overline{x}_{j_2} \neq a_{j_1j_21}, a_{j_2j_21}$ is

$$\Delta\dot{x}_{j_1} = a_{j_1j_10}(\overline{x}_{j_2} - a_{j_1j_21})(\Delta x_{j_1})^2$$

and

$$\dot{x}_{j_2} = a_{j_2j_20}(\overline{x}_{j_2} - a_{j_2j_21})^2 \neq 0.$$

Thus, the equilibrium of $x_{j_2}^* = a_{j_1j_21}$ with $\overline{x}_{j_1} \neq a_{j_1j_11}$ has the following properties.

- For $a_{j_1j_10}(\overline{x}_{j_2} - a_{j_1j_21}) > 0$ and $a_{j_2j_20}(\overline{x}_{j_2} - a_{j_2j_21})^2 > 0$,

$$(a_{j_1j_11}, \dot{x}_{j_2}) = \underbrace{(\mathrm{US}, \mathrm{pF})}_{\text{upper-saddle flow }(+)}.$$

- For $a_{j_1j_10}(\overline{x}_{j_2} - a_{j_1j_21}) < 0$ and $a_{j_2j_20}(\overline{x}_{j_2} - a_{j_2j_21})^2 > 0$,

$$(a_{j_1j_11}, \dot{x}_{j_2}) = \underbrace{(\mathrm{LS}, \mathrm{pF})}_{\text{lower-saddle flow }(+)}.$$

- For $a_{j_1j_10}(\overline{x}_{j_2} - a_{j_1j_21}) > 0$ and $a_{j_2j_20}(\overline{x}_{j_2} - a_{j_2j_21})^2 < 0$,

$$(a_{j_1j_11}, \dot{x}_{j_2}) = \underbrace{(\mathrm{US}, \mathrm{nF})}_{\text{upper-saddle flow }(-)}.$$

- For $a_{j_1j_10}(\overline{x}_{j_2} - a_{j_1j_21}) < 0$ and $a_{j_2j_20}(\overline{x}_{j_2} - a_{j_2j_21})^2 < 0$,

$$(a_{j_1j_1 1}, \dot{x}_{j_2}) = \underbrace{(\text{LS}, \text{nF})}_{\text{lower-saddle flow }(-)}.$$

From the above analysis, let $\bar{x}_{j_2} = a_{j_2j_2 1}$. Therefore, the equilibrium of $(x^*_{j_1}, x^*_{j_2}) = (a_{j_1j_1 1}, a_{j_1j_2 1})$ with $a_{j_2j_2 1} \neq a_{j_1j_2 1}$ has the following properties as in Eqs. (1.194)–(1.197).

- For $a_{j_1j_1 0} > 0$ and $a_{j_2j_2 0}(a_{j_1j_2 1} - a_{j_2j_2 1})^2 > 0$,

$$(a_{j_1j_1 1}, a_{j_1j_2 1}) = \underbrace{(\text{UP}:\text{UP}, \text{pF})}_{\text{hyperbolic-secant to hyperbolic flow }(+)}.$$

  The equilibrium of $(x^*_{j_1}, x^*_{j_2}) = (a_{j_1j_1 1}, a_{j_1j_2 1})$ is a (UP:UP, pF)-positive hyperbolic-secant to hyperbolic flow.
- For $a_{j_1j_1 0} < 0$ and $a_{j_2j_2 0}(a_{j_1j_2 1} - a_{j_2j_2 1})^2 > 0$,

$$(a_{j_1j_1 1}, a_{j_1j_2 1}) = \underbrace{(\text{DP}:\text{DP}, \text{pF})}_{\text{hyperbolic to hyperbolic-secant flow }(+)}.$$

  The equilibrium of $(x^*_{j_1}, x^*_{j_2}) = (a_{j_1j_1 1}, a_{j_1j_2 1})$ is a (DP:DP, pF)-positive hyperbolic to hyperbolic-secant flow.
- For $a_{j_1j_1 0} > 0$ and $a_{j_2j_2 0}(a_{j_1j_2 1} - a_{j_2j_2 1})^2 < 0$,

$$(a_{j_1j_1 1}, a_{j_1j_2 1}) = \underbrace{(\text{DP}:\text{DP}, \text{nF})}_{\text{hyperbolic to hyperbolic-secant flow }(-)}.$$

  The equilibrium of $(x^*_{j_1}, x^*_{j_2}) = (a_{j_1j_1 1}, a_{j_1j_2 1})$ is an (DP:DP,nF)-negative hyperbolic to hyperbolic-secant flow.
- For $a_{j_1j_1 0} < 0$ and $a_{j_2j_2 0}(a_{j_1j_2 1} - a_{j_2j_2 1})^2 < 0$,

$$(a_{j_1j_1 1}, a_{j_1j_2 1}) = \underbrace{(\text{UP}:\text{UP}, \text{nF})}_{\text{hyperbolic-secant to hyperbolic flow }(-)}.$$

  The equilibrium of $(x^*_{j_1}, x^*_{j_2}) = (a_{j_1j_1 1}, a_{j_1j_2 1})$ is a (UP:UP, nF)-negative hyperbolic-secant to hyperbolic flow.

(v_2) For $a_{j_2j_2 1} = a_{j_1j_2 1}$ in phase space,

$$\frac{dx_{j_1}}{dx_{j_2}} = \frac{a_{j_1j_1 0}}{a_{j_2j_2 0}} \frac{(x_{j_1} - a_{j_1j_1 1})^2 (x_{j_2} - a_{j_1j_2 1})}{(x_{j_2} - a_{j_2j_2 1})^2} = \frac{a_{j_1j_1 0}}{a_{j_2j_2 0}} \frac{(x_{j_1} - a_{j_1j_1 1})^2}{(x_{j_2} - a_{j_2j_2 1})},$$

and the deformation of the foregoing equation is

$$\frac{dx_{j_1}}{(x_{j_1}-a_{j_1j_11})^2}=\frac{a_{j_1j_10}}{a_{j_2j_20}}\frac{dx_{j_2}}{(x_{j_2}-a_{j_2j_21})}.$$

With an initial condition of (x_{j_10},x_{j_20}) at $t=t_0$, the integration of the above equation gives the first integral manifold for $a_{j_2j_21}=a_{j_1j_21}$ as

$$-(\frac{1}{x_{j_1}-a_{j_1j_11}}-\frac{1}{x_{j_11}-a_{j_1j_11}})=\frac{a_{j_1j_10}}{a_{j_2j_20}}\ln\frac{|x_{j_2}-a_{j_2j_21}|}{|x_{j_20}-a_{j_2j_21}|}.$$

(v_{2a}) At $x^*_{j_2}=a_{j_1j_21}=a_{j_2j_21}$ with $\bar{x}_{j_1}\neq a_{j_1j_11}$, in phase space, for a small $\varepsilon>0$,

$$\frac{dx_{j_1}}{dx_{j_2}}\bigg|_{x^*_{j_2}=a_{j_2j_21}\pm\varepsilon}=\pm\frac{a_{j_1j_10}}{a_{j_2j_20}}\frac{(\bar{x}_{j_1}-a_{j_1j_11})^2}{\varepsilon}.$$

If

$$\frac{a_{j_1j_10}}{a_{j_2j_20}}(\bar{x}_{j_1}-a_{j_1j_11})^2>0,$$

there is a down-up flow. If

$$\frac{a_{j_1j_10}}{a_{j_2j_20}}(\bar{x}_{j_1}-a_{j_1j_11})^2<0,$$

there is an up-down flow. The variational equation is

$$\Delta\dot{x}_{j_2}=a_{j_2j_20}(\Delta x_{j_2})^2.$$

The down-up and up-down flows are upper-saddle and lower-saddle for $a_{j_2j_20}>0$ and $a_{j_2j_20}<0$, respectively.

Therefore, the infinite-equilibrium of $x^*_{j_2}=a_{j_1j_21}=a_{j_2j_21}$ with $\bar{x}_{j_1}\neq a_{j_1j_11}$ has the following properties as in Eqs. (1.199)–(1.202).

- For $a_{j_1j_10}(\bar{x}_{j_1}-a_{j_1j_11})^2>0$ and $a_{j_2j_20}>0$,

$$(\bar{x}_{j_1},a_{j_1j_21})=\underbrace{(\text{DU},\text{US})}_{\text{down-up upper-saddle}}.$$

  The infinite-equilibrium of $x^*_{j_2}=a_{j_1j_21}=a_{j_2j_21}$ is an (DU,US)-down-up upper-saddle.

- For $a_{j_1j_10}(\bar{x}_{j_1}-a_{j_1j_11})^2<0$ and $a_{j_2j_20}>0$,

$$(\bar{x}_{j_1}, a_{j_1j_2 1}) = \underbrace{(\text{UD}, \text{US})}_{\text{up-down upper-saddle}}.$$

The infinite-equilibrium of $x^*_{j_2} = a_{j_1j_2 1} = a_{j_2j_2 1}$ is a (UD,US)-up-down upper-saddle.

- For $a_{j_1j_1 0}(\bar{x}_{j_1} - a_{j_1j_1 1})^2 > 0$ and $a_{j_2j_2 0} < 0$,

$$(\bar{x}_{j_1}, a_{j_1j_2 1}) = \underbrace{(\text{UD}, \text{LS})}_{\text{up-down lower-saddle}}.$$

The infinite-equilibrium of $x^*_{j_2} = a_{j_1j_2 1} = a_{j_2j_2 1}$ is a (UD,LS)-up-down lower-saddle.

- For $a_{j_1j_1 0}(\bar{x}_{j_1} - a_{j_1j_1 1})^2 < 0$ and $a_{j_2j_2 0} < 0$,

$$(\bar{x}_{j_1}, a_{j_1j_2 1}) = \underbrace{(\text{DU}, \text{LS})}_{\text{down-up lower-saddle}}.$$

The infinite-equilibrium of $x^*_{j_2} = a_{j_1j_2 1} = a_{j_2j_2 1}$ is a (DU,LS)-down-up lower-saddle.

(v_{2b}) From case (v_{2a}), let $\bar{x}_{j_1} = a_{j_1j_1 1}$. Thus, with $a_{j_2j_2 1} = a_{j_1j_2 1}$, the equilibrium of $(x^*_{j_1}, x^*_{j_2}) = (a_{j_1j_1 1}, a_{j_1j_2 1})$ has the following properties as in Eqs. (1.203)–(1.206).

- For $a_{j_1j_1 0} > 0$ and $a_{j_2j_2 0} > 0$,

$$(a_{j_1j_1 1}, a_{j_1j_2 1}) = \underbrace{(\text{DU} : \text{DU}, \text{US})}_{\text{down-up upper-saddle}}.$$

The equilibrium of $(x^*_{j_1}, x^*_{j_2}) = (a_{j_1j_1 1}, a_{j_1j_2 1})$ is a (DU:DU,US)-down-up upper-saddle.

- For $a_{j_1j_1 0} < 0$ and $a_{j_2j_2 0} > 0$,

$$(a_{j_1j_1 1}, a_{j_1j_2 1}) = \underbrace{(\text{UD} : \text{UD}, \text{US})}_{\text{up-down upper-saddle}}.$$

The equilibrium of $(x^*_{j_1}, x^*_{j_2}) = (a_{j_1j_1 1}, a_{j_1j_2 1})$ is a (UD:UD,US)-up-down upper-saddle.

- For $a_{j_1j_1 0} > 0$ and $a_{j_2j_2 0} < 0$,

$$(a_{j_1j_1 1}, a_{j_1j_2 1}) = \underbrace{(\text{UD} : \text{UD}, \text{LS})}_{\text{up-down lower-saddle}}.$$

The equilibrium of $(x_{j_1}^*, x_{j_2}^*) = (a_{j_1j_11}, a_{j_1j_21})$ is a (UD:UD,LS)-up-down lower-saddle.

- For $a_{j_1j_10} < 0$ and $a_{j_2j_20} < 0$,

$$(a_{j_1j_11}, a_{j_1j_21}) = \underbrace{(\mathrm{DU} : \mathrm{DU}, \mathrm{US})}_{\text{down-up upper-saddle}}.$$

The equilibrium of $(x_{j_1}^*, x_{j_2}^*) = (a_{j_1j_11}, a_{j_1j_21})$ is a (DU:DU,LS)-down-up lower-saddle.

(vi) For $\Delta_{j_1j_1} = 0$ and $\Delta_{j_2j_2} > 0$, the standard form is

$$\begin{aligned} \dot{x}_{j_2} &= a_{j_2j_20}(x_{j_2} - a_{j_2j_21})(x_{j_2} - a_{j_2j_22}), \\ \dot{x}_{j_1} &= a_{j_1j_10}(x_{j_1} - a_{j_1j_11})^2(x_{j_2} - a_{j_1j_21}) \end{aligned}$$

where

$$\begin{aligned} &a_{j_1j_11} = -\frac{1}{2}B_{j_1j_1}, \\ &b_{j_2j_21}, b_{j_2j_22} = -\frac{1}{2}(B_{j_2j_2} \pm \sqrt{\Delta_{j_2j_2}}), \\ &\{a_{j_2j_21}, a_{j_2j_22}\} = \mathrm{sort}\{b_{j_2j_21}, b_{j_2j_22}\}, a_{j_2j_21} < a_{j_2j_22}. \end{aligned}$$

(vi_1) In phase space, $a_{j_2j_2s_1} \neq a_{j_1j_21}$, we have

$$\frac{dx_{j_1}}{dx_{j_2}} = \frac{a_{j_1j_10}}{a_{j_2j_20}} \frac{(x_{j_1} - a_{j_1j_11})^2(x_{j_2} - a_{j_1j_21})}{(x_{j_2} - a_{j_2j_21})(x_{j_2} - a_{j_2j_22})},$$

and the deformation of the foregoing equation is

$$\begin{aligned} \frac{1}{(x_{j_1} - a_{j_1j_11})^2} dx_{j_1} = \frac{a_{j_1j_10}}{a_{j_2j_20}} &\left[\frac{a_{j_2j_21} - a_{j_1j_21}}{(a_{j_2j_21} - a_{j_2j_22})(x_{j_2} - a_{j_2j_21})} \right. \\ &\left. + \frac{a_{j_2j_22} - a_{j_1j_21}}{(a_{j_2j_22} - a_{j_2j_21})(x_{j_2} - a_{j_2j_22})} \right] dx_{j_2}. \end{aligned}$$

With an initial condition of (x_{j_10}, x_{j_20}) at $t = t_0$, the integration of the above equation gives the first integral manifold $(a_{j_2j_2s_1} \neq a_{j_1j_21})$ as

$$-\left(\frac{1}{x_{j_1} - a_{j_1j_11}} - \frac{1}{x_{j_10} - a_{j_1j_11}}\right) = \frac{a_{j_1j_10}}{a_{j_2j_20}} \sum\nolimits_{s_1=1}^{2} \frac{(a_{j_2j_2s_1} - a_{j_1j_21})}{(a_{j_2j_2s_1} - a_{j_2j_2s_2})} \ln \frac{|x_{j_2} - a_{j_2j_2s_1}|}{|x_{j_20} - a_{j_2j_2s_1}|}.$$

(vi_{1a}) At $(x_{j_1}^*, x_{j_2}^*) = (a_{j_1j_11}, a_{j_2j_2s_1})$, the variational equations are

$$\Delta \dot{x}_{j_2} = a_{j_2j_20}(a_{j_2j_2s_1} - a_{j_2j_2s_2})\Delta x_{j_2},$$
$$\Delta \dot{x}_{j_1} = a_{j_1j_10}(a_{j_2j_2s_1} - a_{j_1j_21})(\Delta x_{j_1})^2.$$

The flows $(x^*_{j_1}, x^*_{j_2}) = (a_{j_1j_11}, a_{j_2j_2s_1})$ in the x_{j_1}-direction are upper-saddle and lower-saddle for $a_{j_1j_10}(a_{j_2j_2s_1} - a_{j_1j_21}) > 0$ and $a_{j_1j_10}(a_{j_2j_2s_1} - a_{j_1j_21}) < 0$, respectively. The flows $(x^*_{j_1}, x^*_{j_2}) = (a_{j_1j_11}, a_{j_2j_2s_1})$ are source and sink for $a_{j_2j_20}(a_{j_2j_2s_1} - a_{j_2j_2s_2}) > 0$ and $a_{j_2j_20}(a_{j_2j_2s_1} - a_{j_2j_2s_2}) < 0$ in the x_{j_2}-direction, respectively.

Therefore, the equilibrium of $(x^*_{j_1}, x^*_{j_2}) = (a_{j_1j_11}, a_{j_2j_2s_1})$ ($s_1, s_2 \in \{1, 2\}, s_1 \neq s_2$) has the following properties as in Eqs. (1.210)–(1.213).

- For $a_{j_1j_10}(a_{j_2j_2s_1} - a_{j_1j_21}) > 0$ and $a_{j_2j_20}(a_{j_2j_2s_1} - a_{j_2j_2s_2}) > 0$,

$$(a_{j_1j_11}, a_{j_2j_2s_1}) = \underbrace{(\text{US, SO})}_{\text{upper-saddle source}}.$$

  The equilibrium of $(x^*_{j_1}, x^*_{j_2}) = (a_{j_1j_11}, a_{j_2j_2s_1})$ is a (US,SO)-upper-saddle source.
- For $a_{j_1j_10}(a_{j_2j_2s_1} - a_{j_1j_21}) < 0$ and $a_{j_2j_20}(a_{j_2j_2s_1} - a_{j_2j_2s_2}) > 0$,

$$(a_{j_1j_11}, a_{j_2j_2s_1}) = \underbrace{(\text{LS, SO})}_{\text{lower-saddle source}}.$$

  The equilibrium of $(x^*_{j_1}, x^*_{j_2}) = (a_{j_1j_11}, a_{j_2j_2s_1})$ is an (LS,SO)-lower-saddle source.
- For $a_{j_1j_10}(a_{j_2j_2s_1} - a_{j_1j_21}) > 0$ and $a_{j_2j_20}(a_{j_2j_2s_1} - a_{j_2j_2s_2}) < 0$,

$$(a_{j_1j_11}, a_{j_2j_2s_1}) = \underbrace{(\text{US, SI})}_{\text{upper-saddle sink}}.$$

  The equilibrium of $(x^*_{j_1}, x^*_{j_2}) = (a_{j_1j_11}, a_{j_2j_2s_1})$ is a (US,SI)-upper-saddle sink.
- For $a_{j_1j_10}(a_{j_2j_2s_1} - a_{j_1j_21}) < 0$ and $a_{j_2j_20}(a_{j_2j_2s_1} - a_{j_2j_2s_2}) < 0$,

$$(a_{j_1j_11}, a_{j_2j_2s_1}) = \underbrace{(\text{LS, SI})}_{\text{lower-saddle sink}}.$$

  The equilibrium of $(x^*_{j_1}, x^*_{j_2}) = (a_{j_1j_11}, a_{j_2j_2s_1})$ is an (LS,SI)-lower-saddle sink.

(vi$_{1b}$) In phase space, at $x^*_{j_2} = a_{j_1j_21}$ with $\bar{x}_{j_1} \neq a_{j_1j_11}$,

$$\frac{dx_{j_1}}{dx_{j_2}}\bigg|_{x^*_{j_2} = a_{j_1j_21}} = \frac{a_{j_1j_10}}{a_{j_2j_20}} \frac{(\bar{x}_{j_1} - a_{j_1j_11})^2(x_{j_2} - a_{j_1j_21})}{(x_{j_2} - a_{j_2j_21})(x_{j_2} - a_{j_2j_22})}\bigg|_{x^*_{j_2} = a_{j_1j_21}} = 0.$$

If

$$\frac{d^2x_{j_1}}{dx_{j_2}^2}\bigg|_{x_{j_2}^*=a_{j_1j_2 1}} = \frac{a_{j_1j_1 0}}{a_{j_2j_2 0}} \frac{(\overline{x}_{j_1} - a_{j_1j_1 1})^2}{(a_{j_1j_2 1} - a_{j_2j_2 1})(a_{j_1j_2 1} - a_{j_2j_2 2})} > 0,$$

there is an up-parabola flow at $x_{j_2}^* = a_{j_1j_2 1}$ with $\overline{x}_{j_1} \neq a_{j_1j_1 1}$. If

$$\frac{d^2x_{j_1}}{dx_{j_2}^2}\bigg|_{x_{j_2}^*=a_{j_1j_2 1}} = \frac{a_{j_1j_1 0}}{a_{j_2j_2 0}} \frac{(\overline{x}_{j_1} - a_{j_1j_1 1})^2}{(a_{j_1j_2 1} - a_{j_2j_2 1})(a_{j_1j_2 1} - a_{j_2j_2 2})} < 0,$$

there is a down-parabola flow at $x_{j_2}^* = a_{j_1j_2 1}$ with $\overline{x}_{j_1} \neq a_{j_1j_1 1}$. Let

$$\begin{aligned} \dot{x}_{j_2} &= a_{j_2j_2 0}(a_{j_1j_2 1} - a_{j_2j_2 1})(a_{j_1j_2 1} - a_{j_2j_2 2}) \\ &= a_{j_2j_2 0}\prod\nolimits_{s_1=1}^{2}(a_{j_1j_2 1} - a_{j_2j_2 s_1}) \neq 0. \end{aligned}$$

The parabola flows at $x_{j_2}^* = a_{j_1j_2 1}$ with $\overline{x}_{j_1} \neq a_{j_1j_1 1}$ are positive and negative for $\dot{x}_{j_2} > 0$ and $\dot{x}_{j_2} < 0$, respectively.

Thus, the equilibrium of $x_{j_2}^* = a_{j_1j_2 1}$ with $\overline{x}_{j_1} \neq a_{j_1j_1 1}$ has the following properties.

- For $a_{j_1j_1 0}(\overline{x}_{j_1} - a_{j_1j_1 1})^2 > 0$ and $a_{j_2j_2 0}\prod_{s_1=1}^{2}(a_{j_1j_2 1} - a_{j_2j_2 s_1}) > 0$,

$$(a_{j_1j_2 1}, \dot{x}_{j_2}) = \underbrace{(\text{UP}, \text{pF})}_{\text{up-parabola flow }(+)}.$$

- For $a_{j_1j_1 0}(\overline{x}_{j_1} - a_{j_1j_1 1})^2 < 0$ and $a_{j_2j_2 0}\prod_{s_1=1}^{2}(a_{j_1j_2 1} - a_{j_2j_2 s_1}) > 0$,

$$(a_{j_1j_2 1}, \dot{x}_{j_2}) = \underbrace{(\text{DP}, \text{pF})}_{\text{down-parabola flow }(+)}.$$

- For $a_{j_1j_1 0}(\overline{x}_{j_1} - a_{j_1j_1 1})^2 > 0$ and $a_{j_2j_2 0}\prod_{s_1=1}^{2}(a_{j_1j_2 1} - a_{j_2j_2 s_1}) < 0$,

$$(a_{j_1j_2 1}, \dot{x}_{j_2}) = \underbrace{(\text{DP}, \text{nF})}_{\text{down-parabola flow }(-)}.$$

- For $a_{j_1j_1 0}(\overline{x}_{j_1} - a_{j_1j_1 1})^2 < 0$ and $a_{j_2j_2 0}\prod_{s_1=1}^{2}(a_{j_1j_2 1} - a_{j_2j_2 s_1}) < 0$,

$$(a_{j_1j_2 1}, \dot{x}_{j_2}) = \underbrace{(\text{UP}, \text{nF})}_{\text{up-parabola flow }(-)}.$$

The variational equation at $x_{j_1}^* = a_{j_1j_1 1}$ with $\overline{x}_{j_2} \neq a_{j_1j_2 1}$ is

$$\Delta\dot{x}_{j_1} = a_{j_1j_10}(\overline{x}_{j_2} - a_{j_1j_21})(\Delta x_{j_1})^2$$

and

$$\dot{x}_{j_2} = a_{j_2j_20}\prod\nolimits_{s_1=1}^{2}(\overline{x}_{j_2} - a_{j_2j_2s_1}).$$

The parabola flows are upper-saddle and lower-saddle for $a_{j_1j_10}(\overline{x}_{j_2} - a_{j_1j_21}) > 0$ and $a_{j_1j_10}(\overline{x}_{j_2} - a_{j_1j_21}) < 0$, respectively.

Thus, the equilibrium of $x^*_{j_2} = a_{j_1j_21}$ with $\overline{x}_{j_1} \neq a_{j_1j_11}$ has the following properties.

- For $a_{j_1j_10}(\overline{x}_{j_2} - a_{j_1j_21}) > 0$ and $a_{j_2j_20}\prod_{s_1=1}^{2}(\overline{x}_{j_2} - a_{j_2j_2s_1}) > 0$,

$$(a_{j_1j_11}, \dot{x}_{j_2}) = \underbrace{(\text{US}, \text{pF})}_{\text{upper-saddle flow }(+)}.$$

- For $a_{j_1j_10}(\overline{x}_{j_2} - a_{j_1j_21}) < 0$ and $a_{j_2j_20}\prod_{s_1=1}^{2}(\overline{x}_{j_2} - a_{j_2j_2s_1}) > 0$,

$$(a_{j_1j_11}, \dot{x}_{j_2}) = \underbrace{(\text{LS}, \text{pF})}_{\text{lower-saddle flow }(+)}.$$

- For $a_{j_1j_10}(\overline{x}_{j_2} - a_{j_1j_21}) > 0$ and $a_{j_2j_20}\prod_{s_1=1}^{2}(\overline{x}_{j_2} - a_{j_2j_2s_1}) < 0$,

$$(a_{j_1j_11}, \dot{x}_{j_2}) = \underbrace{(\text{US}, \text{nF})}_{\text{upper-saddle flow }(-)}.$$

- For $a_{j_1j_10}(\overline{x}_{j_2} - a_{j_1j_21}) < 0$ and $a_{j_2j_20}\prod_{s_1=1}^{2}(\overline{x}_{j_2} - a_{j_2j_2s_1}) < 0$,

$$(a_{j_1j_11}, \dot{x}_{j_2}) = \underbrace{(\text{LS}, \text{nF})}_{\text{lower-saddle flow }(-)}.$$

From the above analysis, the equilibrium of $(x^*_{j_1}, x^*_{j_2}) = (a_{j_1j_11}, a_{j_1j_21})$ with $a_{j_1j_21} \neq a_{j_2j_21}, a_{j_2j_22}$ has the following properties as in Eqs. (1.214)–(1.217).

- For $a_{j_1j_10} > 0$ and $a_{j_2j_20}\prod_{s_1=1}^{2}(a_{j_1j_21} - a_{j_2j_2s_1}) > 0$,

$$(a_{j_1j_11}, a_{j_1j_21}) = \underbrace{(\text{UP}:\text{UP}, \text{pF})}_{\text{hyperbolic-secant-to-hyperbolic flow }(+)}.$$

The equilibrium of $(x^*_{j_1}, x^*_{j_2}) = (a_{j_1j_11}, a_{j_1j_21})$ is a (UP:UP, pF)-positive hyperbolic-secant-to-hyperbolic flow.

- For $a_{j_1j_10}<0$ and $a_{j_2j_20}\prod_{s_1=1}^{2}(a_{j_1j_21}-a_{j_2j_2s_1})>0$,

$$(a_{j_1j_11},a_{j_1j_21}) = \underbrace{(\mathrm{DP}:\mathrm{DP},\mathrm{pF})}_{\text{hyperbolic-to-hyperbolic-secant flow }(+)}.$$

The equilibrium of $(x_{j_1}^*,x_{j_2}^*)=(a_{j_1j_11},a_{j_1j_21})$ is a (DP:DP, pF)-positive hyperbolic-to-hyperbolic-secant flow.

- For $a_{j_1j_10}>0$ and $a_{j_2j_20}\prod_{s_1=1}^{2}(a_{j_1j_21}-a_{j_2j_2s_1})<0$,

$$(a_{j_1j_11},a_{j_1j_21}) = \underbrace{(\mathrm{DP}:\mathrm{DP},\mathrm{nF})}_{\text{hyperbolic-to-hyperbolic-secant flow }(-)}.$$

The equilibrium of $(x_{j_1}^*,x_{j_2}^*)=(a_{j_1j_11},a_{j_1j_21})$ is a (DP:DP, nF)-negative hyperbolic-to-hyperbolic-secant flow.

- For $a_{j_1j_10}<0$ and $a_{j_2j_20}\prod_{s_1=1}^{2}(a_{j_1j_21}-a_{j_2j_2s_1})<0$,

$$(a_{j_1j_11},a_{j_1j_21}) = \underbrace{(\mathrm{UP}:\mathrm{UP},\mathrm{nF})}_{\text{hyperbolic-secant-to-hyperbolic flow }(-)}.$$

The equilibrium of $(x_{j_1}^*,x_{j_2}^*)=(a_{j_1j_11},a_{j_1j_21})$ is a (UP:UP, nF)-negative hyperbolic-secant-to-hyperbolic flow.

(vi$_2$) For $a_{j_2j_2s_1}=a_{j_1j_21}$, we have

$$\frac{dx_{j_1}}{dx_{j_2}}=\frac{a_{j_1j_10}}{a_{j_2j_20}}\frac{(x_{j_1}-a_{j_1j_11})^2}{(x_{j_2}-a_{j_2j_2s_2})},$$

and the deformation of the foregoing equation is

$$\frac{dx_{j_1}}{(x_{j_1}-a_{j_1j_11})^2}=\frac{a_{j_1j_10}}{a_{j_2j_20}}\frac{dx_{j_2}}{(x_{j_2}-a_{j_2j_2s_1})}.$$

With an initial condition of (x_{j_10},x_{j_20}) at $t=t_0$, the integration of the above equation gives the first integral manifold $(a_{j_2j_2s_1}=a_{j_1j_21})$ as

$$-\left(\frac{1}{x_{j_1}-a_{j_1j_11}}-\frac{1}{x_{j_1}1-a_{j_1j_11}}\right)=\frac{a_{j_1j_10}}{a_{j_2j_20}}\ln\frac{|x_{j_2}-a_{j_2j_2s_2}|}{|x_{j_2}-a_{j_2j_2s_2}|}.$$

(vi$_{2a}$) At $x_{j_2}^* = a_{j_1j_21} = a_{j_2j_2s_1}$ with $\bar{x}_{j_1} \neq a_{j_1j_11}$, if

$$\frac{dx_{j_1}}{dx_{j_2}}\bigg|_{x_{j_2}^* = a_{j_1j_21}} = \frac{a_{j_1j_10}}{a_{j_2j_20}} \frac{(\bar{x}_{j_1} - a_{j_1j_11})^2}{(a_{j_2j_2s_1} - a_{j_2j_2s_2})} > 0,$$

there is an increasing-inflection flow, and if

$$\frac{dx_{j_1}}{dx_{j_2}}\bigg|_{x_{j_2}^* = a_{j_1j_21}} = \frac{a_{j_1j_10}}{a_{j_2j_20}} \frac{(\bar{x}_{j_1} - a_{j_1j_11})^2}{(a_{j_2j_2s_1} - a_{j_2j_2s_2})} < 0,$$

there is a decreasing-inflection flow. The corresponding variational equation is

$$\Delta\dot{x}_{j_2} = a_{j_2j_20}(a_{j_2j_2s_1} - a_{j_2j_2s_2})\Delta x_{j_2},$$

The parabola flow at $x_{j_2}^* = a_{j_1j_21} = a_{j_2j_2s_1}$ with $\bar{x}_{j_1} \neq a_{j_1j_11}$ are source and sink flows for $a_{j_2j_20}(a_{j_2j_2s_1} - a_{j_2j_2s_2}) > 0$ and $a_{j_2j_20}(a_{j_2j_2s_1} - a_{j_2j_2s_2}) < 0$, respectively.

Therefore, the infinite-equilibrium of $x_{j_2}^* = a_{j_1j_21} = a_{j_2j_2s_1}$ with $\bar{x}_{j_1} \neq a_{j_1j_11}$ has the following properties as in Eqs. (1.219)–(1.222).

- For $a_{j_1j_10}(\bar{x}_{j_1} - a_{j_1j_11})^2 > 0$ and $a_{j_2j_20}(a_{j_2j_2s_1} - a_{j_2j_2s_2}) > 0$,

$$(\bar{x}_{j_1}, a_{j_2j_2s_1}) = \underbrace{(\text{II, SO})}_{\text{increasing-inflection source}}.$$

The infinite-equilibrium of $x_{j_2}^* = a_{j_1j_21} = a_{j_2j_2s_1}$ is an (II,SO)-increasing-inflection source.

- For $a_{j_1j_10}(\bar{x}_{j_1} - a_{j_1j_11}) < 0$ and $a_{j_2j_20}(a_{j_2j_2s_1} - a_{j_2j_2s_2}) > 0$,

$$(\bar{x}_{j_1}, a_{j_2j_2s_1}) = \underbrace{(\text{DI, SO})}_{\text{decreasing-inflection source}}.$$

The infinite-equilibrium of $x_{j_2}^* = a_{j_1j_21} = a_{j_2j_2s_1}$ is a (DI,SO)-decreasing-inflection source.

- For $a_{j_1j_10}(\bar{x}_{j_1} - a_{j_1j_11}) > 0$ and $a_{j_2j_20}(a_{j_2j_2s_1} - a_{j_2j_2s_2}) < 0$,

$$(\bar{x}_{j_1}, a_{j_2j_2s_1}) = \underbrace{(\text{DI, SI})}_{\text{decreasing-inflection sink}}.$$

The infinite-equilibrium of $x_{j_2}^* = a_{j_1j_21} = a_{j_2j_2s_1}$ is a (DI,SI)-decreasing-inflection sink.

- For $a_{j_1j_10}(\bar{x}_{j_1} - a_{j_1j_11}) < 0$ and $a_{j_2j_20}(a_{j_2j_2s_1} - a_{j_2j_2s_2}) < 0$,

$$(\bar{x}_{j_1}, a_{j_2j_2s_1}) = \underbrace{(\text{II, SI})}_{\text{increasing-inflection sink}}.$$

The infinite-equilibrium of $x^*_{j_2} = a_{j_1j_21} = a_{j_2j_2s_1}$ is an (II,SI)-increasing-inflection sink.

(vi_{2b}) From case (vi_{2b}), for $\bar{x}_{j_1} = a_{j_1j_11}$, the equilibrium of $(x^*_{j_1}, x^*_{j_2}) = (a_{j_1j_11}, a_{j_1j_2s_2})$ with $a_{j_2j_2s_1} = a_{j_1j_21}$ has the following properties as in Eqs. (1.223)–(1.226).

- For $a_{j_1j_10} > 0$ and $a_{j_2j_20}(a_{j_2j_2s_1} - a_{j_2j_2s_2}) > 0$,

$$(a_{j_1j_11}, a_{j_2j_2s_1}) = \underbrace{(\text{II : II, SO})}_{\text{inflection source}}.$$

The equilibrium of $(x^*_{j_1}, x^*_{j_2}) = (a_{j_1j_11}, a_{j_2j_2s_1})$ is an (II:II,SO)-increasing-inflection source.

- For $a_{j_1j_10} < 0$ and $a_{j_2j_20}(a_{j_2j_2s_1} - a_{j_2j_2s_2}) > 0$,

$$(a_{j_1j_11}, a_{j_2j_2s_1}) = \underbrace{(\text{DI : DI, SO})}_{\text{inflection source}}.$$

The equilibrium of $(x^*_{j_1}, x^*_{j_2}) = (a_{j_1j_11}, a_{j_2j_2s_1})$ is a (DI:DI,SO)-decreasing-inflection source.

- For $a_{j_1j_10} > 0$ and $a_{j_2j_20}(a_{j_2j_2s_1} - a_{j_2j_2s_2}) < 0$,

$$(a_{j_1j_11}, a_{j_2j_2s_1}) = \underbrace{(\text{DI : DI, SI})}_{\text{inflection sink}}.$$

The equilibrium of $(x^*_{j_1}, x^*_{j_2}) = (a_{j_1j_11}, a_{j_2j_2s_1})$ is a (DI:DI,SI)-decreasing-inflection sink.

- For $a_{j_1j_10} < 0$ and $a_{j_2j_20}(a_{j_2j_2s_1} - a_{j_2j_2s_2}) < 0$,

$$(a_{j_1j_11}, a_{j_2j_2s_1}) = \underbrace{(\text{II : II, SI})}_{\text{inflection sink}}.$$

The equilibrium of $(x^*_{j_1}, x^*_{j_2}) = (a_{j_1j_11}, a_{j_2j_2s_1})$ is an (II:II,SI)-increasing-inflection sink.

(vi_{2c}) At $(x^*_{j_1}, x^*_{j_2}) = (a_{j_1j_11}, a_{j_2j_2s_2})$, the variational equations are

$$\Delta\dot{x}_{j_2} = a_{j_2j_20}(a_{j_2j_2s_2} - a_{j_2j_2s_1})\Delta x_{j_2},$$
$$\Delta\dot{x}_{j_1} = a_{j_1j_10}(a_{j_2j_2s_2} - a_{j_1j_21})(\Delta x_{j_1})^2$$

The flow for $a_{j_1j_10}(a_{j_2j_2s_2} - a_{j_1j_21}) > 0$ and $a_{j_1j_10}(a_{j_2j_2s_2} - a_{j_1j_21}) < 0$ are upper-saddle and lower saddle in the x_{j_1}-direction, respectively. The flow are source and sink for $a_{j_2j_20}(a_{j_2j_2s_2} - a_{j_2j_2s_1}) > 0$ and $a_{j_2j_20}(a_{j_2j_2s_2} - a_{j_2j_2s_1}) < 0$ in the x_{j_2}-direction, respectively.

Therefore, the equilibrium of $(x^*_{j_1}, x^*_{j_2}) = (a_{j_1j_11}, a_{j_2j_2s_2})$ with $a_{j_2j_2s_1} = a_{j_1j_21}$ has the following properties as in Eqs. (1.227)–(1.230).

- For $a_{j_1j_10}(a_{j_2j_2s_2} - a_{j_1j_21}) > 0$ and $a_{j_2j_20}(a_{j_2j_2s_2} - a_{j_2j_2s_1}) > 0$,

$$(a_{j_1j_11}, a_{j_2j_2s_2}) = \underbrace{(\text{US}, \text{SO})}_{\text{upper-saddle source}}.$$

The equilibrium of $(x^*_{j_1}, x^*_{j_2}) = (a_{j_1j_11}, a_{j_2j_2s_2})$ is a (US,SO)-upper-saddle source.

- For $a_{j_1j_10}(a_{j_2j_2s_2} - a_{j_1j_21}) < 0$ and $a_{j_2j_20}(a_{j_2j_2s_2} - a_{j_2j_2s_1}) > 0$,

$$(a_{j_1j_11}, a_{j_2j_2s_2}) = \underbrace{(\text{LS}, \text{SO})}_{\text{lower-saddle source}}.$$

The equilibrium of $(x^*_{j_1}, x^*_{j_2}) = (a_{j_1j_11}, a_{j_2j_2s_2})$ is an (LS,SO)-lower-saddle source.

- For $a_{j_1j_10}(a_{j_2j_2s_2} - a_{j_1j_21}) > 0$ and $a_{j_2j_20}(a_{j_2j_2s_2} - a_{j_2j_2s_1}) < 0$,

$$(a_{j_1j_11}, a_{j_2j_2s_2}) = \underbrace{(\text{US}, \text{SI})}_{\text{upper-saddle sink}}.$$

The equilibrium of $(x^*_{j_1}, x^*_{j_2}) = (a_{j_1j_11}, a_{j_2j_2s_2})$ is a (US,SI)-upper-saddle sink.

- For $a_{j_1j_10}(a_{j_2j_2s_2} - a_{j_1j_21}) < 0$ and $a_{j_2j_20}(a_{j_2j_2s_2} - a_{j_2j_2s_1}) < 0$,

$$(a_{j_1j_11}, a_{j_2j_2s_2}) = \underbrace{(\text{LS}, \text{SI})}_{\text{lower-saddle sink}}.$$

The equilibrium of $(x^*_{j_1}, x^*_{j_2}) = (a_{j_1j_11}, a_{j_2j_2s_2})$ is an (LS,SI)-lower-saddle sink.

(vii) For $\Delta_{j_1j_2} > 0$ and $\Delta_{j_2j_2} < 0$, the standard form is

$$\dot{x}_{j_2} = a_{j_2j_20}\left[(x_{j_2} - a_{j_2j_2})^2 + b_{j_2j_2}\right],$$
$$\dot{x}_{j_1} = a_{j_1j_10}(x_{j_1} - a_{j_1j_11})(x_{j_1} - a_{j_1j_12})(x_{j_2} - a_{j_1j_21})$$

where

$$\begin{aligned}
&b_{j_1j_11}, b_{j_1j_12} = -\frac{1}{2}(B_{j_1j_1} \pm \sqrt{\Delta_{j_1j_1}}),\\
&\{a_{j_1j_11}, a_{j_1j_12}\} = \mathrm{sort}\{b_{j_1j_11}, b_{j_1j_12}\}, a_{j_1j_11} < a_{j_1j_12};\\
&a_{j_2j_2} = -\frac{1}{2}B_{j_2j_2}, b_{j_2j_2} = \frac{1}{4}(-\Delta_{j_2j_2}).
\end{aligned}$$

In phase space,

$$\frac{dx_{j_1}}{dx_{j_2}} = \frac{a_{j_1j_10}}{a_{j_2j_20}} \frac{(x_{j_1} - a_{j_1j_11})(x_{j_1} - a_{j_1j_12})(x_{j_2} - a_{j_1j_21})}{(x_{j_2} - a_{j_2j_2})^2 + b_{j_2j_2}},$$

and the deformation of the foregoing equation is

$$\frac{1}{(x_{j_1} - a_{j_1j_11})(x_{j_1} - a_{j_1j_12})} dx_{j_1} = \frac{a_{j_1j_10}}{a_{j_2j_20}} \left[\frac{(x_{j_2} - a_{j_2j_2})}{(x_{j_2} - a_{j_2j_2})^2 + b_{j_2j_2}} + \frac{(a_{j_2j_2} - a_{j_1j_21})}{(x_{j_2} - a_{j_2j_2})^2 + b_{j_2j_2}} \right] dx_{j_2}.$$

With an initial condition of (x_{j_10}, x_{j_20}) at $t = t_0$, the integration of the above equation gives the first integral manifold as

$$\begin{aligned}
&\frac{1}{a_{j_1j_12} - a_{j_1j_11}} \left(\ln \frac{|x_{j_1} - a_{j_1j_12}|}{|x_{j_10} - a_{j_1j_12}|} - \ln \frac{|x_{j_1} - a_{j_1j_11}|}{|x_{j_10} - a_{j_1j_11}|} \right)\\
&= \frac{a_{j_1j_10}}{a_{j_2j_20}} \left[\frac{1}{2} \ln \frac{|(x_{j_2} - a_{j_2j_2})^2 + b_{j_2j_2}|}{|(x_{j_20} - a_{j_2j_2})^2 + b_{j_2j_2}|} \right.\\
&\quad \left. + \frac{a_{j_2j_2} - a_{j_1j_21}}{\sqrt{b_{j_2j_2}}} \left(\arctan \frac{x_{j_2} - a_{j_2j_2}}{\sqrt{b_{j_2j_2}}} - \arctan \frac{x_{j_20} - a_{j_2j_2}}{\sqrt{b_{j_2j_2}}} \right) \right].
\end{aligned}$$

For $x^*_{j_2} = a_{j_1j_2s_1}$ and $\overline{x}_{j_1} \neq a_{j_1j_11}$ ($s_1, s_2 \in \{1, 2\}, s_1 \neq s_2$), in phase space,

$$\left.\frac{dx_{j_1}}{dx_{j_2}}\right|_{x^*_{j_2} = a_{j_1j_21}} = \left.\frac{a_{j_1j_10}}{a_{j_2j_20}} \frac{\prod_{s_1=1}^2 (\overline{x}_{j_1} - a_{j_1j_1s_1})(x_{j_2} - a_{j_1j_21})}{(x_{j_2} - a_{j_2j_2})^2 + b_{j_2j_2}}\right|_{x^*_{j_2} = a_{j_1j_21}} = 0.$$

If

$$\left.\frac{d^2x_{j_1}}{dx^2_{j_2}}\right|_{x^*_{j_2} = a_{j_1j_21}} = \frac{a_{j_1j_10}}{a_{j_2j_20}} \frac{\prod_{s_1=1}^2 (\overline{x}_{j_1} - a_{j_1j_1s_1})}{(a_{j_1j_21} - a_{j_2j_2})^2 + b_{j_2j_2}} > 0,$$

there is an up-parabola flow at $x^*_{j_2} = a_{j_1j_2s_1}$ with $\overline{x}_{j_1} \neq a_{j_1j_11}$. If

$$\frac{d^2 x_{j_1}}{dx_{j_2}^2}\Big|_{x_{j_2}^* = a_{j_1 j_2 1}} = \frac{a_{j_1 j_1 0}}{a_{j_2 j_2 0}} \frac{\prod_{s_1=1}^{2}(\overline{x}_{j_1} - a_{j_1 j_1 s_1})}{(a_{j_1 j_2 1} - a_{j_2 j_2})^2 + b_{j_2 j_2}} < 0,$$

there is a down-parabola flow at $x_{j_2}^* = a_{j_1 j_2 s_1}$ with $\overline{x}_{j_1} \neq a_{j_1 j_1 1}$. Let

$$\dot{x}_{j_2} = a_{j_2 j_2 0}\left[(a_{j_1 j_2 1} - a_{j_2 j_2})^2 + b_{j_2 j_2}\right],$$

with

$$(a_{j_1 j_2 1} - a_{j_2 j_2})^2 + b_{j_2 j_2} > 0.$$

The inflection flows at $x_{j_2}^* = a_{j_1 j_2 1}$ with $\overline{x}_{j_1} \neq a_{j_1 j_1 1}$ are positive and negative for $a_{j_2 j_2 0} > 0$ and $a_{j_2 j_2 0} < 0$, respectively.

- For $a_{j_1 j_1 0}\prod_{s_1=1}^{2}(\overline{x}_{j_1} - a_{j_1 j_1 s_1}) > 0$ and $a_{j_2 j_2 0} > 0$,

$$(a_{j_1 j_2 s_1}, \dot{x}_{j_2}) = \underbrace{(\text{UP}, \text{pF})}_{\text{up-parabola flow }(+)}.$$

- For $a_{j_1 j_1 0}\prod_{s_1=1}^{2}(\overline{x}_{j_1} - a_{j_1 j_1 s_1}) < 0$ and $a_{j_2 j_2 0} > 0$,

$$(a_{j_1 j_2 s_1}, \dot{x}_{j_2}) = \underbrace{(\text{DP}, \text{pF})}_{\text{down-parabola flow }(+)}.$$

- For $a_{j_1 j_1 0}\prod_{s_1=1}^{2}(\overline{x}_{j_1} - a_{j_1 j_1 s_1}) > 0$ and $a_{j_2 j_2 0} < 0$,

$$(a_{j_1 j_2 s_1}, \dot{x}_{j_2}) = \underbrace{(\text{DP}, \text{nF})}_{\text{down-parabola flow }(-)}.$$

- For $a_{j_1 j_1 0}\prod_{s_1=1}^{2}(\overline{x}_{j_1} - a_{j_1 j_1 s_1}) < 0$ and $a_{j_2 j_2 0} < 0$,

$$(a_{j_1 j_2 s_1}, \dot{x}_{j_2}) = \underbrace{(\text{UP}, \text{nF})}_{\text{up-parabola flow }(-)}.$$

From the above analysis, thus, the equilibrium of $(x_{j_1}^*, x_{j_2}^*) = (a_{j_1 j_1 s_1}, a_{j_1 j_2 1})$ $(s_1, s_2 \in \{1, 2\}, s_1 \neq s_2)$ has the following properties as in Eqs. (1.234)–(1.237).

- For $a_{j_1 j_1 0}(a_{j_1 j_1 s_1} - a_{j_1 j_1 s_2}) > 0$ and $a_{j_2 j_2 0} > 0$,

$$(a_{j_1j_1s_1}, a_{j_1j_2 1}) = \underbrace{(\text{DP}:\text{UP},\text{pF})}_{\text{hyperbolic flow }(+)}.$$

The equilibrium of $(x^*_{j_1}, x^*_{j_2}) = (a_{j_1j_1s_1}, a_{j_1j_2 1})$ is a (DP:UP,pF)-positive hyperbolic flow.

- For $a_{j_1j_1 0}(a_{j_1j_1s_1} - a_{j_1j_1s_2}) < 0$ and $a_{j_2j_2 0} > 0$,

$$(a_{j_1j_1s_1}, a_{j_1j_2 1}) = \underbrace{(\text{UP}:\text{DP},\text{pF})}_{\text{hyperbolic-secant flow }(+)}.$$

The equilibrium of $(x^*_{j_1}, x^*_{j_2}) = (a_{j_1j_1s_1}, a_{j_1j_2 1})$ is a (UP:DP,pF)-positive hyperbolic-secant flow.

- For $a_{j_1j_1 0}(a_{j_1j_1s_1} - a_{j_1j_1s_2}) > 0$ and $a_{j_2j_2 0} < 0$,

$$(a_{j_1j_1s_1}, a_{j_1j_2 1}) = \underbrace{(\text{UP}:\text{DP},\text{nF})}_{\text{hyperbolic-secant flow }(-)}.$$

The equilibrium of $(x^*_{j_1}, x^*_{j_2}) = (a_{j_1j_1s_1}, a_{j_1j_2 1})$ is a (UP:DP,nF)-negative hyperbolic-secant flow.

- For $a_{j_1j_1 0}(a_{j_1j_1s_1} - a_{j_1j_1s_2}) < 0$ and $a_{j_2j_2 0} < 0$,

$$(a_{j_1j_1s_1}, a_{j_1j_2 1}) = \underbrace{(\text{DP}:\text{UP},\text{nF})}_{\text{hyperbolic flow }(-)}.$$

The equilibrium of $(x^*_{j_1}, x^*_{j_2}) = (a_{j_1j_1s_1}, a_{j_1j_2 1})$ is a (DP:UP,nF)-negative hyperbolic flow.

(viii) For $\Delta_{j_1j_2} > 0$ and $\Delta_{j_2j_2} = 0$, the standard form is

$$\begin{aligned}
\dot{x}_{j_2} &= a_{j_2j_2 0}(x_{j_2} - a_{j_2j_2 1})^2,\\
\dot{x}_{j_1} &= a_{j_1j_1 0}(x_{j_1} - a_{j_1j_1 1})(x_{j_1} - a_{j_1j_1 2})(x_{j_2} - a_{j_1j_2 1})
\end{aligned}$$

where

$$\begin{aligned}
& b_{j_1j_1 1}, b_{j_1j_1 2} = -\frac{1}{2}(B_{j_1j_1} \pm \sqrt{\Delta_{j_1j_1}}),\\
& \{a_{j_1j_1 1}, a_{j_1j_1 2}\} = \text{sort}\{b_{j_1j_1 1}, b_{j_1j_1 2}\}, a_{j_1j_1 1} < a_{j_1j_1 2};\\
& a_{j_2j_2 1} = -\frac{1}{2}B_{j_2j_2}.
\end{aligned}$$

(viii$_1$) For $a_{j_2j_2 1} \neq a_{j_1j_2 1}, a_{j_1j_2 2}$, in phase space,

$$\frac{dx_{j_1}}{dx_{j_2}} = \frac{a_{j_1 j_1 0}}{a_{j_2 j_2 0}} \frac{(x_{j_1} - a_{j_1 j_1 1})(x_{j_1} - a_{j_1 j_1 2})(x_{j_2} - a_{j_1 j_2 1})}{(x_{j_2} - a_{j_2 j_2 1})^2},$$

and the deformation of the foregoing equation is

$$\frac{1}{(x_{j_1} - a_{j_1 j_1 1})(x_{j_1} - a_{j_1 j_1 2})} dx_{j_1} = \frac{a_{j_1 j_1 0}}{a_{j_2 j_2 0}} \left[\frac{1}{(x_{j_2} - a_{j_2 j_2 1})} + \frac{(a_{j_2 j_2 1} - a_{j_1 j_2 1})}{(x_{j_2} - a_{j_2 j_2 1})^2} \right] dx_{j_2}.$$

With an initial condition of $(x_{j_1 0}, x_{j_2 0})$ at $t = t_0$, the integration of the above equation gives the first integral manifold for $a_{j_2 j_2 1} \neq a_{j_1 j_2 1}, a_{j_1 j_2 2}$ as

$$\begin{aligned} &\frac{1}{a_{j_1 j_1 2} - a_{j_1 j_1 1}} \left(\ln \frac{|x_{j_1} - a_{j_1 j_1 2}|}{|x_{j_1 0} - a_{j_1 j_1 2}|} - \ln \frac{|x_{j_1} - a_{j_1 j_1 1}|}{|x_{j_1 0} - a_{j_1 j_1 1}|} \right) \\ &= \frac{a_{j_1 j_1 0}}{a_{j_2 j_2 0}} \left[\ln \frac{|x_{j_2} - a_{j_2 j_2 1}|}{|x_{j_2 0} - a_{j_2 j_2 1}|} - (a_{j_2 j_2 1} - a_{j_1 j_2 1}) \left(\frac{1}{x_{j_2} - a_{j_2 j_2 1}} - \frac{1}{x_{j_2 0} - a_{j_2 j_2 1}} \right) \right]. \end{aligned}$$

(viii$_{1a}$) At $(x^*_{j_1}, x^*_{j_2}) = (a_{j_1 j_1 s_1}, a_{j_2 j_2 1})$, the variational equations are

$$\begin{aligned} \Delta\dot{x}_{j_2} &= a_{j_2 j_2 0}(\Delta x_{j_2})^2, \\ \Delta\dot{x}_{j_1} &= a_{j_1 j_1 0}(a_{j_2 j_2 1} - a_{j_1 j_2 1})(a_{j_1 j_1 s_1} - a_{j_1 j_1 s_2})\Delta x_{j_1}. \end{aligned}$$

The flows at $(x^*_{j_1}, x^*_{j_2}) = (a_{j_1 j_1 1}, a_{j_2 j_2 1})$ in the x_{j_1}-direction are source and sink for $a_{j_1 j_1 0}(a_{j_2 j_2 1} - a_{j_1 j_2 1})(a_{j_1 j_1 s_1} - a_{j_1 j_1 s_2}) > 0$ and $a_{j_1 j_1 0}(a_{j_2 j_2 1} - a_{j_1 j_2 1})$ $(a_{j_1 j_1 s_1} - a_{j_1 j_1 s_2}) < 0$, respectively. The flows at $(x^*_{j_1}, x^*_{j_2}) = (a_{j_1 j_1 s_1}, a_{j_2 j_2 1})$ in the x_{j_2}-direction are upper-saddle and lower-saddle for $a_{j_2 j_2 0} > 0$ and $a_{j_2 j_2 0} < 0$, respectively.

Therefore, the equilibrium of $(x^*_{j_1}, x^*_{j_2}) = (a_{j_1 j_1 s_1}, a_{j_2 j_2 1})$ has the following properties as in Eqs. (1.241)–(1.244).

- For $a_{j_1 j_1 0}(a_{j_1 j_1 s_1} - a_{j_1 j_1 s_2})(a_{j_2 j_2 1} - a_{j_1 j_2 1}) > 0$ and $a_{j_2 j_2 0} > 0$,

$$(a_{j_1 j_1 s_1}, a_{j_2 j_2 1}) = \underbrace{(\text{SO}, \text{US})}_{\text{upper-saddle source}}.$$

The equilibrium of $(x^*_{j_1}, x^*_{j_2}) = (a_{j_1 j_1 s_1}, a_{j_2 j_2 1})$ is an (SO,US)-upper-saddle source.

- For $a_{j_1 j_1 0}(a_{j_1 j_1 s_1} - a_{j_1 j_1 s_2})(a_{j_2 j_2 1} - a_{j_1 j_2 1}) < 0$ and $a_{j_2 j_2 0} > 0$,

$$(a_{j_1 j_1 s_1}, a_{j_2 j_2 1}) = \underbrace{(\text{SI}, \text{US})}_{\text{upper-saddle-sink}}.$$

The equilibrium of $(x^*_{j_1}, x^*_{j_2}) = (a_{j_1 j_1 s_1}, a_{j_2 j_2 1})$ is an (SI,US)-upper-saddle sink.

- For $a_{j_1j_10}(a_{j_1j_1s_1}-a_{j_1j_1s_2})(a_{j_2j_21}-a_{j_1j_21})>0$ and $a_{j_2j_20}<0$,

$$(a_{j_1j_1s_1},a_{j_2j_21})=\underbrace{(\mathrm{SO},\mathrm{LS})}_{\text{lower-saddle source}}.$$

The equilibrium of $(x^*_{j_1},x^*_{j_2})=(a_{j_1j_1s_1},a_{j_2j_21})$ is an (SO,LS)-lower-saddle source.

- For $a_{j_1j_10}(a_{j_1j_1s_1}-a_{j_1j_1s_2})(a_{j_2j_21}-a_{j_1j_21})<0$ and $a_{j_2j_20}<0$,

$$(a_{j_1j_1s_1},a_{j_2j_21})=\underbrace{(\mathrm{SI},\mathrm{LS})}_{\text{lower-saddle source}}.$$

The equilibrium of $(x^*_{j_1},x^*_{j_2})=(a_{j_1j_1s_1},a_{j_2j_21})$ is an (SI,LS)-lower-saddle sink.

– The saddle-source and saddle-sink are the appearing bifurcations of saddle-source and saddle-sink.

(viii$_{1b}$) For $x^*_{j_2}=a_{j_1j_21}\neq a_{j_2j_21}$ and $\bar{x}_{j_1}\neq a_{j_1j_1s_1}$ ($s_1,s_2\in\{1,2\}$, $s_1\neq s_2$), in phase space,

$$\frac{dx_{j_1}}{dx_{j_2}}\Big|_{x^*_{j_2}=a_{j_1j_21}}=\frac{a_{j_1j_10}}{a_{j_2j_20}}\frac{\prod_{s_1=1}^{2}(\bar{x}_{j_1}-a_{j_1j_1s_1})(x_{j_2}-a_{j_1j_21})}{(x_{j_2}-a_{j_2j_21})^2}\Big|_{x^*_{j_2}=a_{j_1j_21}}=0.$$

If

$$\frac{d^2x_{j_1}}{dx^2_{j_2}}\Big|_{x^*_{j_2}=a_{j_1j_21}}=\frac{a_{j_1j_10}}{a_{j_2j_20}}\frac{\prod_{s_1=1}^{2}(\bar{x}_{j_1}-a_{j_1j_1s_1})}{(a_{j_1j_21}-a_{j_2j_21})^2}>0,$$

there is an up-parabola flow at $x^*_{j_2}=a_{j_1j_2s_1}\neq a_{j_2j_21}$ with $\bar{x}_{j_1}\neq a_{j_1j_11}$. If

$$\frac{d^2x_{j_1}}{dx^2_{j_2}}\Big|_{x^*_{j_2}=a_{j_1j_21}}=\frac{a_{j_1j_10}}{a_{j_2j_20}}\frac{\prod_{s_1=1}^{2}(\bar{x}_{j_1}-a_{j_1j_1s_1})}{(a_{j_1j_21}-a_{j_2j_21})^2}<0,$$

there is a down-parabola flow at $x^*_{j_2}=a_{j_1j_21}$ with $\bar{x}_{j_1}\neq a_{j_1j_1s_1},a_{j_1j_1s_2}$. Let

$$\dot{x}_{j_2}=a_{j_2j_20}(a_{j_1j_21}-a_{j_2j_21})^2\neq 0.$$

The parabola flows at $x^*_{j_2}=a_{j_1j_21}\neq a_{j_2j_21}$ with $\bar{x}_{j_1}\neq a_{j_1j_1s_1},a_{j_1j_1s_2}$ are positive and negative for $a_{j_2j_20}>0$ and $a_{j_2j_20}<0$, respectively.

- For $a_{j_1j_10}\prod_{s_1=1}^{2}(\overline{x}_{j_1} - a_{j_1j_1s_1}) > 0$ and $a_{j_2j_20}(a_{j_1j_21} - a_{j_2j_21})^2 > 0$,

$$(a_{j_1j_21}, \dot{x}_{j_2}) = \underbrace{(\text{UP}, \text{pF})}_{\text{up-parabola flow } (+)}.$$

- For $a_{j_1j_10}\prod_{s_1=1}^{2}(\overline{x}_{j_1} - a_{j_1j_1s_1}) < 0$ and $a_{j_2j_20}(a_{j_1j_21} - a_{j_2j_21})^2 > 0$,

$$(a_{j_1j_2s_1}, \dot{x}_{j_2}) = \underbrace{(\text{DP}, \text{pF})}_{\text{down-parabola flow } (+)}.$$

- For $a_{j_1j_10}\prod_{s_1=1}^{2}(\overline{x}_{j_1} - a_{j_1j_1s_1}) > 0$ and $a_{j_2j_20}(a_{j_1j_21} - a_{j_2j_21})^2 < 0$,

$$(a_{j_1j_2s_1}, \dot{x}_{j_2}) = \underbrace{(\text{DP}, \text{nF})}_{\text{down-parabola flow } (-)}.$$

- For $a_{j_1j_10}\prod_{s_1=1}^{2}(\overline{x}_{j_1} - a_{j_1j_1s_1}) < 0$ and $a_{j_2j_20}(a_{j_1j_21} - a_{j_2j_21})^2 < 0$,

$$(a_{j_1j_2s_1}, \dot{x}_{j_2}) = \underbrace{(\text{UP}, \text{nF})}_{\text{up-parabola flow } (-)}.$$

From the above analysis, let $\overline{x}_{j_1} = a_{j_1j_1s_1}$ (s_1, $s_2 \in \{1, 2\}$, $s_1 \neq s_2$). Therefore, the equilibrium of $(x^*_{j_1}, x^*_{j_2}) = (a_{j_1j_1s_1}, a_{j_1j_21})$ with $a_{j_1j_21} \neq a_{j_2j_21}$ (s_1, $s_2 \in \{1, 2\}$, $s_1 \neq s_2$) has the following properties as in Eqs. (1.245)–(1.248).

- For $a_{j_1j_10}(a_{j_1j_1s_1} - a_{j_1j_1s_2}) > 0$ and $a_{j_2j_20}(a_{j_1j_21} - a_{j_2j_21})^2 > 0$,

$$(a_{j_1j_1s_1}, a_{j_1j_21}) = \underbrace{(\text{DP} : \text{UP}, \text{pF})}_{\text{hyperbolic flow } (+)}.$$

  The equilibrium of $(x^*_{j_1}, x^*_{j_2}) = (a_{j_1j_1s_1}, a_{j_1j_21})$ is a (DP:UP, pF)-positive hyperbolic flow.
- For $a_{j_1j_10}(a_{j_1j_1s_1} - a_{j_1j_1s_2}) < 0$ and $a_{j_2j_20}(a_{j_1j_21} - a_{j_2j_21})^2 > 0$,

$$(a_{j_1j_1s_1}, a_{j_1j_21}) = \underbrace{(\text{UP} : \text{DP}, \text{pF})}_{\text{hyperbolic-secant flow } (+)}.$$

  The equilibrium of $(x^*_{j_1}, x^*_{j_2}) = (a_{j_1j_1s_1}, a_{j_1j_21})$ is a (UP:DP, pF)-positive hyperbolic-secant flow.
- For $a_{j_1j_10}(a_{j_1j_2s_1} - a_{j_1j_2s_2}) > 0$ and $a_{j_2j_20}(a_{j_1j_2s_1} - a_{j_2j_21})^2 < 0$,

$$(a_{j_1j_1s_1}, a_{j_1j_2 1}) = \underbrace{(\text{UP}:\text{DP},\text{nF})}_{\text{hyperbolic-secant flow }(-)}.$$

The equilibrium of $(x^*_{j_1}, x^*_{j_2}) = (a_{j_1j_1s_1}, a_{j_1j_2 1})$ is a (UP:DP,nF)-negative hyperbolic-secant flow.

- For $a_{j_1j_1 0}(a_{j_1j_2s_1} - a_{j_1j_2s_2}) < 0$ and $a_{j_2j_2 0}(a_{j_1j_2s_1} - a_{j_2j_2 1})^2 < 0$,

$$(a_{j_1j_1s_1}, a_{j_1j_2 1}) = \underbrace{(\text{DP}:\text{UP},\text{nF})}_{\text{hyperbolic flow }(-)}.$$

The equilibrium of $(x^*_{j_1}, x^*_{j_2}) = (a_{j_1j_1s_1}, a_{j_1j_2 1})$ is a (DP:UP,nF)-negative hyperbolic flow.

(viii$_2$) For $a_{j_2j_2 1} = a_{j_1j_2 1}$, we have

$$\begin{aligned}\frac{dx_{j_1}}{dx_{j_2}} &= \frac{a_{j_1j_1 0}}{a_{j_2j_2 0}} \frac{(x_{j_1} - a_{j_1j_1 1})(x_{j_1} - a_{j_1j_1 2})(x_{j_2} - a_{j_1j_2 1})}{(x_{j_2} - a_{j_2j_2 1})^2} \\ &= \frac{a_{j_1j_1 0}}{a_{j_2j_2 0}} \frac{(x_{j_1} - a_{j_1j_1 1})(x_{j_1} - a_{j_1j_1 2})}{(x_{j_2} - a_{j_2j_2 1})},\end{aligned}$$

and the deformation of the foregoing equation is

$$\frac{1}{a_{j_1j_1 2} - a_{j_1j_1 1}}\left(\frac{1}{x_{j_1} - a_{j_1j_1 2}} - \frac{1}{x_{j_1} - a_{j_1j_1 1}}\right)dx_{j_1} = \frac{a_{j_1j_1 0}}{a_{j_2j_2 0}} \frac{dx_{j_2}}{x_{j_2} - a_{j_2j_2 1}}.$$

With an initial condition of $(x_{j_1 0}, x_{j_2 0})$ at $t = t_0$, the integration of the above equation gives the first integral manifold for $a_{j_2j_2 1} = a_{j_1j_2 1}$ is

$$\frac{1}{a_{j_1j_1 2} - a_{j_1j_1 1}}\left(\ln\frac{|x_{j_1} - a_{j_1j_1 2}|}{|x_{j_1 0} - a_{j_1j_1 2}|} - \ln\frac{|x_{j_1} - a_{j_1j_1 1}|}{|x_{j_1 0} - a_{j_1j_1 1}|}\right) = \frac{a_{j_1j_1 0}}{a_{j_2j_2 0}} \ln\frac{|x_{j_2} - a_{j_2j_2 1}|}{|x_{j_2 0} - a_{j_2j_2 1}|}.$$

(viii$_{2a}$) At $x^*_{j_2} = a_{j_1j_2 1} = a_{j_2j_2 1}$ with $\bar{x}_{j_1} \neq a_{j_1j_1 1}, a_{j_1j_1 2}$, for a small $\varepsilon > 0$,

$$\left.\frac{dx_{j_1}}{dx_{j_2}}\right|_{x^*_{j_2} = a_{j_2j_2 1} \pm \varepsilon} = \pm \frac{a_{j_1j_1 0}}{a_{j_2j_2 0}} \frac{(\bar{x}_{j_1} - a_{j_1j_1 1})(\bar{x}_{j_1} - a_{j_1j_1 2})}{\varepsilon}.$$

From the first integral manifold, the flow is an asymptotic flow. If

$$\frac{a_{j_1j_1 0}}{a_{j_2j_2 0}} \prod\nolimits_{s_1=1}^{2} (\bar{x}_{j_1} - a_{j_1j_1s_1}) > 0,$$

the flow at $x^*_{j_2} = a_{j_1j_2 1} = a_{j_2j_2 1}$ is a down-up flow. If

$$\frac{a_{j_1j_10}}{a_{j_2j_20}}\prod\nolimits_{s_1=1}^{2}(\overline{x}_{j_1}-a_{j_1j_1s_1})<0,$$

the flow at $x_{j_2}^*=a_{j_1j_21}=a_{j_2j_21}$ is an up-down flow. The variational equation at $x_{j_2}^*=a_{j_1j_2s_1}=a_{j_2j_21}$ is

$$\Delta\dot{x}_{j_2}=a_{j_2j_20}(\Delta x_{j_2})^2.$$

The asymptotic flows in the x_{j_2}-direction are upper-saddle and lower-saddle for $a_{j_2j_20}>0$ and $a_{j_2j_20}<0$, respectively.

Therefore, the infinite-equilibrium of $x_{j_2}^*=a_{j_1j_21}=a_{j_2j_21}$ with $\overline{x}_{j_1}\neq a_{j_1j_11}$ has the following properties as in Eqs. (1.250)–(1.253).

- For $a_{j_1j_10}\prod_{s_1=1}^{2}(\overline{x}_{j_1}-a_{j_1j_1s_1})>0$ and $a_{j_2j_20}>0$,

$$(\overline{x}_{j_1},a_{j_1j_21})=\underbrace{(\text{DU},\text{US})}_{\text{down-up upper-saddle}}.$$

  The infinite-equilibrium of $x_{j_2}^*=a_{j_1j_21}=a_{j_2j_21}$ is an (DU,US)-down-up upper-saddle.

- For $a_{j_1j_10}\prod_{s_1=1}^{2}(\overline{x}_{j_1}-a_{j_1j_1s_1})<0$ and $a_{j_2j_20}>0$,

$$(\overline{x}_{j_1},a_{j_1j_21})=\underbrace{(\text{UD},\text{US})}_{\text{up-down upper-saddle}}.$$

  The infinite-equilibrium of $x_{j_2}^*=a_{j_1j_21}=a_{j_2j_21}$ is a (UD,US)-up-down upper-saddle.

- For $a_{j_1j_10}\prod_{s_1=1}^{2}(\overline{x}_{j_1}-a_{j_1j_1s_1})>0$ and $a_{j_2j_20}<0$,

$$(\overline{x}_{j_1},a_{j_1j_21})=\underbrace{(\text{UD},\text{LS})}_{\text{up-down lower-saddle}}.$$

  The infinite-equilibrium of $x_{j_2}^*=a_{j_1j_21}=a_{j_2j_21}$ is a (UD,LS)-up-down lower-saddle.

- For $a_{j_1j_10}\prod_{s_1=1}^{2}(\overline{x}_{j_1}-a_{j_1j_1s_1})<0$ and $a_{j_2j_20}<0$,

$$(\overline{x}_{j_1},a_{j_1j_21})=\underbrace{(\text{DU},\text{LS})}_{\text{down-up lower-saddle}}.$$

  The infinite-equilibrium of $x_{j_2}^*=a_{j_1j_21}=a_{j_2j_21}$ is a (DU,LS)-down-up lower-saddle.

(viii$_{2b}$) From case (viii$_{2a}$), the equilibrium of $(x^*_{j_1}, x^*_{j_2}) = (a_{j_1 j_1 s_1}, a_{j_2 j_2 1})$ with $a_{j_2 j_2 1} = a_{j_1 j_2 1}$ has the following properties as in Eqs. (1.254)–(1.257).

- For $a_{j_1 j_1 0}(a_{j_1 j_1 s_1} - a_{j_1 j_1 s_2}) > 0$ and $a_{j_2 j_2 0} > 0$,

$$(a_{j_1 j_1 s_1}, a_{j_2 j_2 1}) = \underbrace{(\mathrm{DU} : \mathrm{UD}, \mathrm{US})}_{\text{down-up-to-up-down upper-saddle}}.$$

  The equilibrium of $(x^*_{j_1}, x^*_{j_2}) = (a_{j_1 j_1 s_1}, a_{j_2 j_2 1})$ is a (DU:UD,US)-down-up-to-up-down upper-saddle.
- For $a_{j_1 j_1 0}(a_{j_1 j_1 s_1} - a_{j_1 j_1 s_2}) < 0$ and $a_{j_2 j_2 0} > 0$,

$$(a_{j_1 j_1 s_1}, a_{j_2 j_2 1}) = \underbrace{(\mathrm{UD} : \mathrm{DU}, \mathrm{US})}_{\text{up-down-to-down-up upper-saddle}}.$$

  The equilibrium of $(x^*_{j_1}, x^*_{j_2}) = (a_{j_1 j_1 s_1}, a_{j_2 j_2 1})$ is a (UD:DU,US)-up-down-to-down-up upper-saddle.
- For $a_{j_1 j_1 0}(a_{j_1 j_1 s_1} - a_{j_1 j_1 s_2}) > 0$ and $a_{j_2 j_2 0} < 0$,

$$(a_{j_1 j_1 s_1}, a_{j_2 j_2 1}) = \underbrace{(\mathrm{UD} : \mathrm{DU}, \mathrm{LS})}_{\text{up-down-to-down-up lower-saddle}}.$$

  The equilibrium of $(x^*_{j_1}, x^*_{j_2}) = (a_{j_1 j_1 s_1}, a_{j_2 j_2 1})$ is a (UD:DU,LS)-up-down-to-down-up lower-saddle.
- For $a_{j_1 j_1 0}(a_{j_1 j_1 s_1} - a_{j_1 j_1 s_2}) < 0$ and $a_{j_2 j_2 0} < 0$,

$$(a_{j_1 j_1 s_1}, a_{j_2 j_2 1}) = \underbrace{(\mathrm{DU} : \mathrm{UD}, \mathrm{LS})}_{\text{down-up-to-up-down lower-saddle}}.$$

  The equilibrium of $(x^*_{j_1}, x^*_{j_2}) = (a_{j_1 j_1 s_1}, a_{j_2 j_2 1})$ is a (DU:UD,LS)-down-up-to-up-down lower-saddle.

(ix) For $\Delta_{j_1 j_2} > 0$ and $\Delta_{j_2 j_2} > 0$, the standard form is

$$\begin{aligned} \dot{x}_{j_2} &= a_{j_2 j_2 0}(x_{j_2} - a_{j_2 j_2 1})(x_{j_2} - a_{j_2 j_2 2}), \\ \dot{x}_{j_1} &= a_{j_1 j_1 0}(x_{j_1} - a_{j_1 j_1 1})(x_{j_1} - a_{j_1 j_1 2})(x_{j_2} - a_{j_1 j_2 1}) \end{aligned}$$

where

$$
\begin{aligned}
&b_{j_1j_11}, b_{j_1j_12} = -\frac{1}{2}(B_{j_1j_1} \pm \sqrt{\Delta_{j_1j_1}}),\\
&\{a_{j_1j_11}, a_{j_1j_12}\} = \text{sort}\{b_{j_1j_11}, b_{j_1j_12}\}, a_{j_1j_11} < a_{j_1j_12};\\
&b_{j_2j_21}, b_{j_2j_22} = -\frac{1}{2}(B_{j_2j_2} \pm \sqrt{\Delta_{j_2j_2}}),\\
&\{a_{j_2j_21}, a_{j_2j_22}\} = \text{sort}\{b_{j_2j_21}, b_{j_2j_22}\}, a_{j_2j_21} < a_{j_2j_22}.
\end{aligned}
$$

(ix_1) For $a_{j_1j_21} \neq a_{j_2j_21}, a_{j_2j_22}$ ($s_1 = 1, 2$), in phase space,

$$
\frac{dx_{j_1}}{dx_{j_2}} = \frac{a_{j_1j_10}}{a_{j_2j_20}} \frac{(x_{j_1} - a_{j_1j_11})(x_{j_1} - a_{j_1j_12})(x_{j_2} - a_{j_1j_21})}{(x_{j_2} - a_{j_2j_21})(x_{j_2} - a_{j_2j_22})},
$$

and the deformation of the foregoing equation is

$$
\frac{dx_{j_1}}{(x_{j_1} - a_{j_1j_11})(x_{j_1} - a_{j_1j_12})} = \frac{a_{j_1j_10}}{a_{j_2j_20}} \sum\nolimits_{l_1=1}^{2} \frac{(a_{j_2j_2l_1} - a_{j_1j_21})}{(a_{j_2j_2l_1} - a_{j_2j_2l_2})} \frac{dx_{j_2}}{(x_{j_2} - a_{j_2j_2l_1})}.
$$

With an initial condition of (x_{j_10}, x_{j_20}) at $t = t_0$, the integration of the above equation gives the first integral manifold as

$$
\begin{aligned}
&\sum\nolimits_{s_1=1}^{2} \frac{1}{(a_{j_1j_1s_1} - a_{j_1j_1s_2})} \ln \frac{|x_{j_1} - a_{j_1j_1s_1}|}{|x_{j_10} - a_{j_1j_1s_1}|}\\
&= \frac{a_{j_1j_10}}{a_{j_2j_20}} \sum\nolimits_{l_1=1}^{2} \frac{(a_{j_2j_2l_1} - a_{j_1j_21})}{(a_{j_2j_2l_1} - a_{j_2j_2l_2})} \ln \frac{|x_{j_2} - a_{j_2j_2l_1}|}{|x_{j_20} - a_{j_2j_2l_1}|}.
\end{aligned}
$$

(ix_{1a}) At $(x_{j_1}^*, x_{j_2}^*) = (a_{j_1j_1s_1}, a_{j_2j_2l_1})$ ($s_1, s_2, l_1, l_2 \in \{1, 2\}, s_1 \neq s_2, l_1 \neq l_2$), the variational equations are

$$
\begin{aligned}
\Delta\dot{x}_{j_2} &= a_{j_2j_20}(a_{j_2j_2l_1} - a_{j_2j_2l_2})\Delta x_{j_2},\\
\Delta\dot{x}_{j_1} &= a_{j_1j_10}(a_{j_1j_1s_1} - a_{j_1j_1s_2})(a_{j_2j_2l_1} - a_{j_1j_21})\Delta x_{j_1}.
\end{aligned}
$$

The flows in the x_{j_1} -direction for $a_{j_1j_10}(a_{j_1j_1s_1} - a_{j_1j_1s_2})(a_{j_2j_2l_1} - a_{j_1j_21}) > 0$ and $a_{j_1j_10}(a_{j_1j_1s_1} - a_{j_1j_1s_2})(a_{j_2j_2l_1} - a_{j_1j_21}) < 0$ are source and sink, respectively. The flows in the x_{j_2} -direction are source and sink for $a_{j_2j_20}(a_{j_2j_2l_1} - a_{j_2j_2l_2}) > 0$ and $a_{j_2j_20}(a_{j_2j_2l_1} - a_{j_2j_2l_2}) < 0$, respectively.

Therefore, the equilibrium of $(x_{j_1}^*, x_{j_2}^*) = (a_{j_1j_1s_1}, a_{j_2j_2l_1})$ (s_1, s_2, l_1, $l_2 \in \{1, 2\}$, $s_1 \neq s_2$, $l_1 \neq l_2$) has the following properties as in Eqs. (1.261)–(1.264).

- For $a_{j_1j_10}(a_{j_2j_2l_1} - a_{j_1j_21})(a_{j_1j_1s_1} - a_{j_1j_1s_2}) > 0$ and $a_{j_2j_20}(a_{j_2j_2l_1} - a_{j_2j_2l_2}) > 0$,

$$(a_{j_1j_1s_1}, a_{j_2j_2l_1}) = \underbrace{(\text{SO}, \text{SO})}_{\text{source}}.$$

The equilibrium of $(x_{j_1}^*, x_{j_2}^*) = (a_{j_1j_1s_1}, a_{j_2j_2l_1})$ is an (SO,SO)-source.

- For $a_{j_1j_1 0}(a_{j_2j_2l_1} - a_{j_1j_2 1})(a_{j_1j_1s_1} - a_{j_1j_1s_2}) < 0$ and $a_{j_2j_2 0}(a_{j_2j_2l_1} - a_{j_2j_2l_2}) > 0$,

$$(a_{j_1j_1 1}, a_{j_2j_2s_1}) = \underbrace{(\text{SI}, \text{SO})}_{\text{saddle}}.$$

The equilibrium of $(x_{j_1}^*, x_{j_2}^*) = (a_{j_1j_1s_1}, a_{j_2j_2l_1})$ is an (SI,SO)-saddle.

- For $a_{j_1j_1 0}(a_{j_2j_2l_1} - a_{j_1j_2 1})(a_{j_1j_1s_1} - a_{j_1j_1s_2}) > 0$ and $a_{j_2j_2 0}(a_{j_2j_2l_1} - a_{j_2j_2l_2}) < 0$,

$$(a_{j_1j_1 1}, a_{j_2j_2s_1}) = \underbrace{(\text{SO}, \text{SI})}_{\text{saddle}}.$$

The equilibrium of $(x_{j_1}^*, x_{j_2}^*) = (a_{j_1j_1s_1}, a_{j_2j_2l_1})$ is an (SO,SI)-saddle.

- For $a_{j_1j_1 0}(a_{j_2j_2l_1} - a_{j_1j_2 1})(a_{j_1j_1s_1} - a_{j_1j_1s_2}) < 0$ and $a_{j_2j_2 0}(a_{j_2j_2l_1} - a_{j_2j_2l_2}) < 0$,

$$(a_{j_1j_1 1}, a_{j_2j_2s_1}) = \underbrace{(\text{SI}, \text{SI})}_{\text{sink}}.$$

The equilibrium of $(x_{j_1}^*, x_{j_2}^*) = (a_{j_1j_1s_1}, a_{j_2j_2l_1})$ is an (SI,SI)-sink.

($\text{ix}_{1\text{b}}$) For $x_{j_2}^* = a_{j_1j_2 1}$ and $x_{j_1}^* = a_{j_1j_1s_1}$ ($s_1 = 1, 2$), in phase space,

$$\frac{dx_{j_1}}{dx_{j_2}}\bigg|_{x_{j_2}^* = a_{j_1j_2 1}} = \frac{a_{j_1j_1 0}}{a_{j_2j_2 0}} \frac{\prod_{s_1=1}^{2}(\overline{x}_{j_1} - a_{j_1j_1s_1})(x_{j_2} - a_{j_1j_2 1})}{\prod_{l_1=1}^{2}(x_{j_2} - a_{j_2j_2l_1})}\bigg|_{x_{j_2}^* = a_{j_1j_2 1}} = 0.$$

If

$$\frac{d^2x_{j_1}}{dx_{j_2}^2}\bigg|_{x_{j_2}^* = a_{j_1j_2 1}} = \frac{a_{j_1j_1 0}}{a_{j_2j_2 0}} \frac{\prod_{s_1=1}^{2}(\overline{x}_{j_1} - a_{j_1j_1s_1})}{\prod_{l_1=1}^{2}(a_{j_1j_2 1} - a_{j_2j_2l_1})} > 0,$$

there is an up-parabola flow at $x_{j_2}^* = a_{j_1j_2 1} \neq a_{j_2j_2 1}, a_{j_2j_2 2}$ with $\overline{x}_{j_1} \neq a_{j_1j_1s_1}, a_{j_1j_1s_2}$. If

$$\frac{d^2x_{j_1}}{dx_{j_2}^2}\bigg|_{x_{j_2}^* = a_{j_1j_2 1}} = \frac{a_{j_1j_1 0}}{a_{j_2j_2 0}} \frac{\prod_{s_1=1}^{2}(\overline{x}_{j_1} - a_{j_1j_1s_1})}{\prod_{l_1=1}^{2}(a_{j_1j_2 1} - a_{j_2j_2l_1})} < 0,$$

there is a down-parabola flow at $x_{j_2}^* = a_{j_1j_2 1} \neq a_{j_2j_2 1}, a_{j_2j_2 2}$ with $\overline{x}_{j_1} \neq a_{j_1j_1s_1}, a_{j_1j_1s_2}$. Let

$$\dot{x}_{j_2} = a_{j_2 j_2 0} \prod\nolimits_{l_1=1}^{2} (a_{j_1 j_2 1} - a_{j_2 j_2 l_1}) \neq 0.$$

The parabola flows at $x_{j_2}^* = a_{j_1 j_2 1} \neq a_{j_2 j_2 1}, a_{j_2 j_2 2}$ with $\bar{x}_{j_1} \neq a_{j_1 j_1 s_1}, a_{j_1 j_1 s_2}$ are positive and negative for $a_{j_2 j_2 0} > 0$ and $a_{j_2 j_2 0} < 0$, respectively.

- For $a_{j_1 j_1 0} \prod_{s_1=1}^{2} (\bar{x}_{j_1} - a_{j_1 j_1 s_1}) > 0$ and $a_{j_2 j_2 0} \prod_{l_1=1}^{2} (a_{j_1 j_2 1} - a_{j_2 j_2 l_1}) > 0$,

$$(a_{j_1 j_2 1}, \dot{x}_{j_2}) = \underbrace{(\text{UP, pF})}_{\text{up-parabola flow } (+)}.$$

- For $a_{j_1 j_1 0} \prod_{s_1=1}^{2} (\bar{x}_{j_1} - a_{j_1 j_1 s_1}) < 0$ and $a_{j_2 j_2 0} \prod_{l_1=1}^{2} (a_{j_1 j_2 1} - a_{j_2 j_2 l_1}) > 0$,

$$(a_{j_1 j_2 s_1}, \dot{x}_{j_2}) = \underbrace{(\text{DP, pF})}_{\text{down-parabola flow } (+)}.$$

- For $a_{j_1 j_1 0} \prod_{s_1=1}^{2} (\bar{x}_{j_1} - a_{j_1 j_1 s_1}) > 0$ and $a_{j_2 j_2 0} \prod_{l_1=1}^{2} (a_{j_1 j_2 1} - a_{j_2 j_2 l_1}) < 0$,

$$(a_{j_1 j_2 s_1}, \dot{x}_{j_2}) = \underbrace{(\text{DP, nF})}_{\text{down-parabola flow } (-)}.$$

- For $a_{j_1 j_1 0} \prod_{s_1=1}^{2} (\bar{x}_{j_1} - a_{j_1 j_1 s_1}) < 0$ and $a_{j_2 j_2 0} \prod_{l_1=1}^{2} (a_{j_1 j_2 1} - a_{j_2 j_2 l_1}) < 0$,

$$(a_{j_1 j_2 s_1}, \dot{x}_{j_2}) = \underbrace{(\text{UP, nF})}_{\text{up-parabola flow } (-)}.$$

From the above analysis, let $\bar{x}_{j_1} = a_{j_1 j_1 s_1}$ $(s_1 \in \{1, 2\}$. Therefore, the equilibrium of $(x_{j_1}^*, x_{j_2}^*) = (a_{j_1 j_1 s_1}, a_{j_1 j_2 1})$ $(s_1,\ s_2 \in \{1, 2\},\ s_1 \neq s_2)$ with $a_{j_1 j_2 1} \neq a_{j_2 j_2 1}, a_{j_2 j_2 2}$ has the following properties as in Eqs. (1.265)–(1.268).

- For $a_{j_1 j_1 0}(a_{j_1 j_1 s_1} - a_{j_1 j_1 s_2}) > 0$ and $a_{j_2 j_2 0} \prod_{l_1=1}^{2} (a_{j_1 j_2 1} - a_{j_2 j_2 l_1}) > 0$,

$$(a_{j_1 j_1 s_1}, a_{j_1 j_2 1}) = \underbrace{(\text{DP : UP, pF})}_{\text{hyperbolic flow } (+)}.$$

  The equilibrium of $(x_{j_1}^*, x_{j_2}^*) = (a_{j_1 j_1 s_1}, a_{j_1 j_2 1})$ is a (DP:UP,pF)-positive hyperbolic flow.

- For $a_{j_1 j_1 0}(a_{j_1 j_1 s_1} - a_{j_1 j_1 s_2}) < 0$ and $a_{j_2 j_2 0} \prod_{l_1=1}^{2} (a_{j_1 j_2 1} - a_{j_2 j_2 l_1}) > 0$,

$$(a_{j_1j_1s_1}, a_{j_1j_21}) = \underbrace{(\text{UP}:\text{DP},\text{pF})}_{\text{hyperbolic-secant flow } (+)}.$$

The equilibrium of $(x_{j_1}^*, x_{j_2}^*) = (a_{j_1j_1s_1}, a_{j_1j_21})$ is a (UP:DP,pF)-positive hyperbolic-secant flow.

- For $a_{j_1j_10}(a_{j_1j_1s_1} - a_{j_1j_1s_2}) > 0$ and $a_{j_2j_20}\prod_{l_1=1}^{2}(a_{j_1j_21} - a_{j_2j_2l_1}) < 0$,

$$(a_{j_1j_1s_1}, a_{j_1j_21}) = \underbrace{(\text{UP}:\text{DP},\text{nF})}_{\text{hyperbolic-secant flow } (-)}.$$

The equilibrium of $(x_{j_1}^*, x_{j_2}^*) = (a_{j_1j_1s_1}, a_{j_1j_21})$ is a (UP:DP,nF)-negative hyperbolic-secant flow

- For $a_{j_1j_10}(a_{j_1j_1s_1} - a_{j_1j_1s_2}) < 0$ and $a_{j_2j_20}\prod_{l_1=1}^{2}(a_{j_1j_21} - a_{j_2j_2l_1}) < 0$,

$$(a_{j_1j_1s_1}, a_{j_1j_21}) = \underbrace{(\text{DP}:\text{UP},\text{nF})}_{\text{hyperbolic flow } (-)}.$$

The equilibrium of $(x_{j_1}^*, x_{j_2}^*) = (a_{j_1j_1s_1}, a_{j_1j_21})$ is a (DP:UP,nF)-negative hyperbolic flow.

(ix$_2$) For $a_{j_1j_21} = a_{j_2j_2l_1}$ ($l_1, l_2 \in \{1,2\}$, $l_1 \neq l_2$), in phase space,

$$\begin{aligned}\frac{dx_{j_1}}{dx_{j_2}} &= \frac{a_{j_1j_10}}{a_{j_2j_20}} \frac{(x_{j_1} - a_{j_1j_1s_1})(x_{j_1} - a_{j_1j_1s_2})(x_{j_2} - a_{j_1j_21})}{(x_{j_2} - a_{j_2j_2l_1})(x_{j_2} - a_{j_2j_2l_2})} \\ &= \frac{a_{j_1j_10}}{a_{j_2j_20}} \frac{(x_{j_1} - a_{j_1j_1s_1})(x_{j_1} - a_{j_1j_1s_2})}{(x_{j_2} - a_{j_2j_2l_2})},\end{aligned}$$

and the deformation of the foregoing equation is

$$\frac{dx_{j_1}}{(x_{j_1} - a_{j_1j_11})(x_{j_1} - a_{j_1j_12})} = \frac{a_{j_1j_10}}{a_{j_2j_20}} \frac{dx_{j_2}}{(x_{j_2} - a_{j_2j_2l_2})}.$$

With an initial condition of (x_{j_10}, x_{j_20}) at $t = t_0$, the integration of the above equation gives the first integral manifold as

$$\frac{1}{a_{j_1j_12} - a_{j_1j_11}} \left(\ln \frac{|x_{j_1} - a_{j_1j_12}|}{|x_{j_10} - a_{j_1j_12}|} - \ln \frac{|x_{j_1} - a_{j_1j_11}|}{|x_{j_10} - a_{j_1j_11}|}\right) = \frac{a_{j_1j_10}}{a_{j_2j_20}} \ln \frac{|x_{j_2} - a_{j_2j_2s_2}|}{|x_{j_2} - a_{j_2j_2s_2}|}.$$

(ix$_{2a}$) For $x_{j_2}^* = a_{j_1j_21} = a_{j_2j_2l_1}$ with $\bar{x}_{j_1} \neq a_{j_1j_1s_1}, a_{j_1j_1s_2}$, if

$$\left.\frac{dx_{j_1}}{dx_{j_2}}\right|_{x^*_{j_2}=a_{j_2j_2l_1}} = \frac{a_{j_1j_10}}{a_{j_2j_20}} \frac{\prod_{s_1=1}^{2}(\overline{x}_{j_1} - a_{j_1j_1s_1})}{(a_{j_2j_2l_1} - a_{j_2j_2l_2})} > 0,$$

there is an increasing-inflection flow, and if

$$\left.\frac{dx_{j_1}}{dx_{j_2}}\right|_{x^*_{j_2}=a_{j_2j_2l_1}} = \frac{a_{j_1j_10}}{a_{j_2j_20}} \frac{\prod_{s_1=1}^{2}(\overline{x}_{j_1} - a_{j_1j_1s_1})}{(a_{j_2j_2l_1} - a_{j_2j_2l_2})} < 0,$$

there is a decreasing-inflection flow. The variational equation at $x^*_{j_2} = a_{j_1j_2s_1} = a_{j_2j_21}$ is

$$\Delta\dot{x}_{j_2} = a_{j_2j_20}(a_{j_2j_2l_1} - a_{j_2j_2l_2})\Delta x_{j_2}.$$

The asymptotic flows for $a_{j_2j_20}(a_{j_2j_2l_1} - a_{j_2j_2l_2}) > 0$ and $a_{j_2j_20}(a_{j_2j_2l_1} - a_{j_2j_2l_2}) < 0$ are source and sink in the x_{j_2}-direction, respectively.

Therefore, the infinite-equilibrium of $x^*_{j_2} = a_{j_1j_21} = a_{j_2j_2l_1}$ with $\overline{x}_{j_1} \neq a_{j_1j_1s_1}, a_{j_1j_1s_2}$ has the following properties as in Eqs. (1.270)–(1.273).

- For $a_{j_1j_10}\prod_{s_1=1}^{2}(\overline{x}_{j_1} - a_{j_1j_1s_1}) > 0$ and $a_{j_2j_20}(a_{j_2j_2l_1} - a_{j_2j_2l_2}) > 0$,

$$(\overline{x}_{j_1}, a_{j_2j_2l_1}) = \underbrace{(\text{II, SO})}_{\text{increasing-inflection source}}.$$

  The infinite-equilibrium of $x^*_{j_2} = a_{j_1j_21} = a_{j_2j_2l_1}$ is an (II,SO)-increasing-inflection source.
- For $a_{j_1j_10}\prod_{s_1=1}^{2}(\overline{x}_{j_1} - a_{j_1j_1s_1}) < 0$ and $a_{j_2j_20}(a_{j_2j_2l_1} - a_{j_2j_2l_2}) > 0$,

$$(\overline{x}_{j_1}, a_{j_2j_2l_1}) = \underbrace{(\text{DI, SO})}_{\text{decreasing-inflection source}}.$$

  The infinite-equilibrium of $x^*_{j_2} = a_{j_1j_21} = a_{j_2j_2l_1}$ is a (DI,SO)-decreasing-inflection source.
- For $a_{j_1j_10}\prod_{s_1=1}^{2}(\overline{x}_{j_1} - a_{j_1j_1s_1}) > 0$ and $a_{j_2j_20}(a_{j_2j_2l_1} - a_{j_2j_2l_2}) < 0$,

$$(\overline{x}_{j_1}, a_{j_2j_2l_1}) = \underbrace{(\text{DI, SI})}_{\text{decreasing-inflection sink}}.$$

  The infinite-equilibrium of $x^*_{j_2} = a_{j_1j_21} = a_{j_2j_2l_1}$ is a (DI,SI)-decreasing-inflection sink.
- For $a_{j_1j_10}\prod_{s_1=1}^{2}(\overline{x}_{j_1} - a_{j_1j_1s_1}) < 0$ and $a_{j_2j_20}(a_{j_2j_2l_1} - a_{j_2j_2l_2}) < 0$,

$$(\bar{x}_{j_1}, a_{j_2j_2l_1}) = \underbrace{(\text{II}, \text{SI})}_{\text{increasing-inflection sink}}.$$

The infinite-equilibrium of $x^*_{j_2} = a_{j_1j_2 1} = a_{j_2j_2l_1}$ is an (II,SI)-increasing-inflection sink.

(ix$_{2b}$) From case (ix$_{2a}$), let $\bar{x}_{j_1} = a_{j_1j_1s_1}$. Therefore, the equilibrium of $(x^*_{j_1}, x^*_{j_2}) = (a_{j_1j_1s_1}, a_{j_2j_2l_1})$ with $a_{j_2j_2s_1} = a_{j_1j_2l_1}$ has the following properties as in Eqs. (1.274)–(1.277).

- For $a_{j_1j_1 0}(a_{j_1j_1s_1} - a_{j_1j_1s_2}) > 0$ and $a_{j_2j_2 0}(a_{j_2j_2l_1} - a_{j_2j_2l_2}) > 0$,

$$(a_{j_1j_1s_1}, a_{j_2j_2l_1}) = \underbrace{(\text{DI} : \text{II}, \text{SO})}_{\text{inflection source}}.$$

The equilibrium of $(x^*_{j_1}, x^*_{j_2}) = (a_{j_1j_1s_1}, a_{j_2j_2l_1})$ is a (DI:II,SO)-inflection source.

- For $a_{j_1j_1 0}(a_{j_1j_1s_1} - a_{j_1j_1s_2}) < 0$ and $a_{j_2j_2 0}(a_{j_2j_2l_1} - a_{j_2j_2l_2}) > 0$,

$$(a_{j_1j_1s_1}, a_{j_2j_2l_1}) = \underbrace{(\text{II} : \text{DI}, \text{SO})}_{\text{inflection source}}.$$

The equilibrium of $(x^*_{j_1}, x^*_{j_2}) = (a_{j_1j_1s_1}, a_{j_2j_2l_1})$ is an (II:DI,SO)-inflection source.

- For $a_{j_1j_1 0}(a_{j_1j_1s_1} - a_{j_1j_1s_2}) > 0$ and $a_{j_2j_2 0}(a_{j_2j_2l_1} - a_{j_2j_2l_2}) < 0$,

$$(a_{j_1j_1s_1}, a_{j_2j_2l_1}) = \underbrace{(\text{II} : \text{DI}, \text{SI})}_{\text{inflection sink}}.$$

The equilibrium of $(x^*_{j_1}, x^*_{j_2}) = (a_{j_1j_1s_1}, a_{j_2j_2l_1})$ is an (II:DI,SI)-inflection sink.

- For $a_{j_1j_1 0}(a_{j_1j_1s_1} - a_{j_1j_1s_2}) < 0$ and $a_{j_2j_2 0}(a_{j_2j_2l_1} - a_{j_2j_2l_2}) < 0$,

$$(a_{j_1j_1s_1}, a_{j_2j_2l_1}) = \underbrace{(\text{DI} : \text{II}, \text{SI})}_{\text{inflection sink}}.$$

The equilibrium of $(x^*_{j_1}, x^*_{j_2}) = (a_{j_1j_1s_1}, a_{j_2j_2l_1})$ is a (DI:II,SI)-inflection sink.

(ix$_{2c}$) For $(x^*_{j_1}, x^*_{j_2}) = (a_{j_1j_1s_1}, a_{j_2j_2l_2})$ with $a_{j_1j_2 1} = a_{j_2j_2l_1}$, the variational equations are

$$\begin{aligned}
\Delta\dot{x}_{j_2} &= a_{j_2j_2 0}(a_{j_2j_2l_2} - a_{j_2j_2l_1})\Delta x_{j_2},\\
\Delta\dot{x}_{j_1} &= a_{j_1j_1 0}(a_{j_1j_1s_2} - a_{j_1j_1s_1})(a_{j_2j_2l_2} - a_{j_1j_2 1})\Delta x_{j_1}.
\end{aligned}$$

The flows in the x_{j_1}-direction for $a_{j_1j_1 0}(a_{j_1j_1s_2} - a_{j_1j_1s_1})(a_{j_2j_2l_2} - a_{j_1j_2 1}) > 0$ and $a_{j_1j_1 0}(a_{j_1j_1s_2} - a_{j_1j_1s_1})(a_{j_2j_2l_2} - a_{j_1j_2 1}) < 0$ are source and sink, respectively. The flows in the x_{j_2} -direction are source and sink for $a_{j_2j_20}(a_{j_2j_2l_2}-a_{j_2j_2l_1})>0$ and $a_{j_2j_20}(a_{j_2j_2l_2}-a_{j_2j_2l_1})<0$, respectively.

Therefore, the equilibrium of $(x_{j_1}^*, x_{j_2}^*)=(a_{j_1j_1s_1}, a_{j_2j_2l_2})$ with $a_{j_1j_21}=a_{j_2j_2l_1}$ has the following properties as in Eqs. (1.278)–(1.281).

- For $a_{j_1j_10}(a_{j_1j_1s_1}-a_{j_1j_1s_2})(a_{j_2j_2l_2}-a_{j_1j_21})>0$ and $a_{j_2j_20}(a_{j_2j_2l_2}-a_{j_2j_2l_1})>0$,
$$(a_{j_1j_1s_1}, a_{j_2j_2l_2})=\underbrace{(\text{SO}, \text{SO})}_{\text{source}}.$$
The equilibrium of $(x_{j_1}^*, x_{j_2}^*)=(a_{j_1j_1s_1}, a_{j_2j_2l_2})$ is an (SO,SO)-source.
- For $a_{j_1j_10}(a_{j_1j_1s_1}-a_{j_1j_1s_2})(a_{j_2j_2l_2}-a_{j_1j_21})<0$ and $a_{j_2j_20}(a_{j_2j_2l_2}-a_{j_2j_2l_1})>0$,
$$(a_{j_1j_1s_1}, a_{j_2j_2l_2})=\underbrace{(\text{SI}, \text{SO})}_{\text{saddle}}.$$
The equilibrium of $(x_{j_1}^*, x_{j_2}^*)=(a_{j_1j_1s_1}, a_{j_2j_2l_2})$ is an (SI,SO)-saddle.
- For $a_{j_1j_10}(a_{j_1j_1s_1}-a_{j_1j_1s_2})(a_{j_2j_2l_2}-a_{j_1j_21})>0$ and $a_{j_2j_20}(a_{j_2j_2l_2}-a_{j_2j_2l_1})<0$,
$$(a_{j_1j_1s_1}, a_{j_2j_2l_2})=\underbrace{(\text{SO}, \text{SI})}_{\text{saddle}}.$$
The equilibrium of $(x_{j_1}^*, x_{j_2}^*)=(a_{j_1j_1s_1}, a_{j_2j_2l_2})$ is an (SO,SI)-saddle.
- For $a_{j_1j_10}(a_{j_1j_1s_1}-a_{j_1j_1s_2})(a_{j_2j_2l_2}-a_{j_1j_21})<0$ and $a_{j_2j_20}(a_{j_2j_2l_2}-a_{j_2j_2l_1})<0$,
$$(a_{j_1j_1s_1}, a_{j_2j_2l_2})=\underbrace{(\text{SI}, \text{SI})}_{\text{sink}}.$$
The equilibrium of $(x_{j_1}^*, x_{j_2}^*)=(a_{j_1j_1s_1}, a_{j_2j_2l_2})$ is an (SI,SI)-sink.

In the end, this theorem is proved. ■

Chapter 2
Saddle-Node Singularity and Bifurcation Dynamics

In this chapter, the appearing and switching bifurcations in self-linear and crossing-quadratic product system with a self-quadratic vector field are discussed, and the corresponding network of equilibriums are presented. The inflection-source (sink) flows and saddle-source (sink) for the appearing bifurcations are presented first. The inflection upper-saddle and lower-saddle on the infinite equilibriums are for inflection-source (sink) flows with connected saddle-source (sink). The inflection-source (sink) flows with connected saddles, source, and sink are presented. The hyperbolic and hyperbolic-secant source (sink) are the switching bifurcations of the inflection-source (sink) flows with connected saddle, source, and sink. The up-down-to-down-up upper-saddle (lower-saddle) are the switching bifurcations of hyperbolic and hyperbolic-secant flows with connected saddle-source (sink). The series of connected hyperbolic and hyperbolic-secant flows with saddle, sink, and source are presented.

2.1 Cubic Systems with Crossing and Self-Singularities

In this section, the crossing and self-singularities existing in the cubic systems will be discussed. The appearing and switching bifurcations will be presented.

2.1.1 Inflection-Source (Sink) Flows and Saddle-Source (Sink)

The inflection-source (sink) flows and saddle-source (sink) are presented for the appearing bifurcations. The inflection-source (sink) flows are the appearing

A. C. J. Luo, *Two-dimensional Product-Cubic Systems, Vol. VII*,
https://doi.org/10.1007/978-3-031-48483-4_2

bifurcations of the connected hyperbolic and hyperbolic-secant flows. The saddle-source (sink) is the appearing bifurcation of saddle and source (sink).

Consider a dynamical system as

$$\begin{aligned}\dot{x}_1 &= a_{110}(x_1 - a_{111})(x_2 - a_{121})^2,\\ \dot{x}_2 &= a_{220}(x_2 - a_{221})^2,\end{aligned} \tag{2.1}$$

and the corresponding first integral manifold is

$$\ln\frac{|x_1 - a_{111}|}{|x_{10} - a_{111}|} = \frac{a_{110}}{a_{220}}\Big\{[(x_2 - a_{221}) - (x_{20} - a_{221})] + 2(a_{221} - a_{121})\ln\frac{|x_2 - a_{221}|}{|x_{20} - a_{221}|} - (a_{221} - a_{121})^2\Big(\frac{1}{x_2 - a_{221}} - \frac{1}{x_{20} - a_{221}}\Big)\Big\}. \tag{2.2}$$

(i) For $a_{121} < a_{221}$, the equilibriums of $x_1^* = a_{111}$ with $x_2^* = a_{121}, a_{221}$ have the following properties, as shown in Fig. 2.1.

$$\left\{\begin{matrix}(a_{111}, a_{221})\\(a_{111}, a_{121})\end{matrix}\right\} = \left\{\begin{matrix}\underbrace{\underbrace{(\text{SO, US})}_{\text{upper-saddle source}}}\\ \underbrace{({}_{\text{DI:II}}\text{SO, pF})}_{\text{inflection source flow }(+)}\end{matrix}\right\} \text{ for } a_{110} > 0 \text{ and } a_{220} > 0; \tag{2.3}$$

$$\left\{\begin{matrix}(a_{111}, a_{221})\\(a_{111}, a_{121})\end{matrix}\right\} = \left\{\begin{matrix}\underbrace{(\text{SI, US})}_{\text{upper-saddle sink}}\\ \underbrace{({}_{\text{II:DI}}\text{SI, pF})}_{\text{inflection sink flow }(+)}\end{matrix}\right\} \text{ for } a_{110} < 0 \text{ and } a_{220} > 0; \tag{2.4}$$

$$\left\{\begin{matrix}(a_{111}, a_{221})\\(a_{111}, a_{121})\end{matrix}\right\} = \left\{\begin{matrix}\underbrace{(\text{SO, LS})}_{\text{lower-saddle source}}\\ \underbrace{({}_{\text{II:DI}}\text{SO, nF})}_{\text{inflection source flow }(-)}\end{matrix}\right\} \text{ for } a_{110} > 0 \text{ and } a_{220} < 0; \tag{2.5}$$

$$\left\{\begin{matrix}(a_{111}, a_{221})\\(a_{111}, a_{121})\end{matrix}\right\} = \left\{\begin{matrix}\underbrace{(\text{SI, LS})}_{\text{lower-saddle sink}}\\ \underbrace{({}_{\text{DI:II}}\text{SI, nF})}_{\text{inflection sink flow }(-)}\end{matrix}\right\} \text{ for } a_{110} < 0 \text{ and } a_{220} < 0. \tag{2.6}$$

The upper-saddle source and sink are the appearing bifurcations of saddle with source and sink, and the lower-saddle source and sink are also the appearing bifurcations of saddle with source and sink. Just the locations of saddle with source

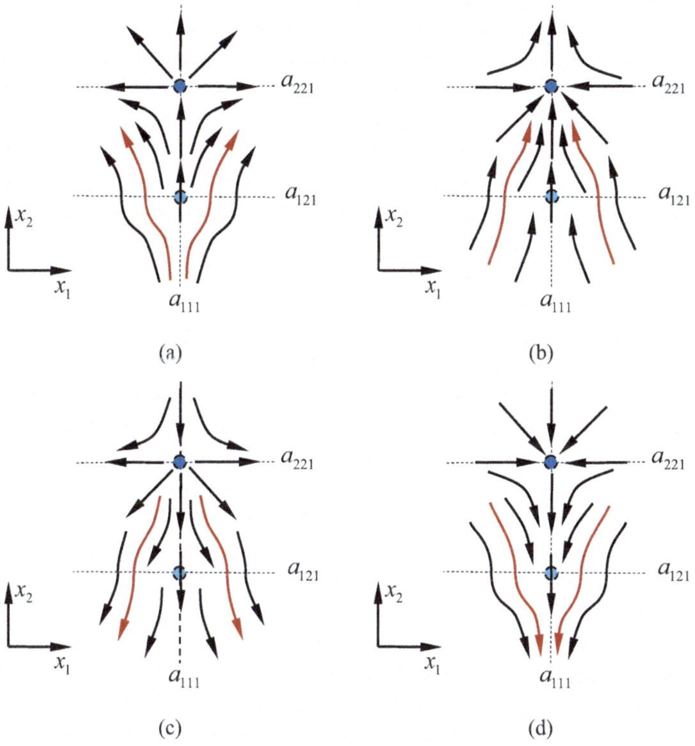

Fig. 2.1 Phase portraits ($a_{121} < a_{221}$) for two-dimensional systems on the x_1-direction with $x_1^* = a_{111}$ and on the x_2-direction with $x_2^* = a_{121}, a_{221}$. The equilibriums flows: (**a**) ($a_{110} > 0$, $a_{220} > 0$), (**b**) ($a_{110} < 0$, $a_{220} > 0$), (**c**) ($a_{110} > 0$, $a_{220} < 0$), (**d**) ($a_{110} < 0$, $a_{220} < 0$)

and sink are switched for the two appearing bifurcations. The third-order inflection source and sink flows are also presented.

(ii) For $a_{221} < a_{121}$, the equilibriums of $x_1^* = a_{111}$ with $x_2^* = a_{121}, a_{221}$ have the properties described as in Eqs. (2.3)–(2.6), and the corresponding phase portraits are also sketched in Fig. 2.2. Compared to the case of $a_{121} < a_{221}$, just the locations of equilibriums are exchanged, and the corresponding dynamical behaviors of the equilibriums are identical.

2.1.2 *Inflection-Saddle Infinite-Equilibriums*

For the above two cases, the switching bifurcations are the inflection-saddle infinite-equilibriums. Consider a dynamical system for $a_{221} = a_{121}$ as

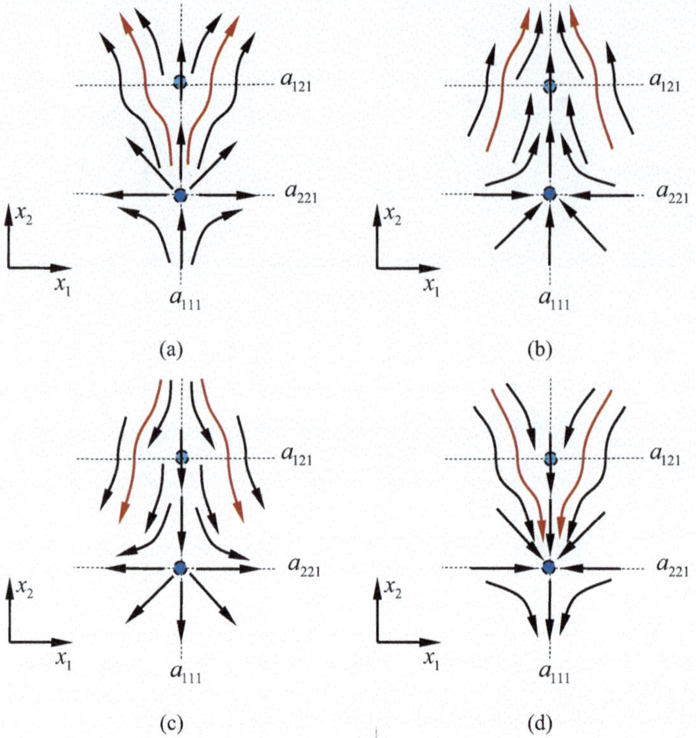

Fig. 2.2 Phase portraits ($a_{221} < a_{121}$) for two-dimensional systems on the x_1-direction with $x_1^* = a_{111}$ and on the x_2-direction with $x_2^* = a_{121}, a_{221}$. The equilibriums flows: (**a**) ($a_{110} > 0$, $a_{220} > 0$), (**b**) ($a_{110} < 0$, $a_{220} > 0$), (**c**) ($a_{110} > 0$, $a_{220} < 0$), (**d**) ($a_{110} < 0$, $a_{220} < 0$)

$$\begin{aligned} \dot{x}_1 &= a_{110}(x_1 - a_{111})(x_2 - a_{121})^2, \\ \dot{x}_2 &= a_{220}(x_2 - a_{221})^2, \end{aligned} \tag{2.7}$$

and the corresponding first integral manifold for $a_{221} = a_{121}$ is

$$\ln \frac{|x_1 - a_{111}|}{|x_{10} - a_{111}|} = \frac{a_{110}}{a_{220}}(x_2 - x_{20}). \tag{2.8}$$

The equilibrium of $(x_1^*, x_2^*) = (a_{111}, a_{221})$ with $a_{221} = a_{121}$ is

$$
\begin{aligned}
(a_{111}, a_{221}) &= \underbrace{(\mathrm{DI}:\mathrm{II}, \mathrm{US})}_{\text{inflection upper-saddle}} \quad \text{for } a_{110} > 0, a_{220} > 0,\\
(a_{111}, a_{221}) &= \underbrace{(\mathrm{II}:\mathrm{DI}, \mathrm{US})}_{\text{inflection upper-saddle}} \quad \text{for } a_{110} < 0, a_{220} > 0,\\
(a_{111}, a_{221}) &= \underbrace{(\mathrm{II}:\mathrm{DI}, \mathrm{LS})}_{\text{inflection lower-saddle}} \quad \text{for } a_{110} > 0, a_{220} < 0,\\
(a_{111}, a_{221}) &= \underbrace{(\mathrm{DI}:\mathrm{II}, \mathrm{LS})}_{\text{inflection lower-saddle}} \quad \text{for } a_{110} < 0, a_{220} < 0.
\end{aligned} \tag{2.9}
$$

and the infinite-equilibrium of $x_2^* = a_{121} = a_{221}$ is summarized as follows.

$$
\left.
\begin{aligned}
(\bar{x}_1, a_{221}) &= \underbrace{(\mathrm{DI}, \mathrm{US})}_{\text{decreasing-infection upper-saddle}} \quad \text{if } \bar{x}_1 \in (-\infty, a_{111}),\\
(\bar{x}_1, a_{221}) &= \underbrace{(\mathrm{II}, \mathrm{US})}_{\text{increasing-inflection upper-saddle}} \quad \text{if } \bar{x}_1 \in (a_{111}, \infty),
\end{aligned}
\right\} \tag{2.10}
$$

for $a_{110} > 0, a_{220} > 0$;

$$
\left.
\begin{aligned}
(\bar{x}_1, a_{221}) &= \underbrace{(\mathrm{II}, \mathrm{US})}_{\text{increasing-infection upper-saddle}} \quad \text{if } \bar{x}_1 \in (-\infty, a_{111}),\\
(\bar{x}_1, a_{221}) &= \underbrace{(\mathrm{DI}, \mathrm{US})}_{\text{decreasing-inflection upper-saddle}} \quad \text{if } \bar{x}_1 \in (a_{111}, \infty),
\end{aligned}
\right\} \tag{2.11}
$$

for $a_{110} < 0, a_{220} > 0$;

$$
\left.
\begin{aligned}
(\bar{x}_1, a_{221}) &= \underbrace{(\mathrm{II}, \mathrm{LS})}_{\text{increasing-infection lower-saddle}} \quad \text{if } \bar{x}_1 \in (-\infty, a_{111}),\\
(\bar{x}_1, a_{221}) &= \underbrace{(\mathrm{DI}, \mathrm{LS})}_{\text{decreasing-inflection lower-saddle}} \quad \text{if } \bar{x}_1 \in (a_{111}, \infty),
\end{aligned}
\right\} \tag{2.12}
$$

for $a_{110} > 0, a_{220} < 0$;

$$
\left.
\begin{aligned}
(\bar{x}_1, a_{221}) &= \underbrace{(\mathrm{DI}, \mathrm{LS})}_{\text{decreasing-infection lower-saddle}} \quad \text{if } \bar{x}_1 \in (-\infty, a_{111}),\\
(\bar{x}_1, a_{221}) &= \underbrace{(\mathrm{II}, \mathrm{LS})}_{\text{increasing-inflection lower-saddle}} \quad \text{if } \bar{x}_1 \in (a_{111}, \infty),
\end{aligned}
\right\} \tag{2.13}
$$

for $a_{110} < 0, a_{220} < 0$.

The infinite-equilibrium of $x_2^* = a_{121} = a_{221}$ is increasing-inflection and decreasing-inflection upper-saddle and lower-saddle, as shown in Fig. 2.3. The infinite-equilibrium is the switching bifurcation for $a_{121} < a_{221}$ and $a_{121} > a_{221}$.

The inflection upper-saddle and lower-saddle on the infinite equilibriums are for inflection-source (sink) flows with connected saddle-source (sink).

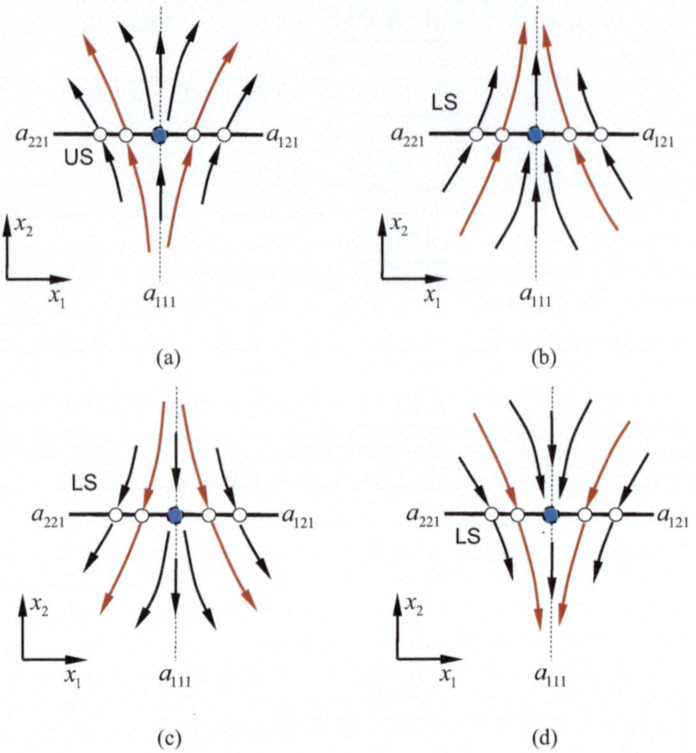

Fig. 2.3 Phase portraits of infinite-equilibrium ($a_{221} = a_{121}$) for two-dimensional systems on the x_1-direction with $x_1^* = a_{111}$ and on the x_2-direction with $x_2^* = a_{121} = a_{221}$. The infinite-equilibriums: (**a**) ($a_{110} > 0$, $a_{220} > 0$), (**b**) ($a_{110} < 0$, $a_{220} > 0$), (**c**) ($a_{110} > 0$, $a_{220} < 0$), (**d**) ($a_{110} < 0$, $a_{220} < 0$)

2.2 Single Crossing-Singular Cubic Systems

In this section, the crossing-singularity in cubic systems will be discussed, and the corresponding appearing and switching bifurcations will be presented.

2.2.1 *Inflection-Source (Sink) Flows with Saddles, Source and Sink*

The inflection-source (sink) flows with connected saddles, source, and sink will be illustrated. The inflection-source (sink) flows are for the appearing bifurcations of the connected hyperbolic and hyperbolic-secant flows.

Consider a dynamical system as

$$\begin{aligned}\dot{x}_1 &= a_{110}(x_1 - a_{111})(x_2 - a_{121})^2,\\ \dot{x}_2 &= a_{220}(x_2 - a_{221})(x_2 - a_{222}),\end{aligned} \tag{2.14}$$

and the corresponding first integral manifold is

$$\begin{aligned}\ln\frac{|x_1 - a_{111}|}{|x_{10} - a_{111}|} = \frac{a_{110}}{a_{220}}\Big\{(x_2 - x_{20}) &+ \frac{(a_{221} - a_{121})^2}{a_{221} - a_{222}}\ln\frac{|x_2 - a_{221}|}{|x_{20} - a_{221}|}\\ &+ \frac{(a_{222} - a_{121})^2}{a_{222} - a_{221}}\ln\frac{|x_2 - a_{222}|}{|x_{20} - a_{222}|}\Big\}.\end{aligned} \tag{2.15}$$

(i) For $a_{121} < a_{221}$ and $a_{222} < a_{121}$, the equilibriums of $x_1^* = a_{111}$ with $x_2^* = a_{121}, a_{221}, a_{222}$ have the following properties, as shown in Figs. 2.4 and 2.5.

$$\left\{\begin{matrix}(a_{111}, a_{222})\\ (a_{111}, a_{221})\\ (a_{111}, a_{121})\end{matrix}\right\} = \left\{\begin{matrix}\underbrace{(\mathrm{SO}, \mathrm{SO})}_{\text{source}}\\ \underbrace{(\mathrm{SO}, \mathrm{SI})}_{\text{saddle}}\\ \underbrace{({}_{\mathrm{DI:II}}\mathrm{SO}, \mathrm{pF})}_{\text{inflection source flow }(+)}\end{matrix}\right\} \text{ for } a_{110} > 0 \text{ and } a_{220} > 0; \tag{2.16}$$

$$\left\{\begin{matrix}(a_{111}, a_{222})\\ (a_{111}, a_{221})\\ (a_{111}, a_{121})\end{matrix}\right\} = \left\{\begin{matrix}\underbrace{(\mathrm{SI}, \mathrm{SO})}_{\text{saddle}}\\ \underbrace{(\mathrm{SI}, \mathrm{SI})}_{\text{sink}}\\ \underbrace{({}_{\mathrm{II:DI}}\mathrm{SI}, \mathrm{pF})}_{\text{inflection sink flow }(+)}\end{matrix}\right\} \text{ for } a_{110} < 0 \text{ and } a_{220} > 0; \tag{2.17}$$

$$\left\{\begin{matrix}(a_{111}, a_{222})\\ (a_{111}, a_{221})\\ (a_{111}, a_{121})\end{matrix}\right\} = \left\{\begin{matrix}\underbrace{(\mathrm{SO}, \mathrm{SI})}_{\text{saddle}}\\ \underbrace{(\mathrm{SO}, \mathrm{SO})}_{\text{source}}\\ \underbrace{({}_{\mathrm{II:DI}}\mathrm{SO}, \mathrm{nF})}_{\text{inflection source flow }(-)}\end{matrix}\right\} \text{ for } a_{110} > 0 \text{ and } a_{220} < 0; \tag{2.18}$$

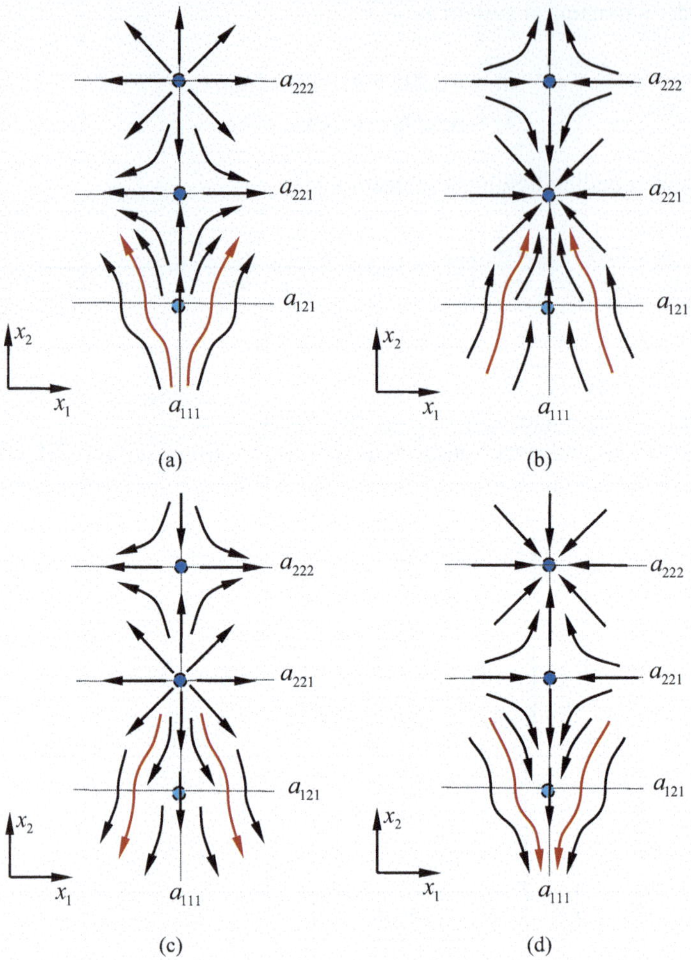

Fig. 2.4 Phase portraits ($a_{121} < a_{221}$) for two-dimensional systems on the x_1-direction with $x_1^* = a_{111}$ and on the x_2-direction with $x_2^* = a_{121}, a_{221}, a_{222}$. The equilibriums flows: (**a**) ($a_{110} > 0$, $a_{220} > 0$), (**b**) ($a_{110} < 0$, $a_{220} > 0$), (**c**) ($a_{110} > 0$, $a_{220} < 0$), (**d**) ($a_{110} < 0$, $a_{220} < 0$)

$$\left\{\begin{array}{l}(a_{111}, a_{222})\\(a_{111}, a_{221})\\(a_{111}, a_{121})\end{array}\right\} = \left\{\begin{array}{c}\underbrace{(\text{SI}, \text{SI})}_{\text{sink}}\\\underbrace{(\text{SI}, \text{SO})}_{\text{source}}\\\underbrace{({}_{\text{DI:II}}\text{SI}, \text{nF})}_{\text{inflection sink flow }(-)}\end{array}\right\} \text{ for } a_{110} < 0 \text{ and } a_{220} < 0. \tag{2.19}$$

For both cases of $a_{121} < a_{221}$ and $a_{222} < a_{121}$, the dynamical behaviors of equilibriums are identical. Just only the locations of equilibriums are switched.

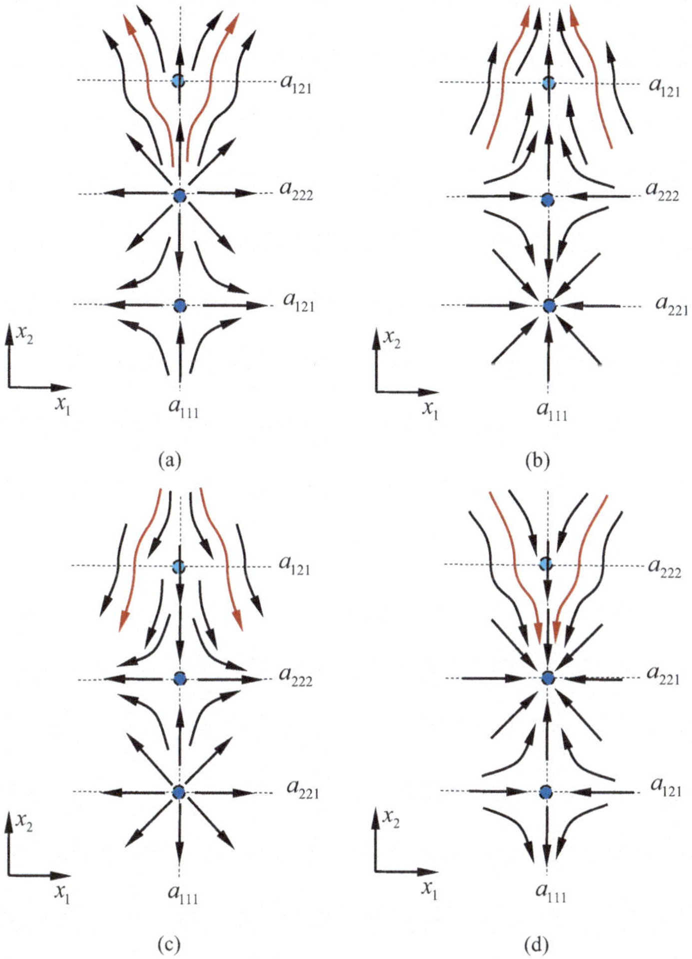

Fig. 2.5 Phase portraits ($a_{222} < a_{121}$) for two-dimensional systems on the x_1-direction with $x_1^* = a_{111}$ and on the x_2-direction with $x_2^* = a_{121}, a_{221}, a_{222}$. The equilibriums flows: (**a**) ($a_{110} > 0$, $a_{220} > 0$), (**b**) ($a_{110} < 0$, $a_{220} > 0$), (**c**) ($a_{110} > 0$, $a_{220} < 0$), (**d**) ($a_{110} < 0$, $a_{220} < 0$)

(ii) For $a_{121} \in (a_{221}, a_{222})$, the equilibriums of $x_1^* = a_{111}$ with $x_2^* = a_{121}, a_{221}, a_{222}$ have the following properties, as shown in Fig. 2.6.

$$\left\{ \begin{matrix} (a_{111}, a_{222}) \\ (a_{111}, a_{121}) \\ (a_{111}, a_{221}) \end{matrix} \right\} = \left\{ \begin{matrix} \underbrace{(\text{SO}, \text{SO})}_{\text{source}} \\ \underbrace{({}_{\text{II:DI}}\text{SO}, \text{nF})}_{\text{inflection source flow } (-)} \\ \underbrace{(\text{SO}, \text{SI})}_{\text{saddle}} \end{matrix} \right\} \text{ for } a_{110} > 0 \text{ and } a_{220} > 0; \tag{2.20}$$

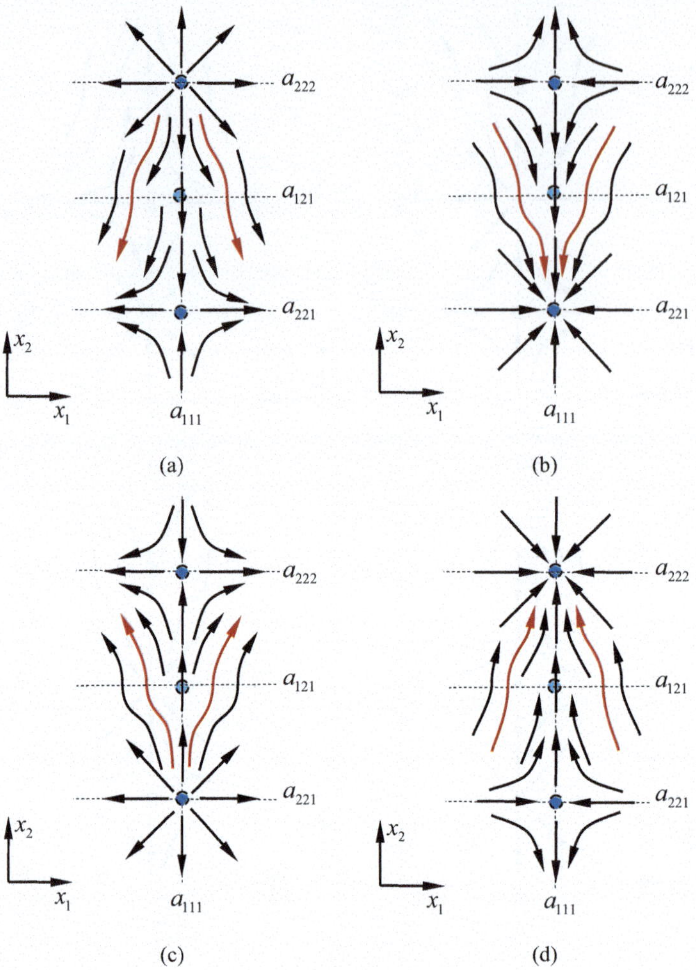

Fig. 2.6 Phase portraits ($a_{121} \in (a_{221}, a_{222})$) for two-dimensional systems on the x_1-direction with $x_1^* = a_{111}$ and on the x_2-direction with $x_2^* = a_{121}, a_{221}, a_{222}$. The equilibriums flows: (**a**) ($a_{110} > 0$, $a_{220} > 0$), (**b**) ($a_{110} < 0$, $a_{220} > 0$), (**c**) ($a_{110} > 0$, $a_{220} < 0$), (**d**) ($a_{110} < 0$, $a_{220} < 0$)

$$\left\{\begin{array}{c}(a_{111}, a_{222})\\(a_{111}, a_{121})\\(a_{111}, a_{221})\end{array}\right\} = \left\{\begin{array}{c}\underbrace{(\text{SI, SO})}_{\text{saddle}}\\\underbrace{({}_{\text{DI:II}}\text{SI, nF})}_{\text{inflection sink flow }(-)}\\\underbrace{(\text{SI, SI})}_{\text{sink}}\end{array}\right\} \text{ for } a_{110} < 0 \text{ and } a_{220} > 0; \tag{2.21}$$

$$\left\{\begin{array}{l}(a_{111}, a_{222})\\(a_{111}, a_{121})\\(a_{111}, a_{221})\end{array}\right\} = \left\{\begin{array}{c}\underbrace{(\mathrm{SO}, \mathrm{SI})}_{\text{saddle}}\\ \underbrace{({}_{\mathrm{DI:II}}\mathrm{SO}, \mathrm{pF})}_{\text{inflection source flow }(+)}\\ \underbrace{(\mathrm{SO}, \mathrm{SO})}_{\text{source}}\end{array}\right\} \text{ for } a_{110} > 0 \text{ and } a_{220} < 0; \tag{2.22}$$

$$\left\{\begin{array}{l}(a_{111}, a_{222})\\(a_{111}, a_{121})\\(a_{111}, a_{221})\end{array}\right\} = \left\{\begin{array}{c}\underbrace{(\mathrm{SI}, \mathrm{SI})}_{\text{sink}}\\ \underbrace{({}_{\mathrm{II:DI}}\mathrm{SI}, \mathrm{pF})}_{\text{inflection sink flow }(+)}\\ \underbrace{(\mathrm{SI}, \mathrm{SO})}_{\text{source}}\end{array}\right\} \text{ for } a_{110} < 0 \text{ and } a_{220} < 0. \tag{2.23}$$

The dynamical behaviors for the equilibriums of (a_{111}, a_{221}) and (a_{111}, a_{222}) are unchanged for comparison of the cases of $a_{121} < a_{221}$ and $a_{222} < a_{121}$. However, the inflection sink and source flows are changed from positive to negative or from negative to positive.

2.2.2 *Parabola-Source (Sink) Infinite-Equilibriums*

For the above two cases, the switching bifurcations will be discussed through the parabola-source (sink) infinite-equilibriums. The hyperbolic and hyperbolic-secant source (sink) are the switching bifurcations of the hyperbolic (hyperbolic-secant) flows with connected saddle, source, and sink.

(i) Consider a dynamical system for $a_{221} = a_{121}$ as

$$\begin{aligned}\dot{x}_1 &= a_{110}(x_1 - a_{111})(x_2 - a_{121})^2,\\ \dot{x}_2 &= a_{220}(x_2 - a_{221})(x_2 - a_{222}),\end{aligned} \tag{2.24}$$

and the corresponding first integral manifold for $a_{221} = a_{121}$ is

$$\ln\frac{|\,x_1 - a_{111}\,|}{|\,x_{10} - a_{111}\,|} = \frac{a_{110}}{a_{220}}\left[(x_2 - x_{20}) + (a_{222} - a_{121})\ln\frac{|\,x_2 - a_{222}\,|}{|\,x_{20} - a_{222}\,|}\right]. \tag{2.25}$$

The equilibriums of $x_2^* = a_{221}, a_{222}, a_{121}$ with $x_1^* = a_{111}$ for$a_{221} = a_{121}$ are

$$\left\{\begin{array}{l}(a_{111},a_{222})\\(a_{111},a_{221})\end{array}\right\}=\left\{\begin{array}{c}\underbrace{(\text{SO},\text{SO})}_{\text{source}}\\\underbrace{(\text{UP}:\text{DP},\text{SI})}_{\text{hyperbolic-secant sink}}\end{array}\right\}\text{ for }a_{110}>0,a_{220}>0,\tag{2.26}$$

$$\left\{\begin{array}{l}(a_{111},a_{222})\\(a_{111},a_{221})\end{array}\right\}=\left\{\begin{array}{c}\underbrace{(\text{SI},\text{SO})}_{\text{saddle}}\\\underbrace{(\text{DP}:\text{UP},\text{SI})}_{\text{hyperbolic sink}}\end{array}\right\}\text{ for }a_{110}<0,a_{220}>0,\tag{2.27}$$

$$\left\{\begin{array}{l}(a_{111},a_{222})\\(a_{111},a_{221})\end{array}\right\}=\left\{\begin{array}{c}\underbrace{(\text{SO},\text{SI})}_{\text{saddle}}\\\underbrace{(\text{DP}:\text{UP},\text{SO})}_{\text{hyperbolic source}}\end{array}\right\}\text{ for }a_{110}>0,a_{220}<0,\tag{2.28}$$

$$\left\{\begin{array}{l}(a_{111},a_{222})\\(a_{111},a_{221})\end{array}\right\}=\left\{\begin{array}{c}\underbrace{(\text{SI},\text{SI})}_{\text{sink}}\\\underbrace{(\text{UP}:\text{DP},\text{SO})}_{\text{hyperbolic-secant source}}\end{array}\right\}\text{ for }a_{110}<0,a_{220}<0;\tag{2.29}$$

and the infinite-equilibrium of $x_2^* = a_{121} = a_{221}$ is summarized as follows.

$$\left.\begin{array}{ll}(\bar{x}_1,a_{221})=\underbrace{(\text{UP},\text{SI})}_{\text{up-parabola sink}} & \text{if }\bar{x}_1\in(-\infty,a_{111}),\\(\bar{x}_1,a_{221})=\underbrace{(\text{DP},\text{SI})}_{\text{down-parabola sink}} & \text{if }\bar{x}_1\in(a_{111},\infty),\\\text{for }a_{110}>0,a_{220}>0;\end{array}\right.\tag{2.30}$$

$$\left.\begin{array}{ll}(\bar{x}_1,a_{221})=\underbrace{(\text{DP},\text{SI})}_{\text{down-parabola sink}} & \text{if }\bar{x}_1\in(-\infty,a_{111}),\\(\bar{x}_1,a_{221})=\underbrace{(\text{UP},\text{SI})}_{\text{up-parabola sink}} & \text{if }\bar{x}_1\in(a_{111},\infty),\\\text{for }a_{110}<0,a_{220}>0;\end{array}\right.\tag{2.31}$$

$$\left.\begin{array}{ll}(\bar{x}_1,a_{221})=\underbrace{(\text{DP},\text{SO})}_{\text{down-parabola source}} & \text{if }\bar{x}_1\in(-\infty,a_{111}),\\(\bar{x}_1,a_{221})=\underbrace{(\text{UP},\text{SO})}_{\text{up-parabola source}} & \text{if }\bar{x}_1\in(a_{111},\infty),\\\text{for }a_{110}>0,a_{220}<0;\end{array}\right.\tag{2.32}$$

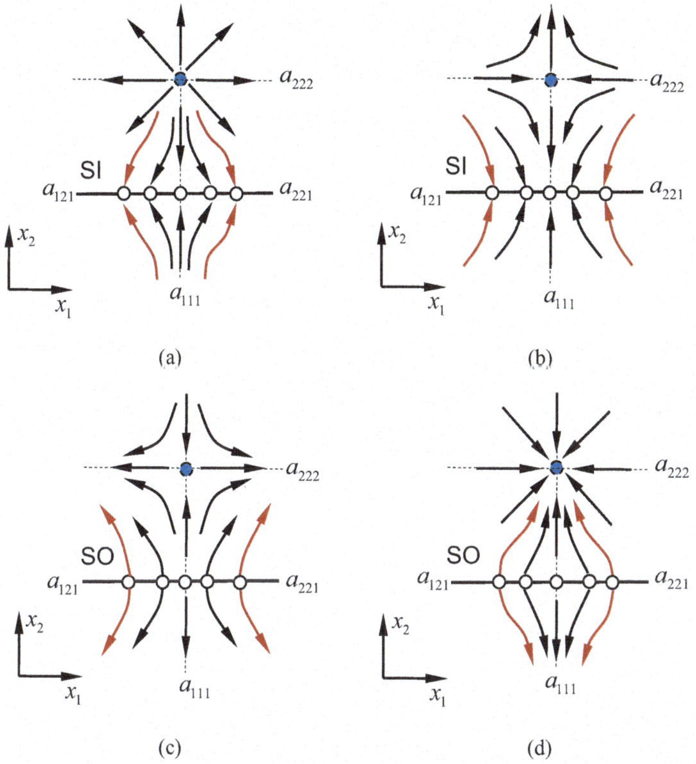

Fig. 2.7 Phase portraits ($a_{121} = a_{221}$) for two-dimensional systems on the x_1-direction with $x_1^* = a_{111}$ and on the x_2-direction with $x_2^* = a_{121}, a_{221}, a_{222}$. The infinite-equilibriums: **(a)** ($a_{110} > 0$, $a_{220} > 0$), **(b)** ($a_{110} < 0$, $a_{220} > 0$), **(c)** ($a_{110} > 0$, $a_{220} < 0$), **(d)** ($a_{110} < 0$, $a_{220} < 0$)

$$
\begin{aligned}
&(\bar{x}_1, a_{221}) = \underbrace{(\mathrm{UP}, \mathrm{SO})}_{\text{up-parabola source}} \quad \text{if } \bar{x}_1 \in (-\infty, a_{111}),\\
&(\bar{x}_1, a_{221}) = \underbrace{(\mathrm{DP}, \mathrm{SO})}_{\text{down-parabola source}} \quad \text{if } \bar{x}_1 \in (a_{111}, \infty),\\
&\text{for } a_{110} < 0, a_{220} < 0.
\end{aligned}
\tag{2.33}
$$

The infinite-equilibriums of $x_2^* = a_{121} = a_{221}$ for up-parabola and down-parabola source and sink are presented in Fig. 2.7. The infinite-equilibriums are for the switching bifurcation for the equilibriums for $a_{121} < a_{221}$ and $a_{221} < a_{121} < a_{222}$.

(ii) Consider a dynamical system for $a_{222} = a_{121}$ as

$$\begin{aligned}\dot{x}_1 &= a_{110}(x_1 - a_{111})(x_2 - a_{121})^2,\\ \dot{x}_2 &= a_{220}(x_2 - a_{221})(x_2 - a_{222}),\end{aligned} \tag{2.34}$$

and the corresponding first integral manifold for $a_{221} = a_{121}$ is

$$\ln \frac{|x_1 - a_{111}|}{|x_{10} - a_{111}|} = \frac{a_{110}}{a_{220}}\left[(x_2 - x_{20}) + (a_{221} - a_{121})\ln \frac{|x_2 - a_{221}|}{|x_{20} - a_{221}|}\right]. \tag{2.35}$$

The infinite-equilibriums of $x_2^* = a_{222} = a_{121}$ are

$$\left\{\begin{matrix}(a_{111}, a_{222})\\(a_{111}, a_{221})\end{matrix}\right\} = \left\{\begin{matrix}\underbrace{(\mathrm{DP}:\mathrm{UP},\mathrm{SO})}_{\text{hyperbolic source}}\\ \underbrace{(\mathrm{SO},\mathrm{SI})}_{\text{saddle}}\end{matrix}\right\} \text{ for } a_{110} > 0, a_{220} > 0, \tag{2.36}$$

$$\left\{\begin{matrix}(a_{111}, a_{222})\\(a_{111}, a_{221})\end{matrix}\right\} = \left\{\begin{matrix}\underbrace{(\mathrm{UP}:\mathrm{DP},\mathrm{SO})}_{\text{hyperbolic-secant source}}\\ \underbrace{(\mathrm{SI},\mathrm{SI})}_{\text{sink}}\end{matrix}\right\} \text{ for } a_{110} < 0, a_{220} > 0, \tag{2.37}$$

$$\left\{\begin{matrix}(a_{111}, a_{222})\\(a_{111}, a_{221})\end{matrix}\right\} = \left\{\begin{matrix}\underbrace{(\mathrm{UP}:\mathrm{DP},\mathrm{SI})}_{\text{hyperbolic-secant sink}}\\ \underbrace{(\mathrm{SO},\mathrm{SO})}_{\text{source}}\end{matrix}\right\} \text{ for } a_{110} > 0, a_{220} < 0, \tag{2.38}$$

$$\left\{\begin{matrix}(a_{111}, a_{222})\\(a_{111}, a_{221})\end{matrix}\right\} = \left\{\begin{matrix}\underbrace{(\mathrm{DP}:\mathrm{UP},\mathrm{SI})}_{\text{hyperbolic sink}}\\ \underbrace{(\mathrm{SI},\mathrm{SO})}_{\text{saddle}}\end{matrix}\right\} \text{ for } a_{110} < 0, a_{220} < 0; \tag{2.39}$$

and the infinite-equilibrium of $x_2^* = a_{121} = a_{222}$ is summarized as follows.

$$\begin{aligned}(\bar{x}_1, a_{222}) &= \underbrace{(\mathrm{DP},\mathrm{SO})}_{\text{down-parabola source}} \quad \text{if } \bar{x}_1 \in (-\infty, a_{111}),\\ (\bar{x}_1, a_{222}) &= \underbrace{(\mathrm{UP},\mathrm{SO})}_{\text{up-parabola source}} \quad \text{if } \bar{x}_1 \in (a_{111}, \infty),\\ &\text{for } a_{110} > 0, a_{220} > 0;\end{aligned} \tag{2.40}$$

$$
\begin{aligned}
(\bar{x}_1, a_{222}) &= \underbrace{(\text{UP}, \text{SO})}_{\text{up-parabola source}} \quad \text{if } \bar{x}_1 \in (-\infty, a_{111}), \\
(\bar{x}_1, a_{222}) &= \underbrace{(\text{DP}, \text{SO})}_{\text{down-parabola source}} \quad \text{if } \bar{x}_1 \in (a_{111}, \infty),
\end{aligned} \tag{2.41}
$$

for $a_{110} < 0, a_{220} > 0$;

$$
\begin{aligned}
(\bar{x}_1, a_{222}) &= \underbrace{(\text{UP}, \text{SI})}_{\text{up-parabola sink}} \quad \text{if } \bar{x}_1 \in (-\infty, a_{111}), \\
(\bar{x}_1, a_{222}) &= \underbrace{(\text{DP}, \text{SI})}_{\text{down-parabola sink}} \quad \text{if } \bar{x}_1 \in (a_{111}, \infty),
\end{aligned} \tag{2.42}
$$

for $a_{110} > 0, a_{220} < 0$;

$$
\begin{aligned}
(\bar{x}_1, a_{222}) &= \underbrace{(\text{DP}, \text{SI})}_{\text{down-parabola sink}} \quad \text{if } \bar{x}_1 \in (-\infty, a_{111}), \\
(\bar{x}_1, a_{222}) &= \underbrace{(\text{UP}, \text{SI})}_{\text{up-parabola sink}} \quad \text{if } \bar{x}_1 \in (a_{111}, \infty),
\end{aligned} \tag{2.43}
$$

for $a_{110} < 0, a_{220} < 0$.

The infinite-equilibrium of $x_2^* = a_{121} = a_{222}$ for up-parabola and down-parabola source and sink is presented in Fig. 2.8. The infinite-equilibrium is for the switching bifurcation for the equilibriums for $a_{121} < a_{222}$ and $a_{121} > a_{222}$.

The hyperbolic and hyperbolic-secant source (sink) are the switching bifurcations for inflection-source (sink) flows with connected saddle, source and sinks.

2.3 Single Self-Singularity Cubic Systems

In this section, the saddle-source and saddle-sink are discussed for the appearing and switching bifurcations, and the up-down and down-up saddle infinite-equilibriums are for the switching bifurcations of the saddle-source and saddle-sink with hyperbolic and hyperbolic-secant flows.

2.3.1 Saddle-Source (Sink) with Connected Hyperbolic Flows

Consider a dynamical system as

$$
\begin{aligned}
\dot{x}_1 &= a_{110}(x_1 - a_{111})(x_2 - a_{121})(x_2 - a_{122}), \\
\dot{x}_2 &= a_{220}(x_2 - a_{221})^2,
\end{aligned} \tag{2.44}
$$

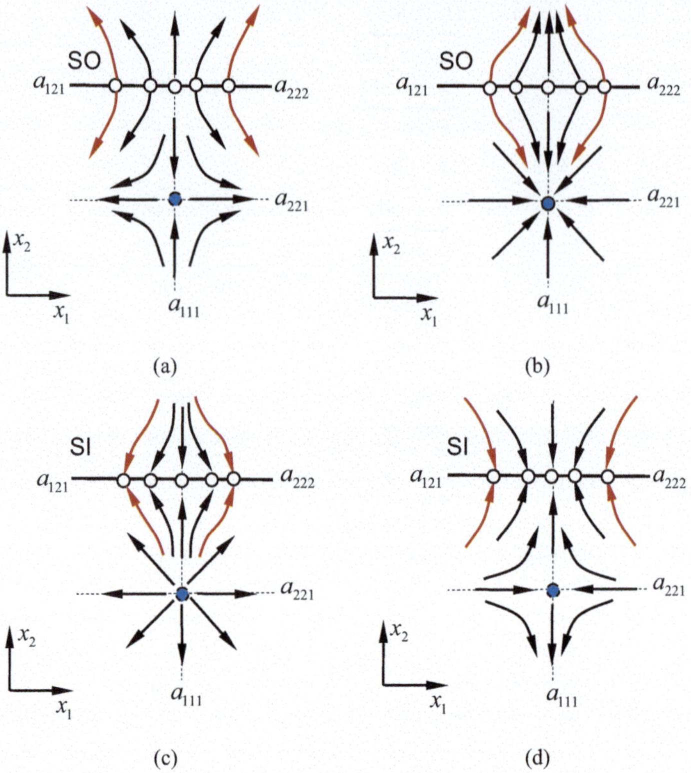

Fig. 2.8 Phase portraits ($a_{121} = a_{222}$) for two-dimensional systems on the x_1-direction with $x_1^* = a_{111}$ and on the x_2-direction with $x_2^* = a_{121}, a_{221}, a_{222}$. The infinite-equilibriums: **(a)** ($a_{110} > 0, a_{220} > 0$), **(b)** ($a_{110} < 0, a_{220} > 0$), **(c)** ($a_{110} > 0, a_{220} < 0$), **(d)** ($a_{110} < 0, a_{220} < 0$)

and the corresponding first integral manifold is

$$\ln \frac{|x_1 - a_{111}|}{|x_{10} - a_{111}|} = \frac{a_{110}}{a_{220}} \left[(x_2 - x_{20}) + (2a_{221} - a_{121} - a_{122}) \ln \frac{|x_2 - a_{221}|}{|x_{20} - a_{221}|} \right. \\ \left. - (a_{221} - a_{121})(a_{221} - a_{122})(\frac{1}{x_2 - a_{221}} - \frac{1}{x_{20} - a_{221}}) \right]. \tag{2.45}$$

(i) For the case of $a_{122} < a_{221}$, the equilibrium series of $x_1^* = a_{111}$ with $x_2^* = a_{121}, a_{122}, a_{221}$ have the following properties, as shown in Fig. 2.9.

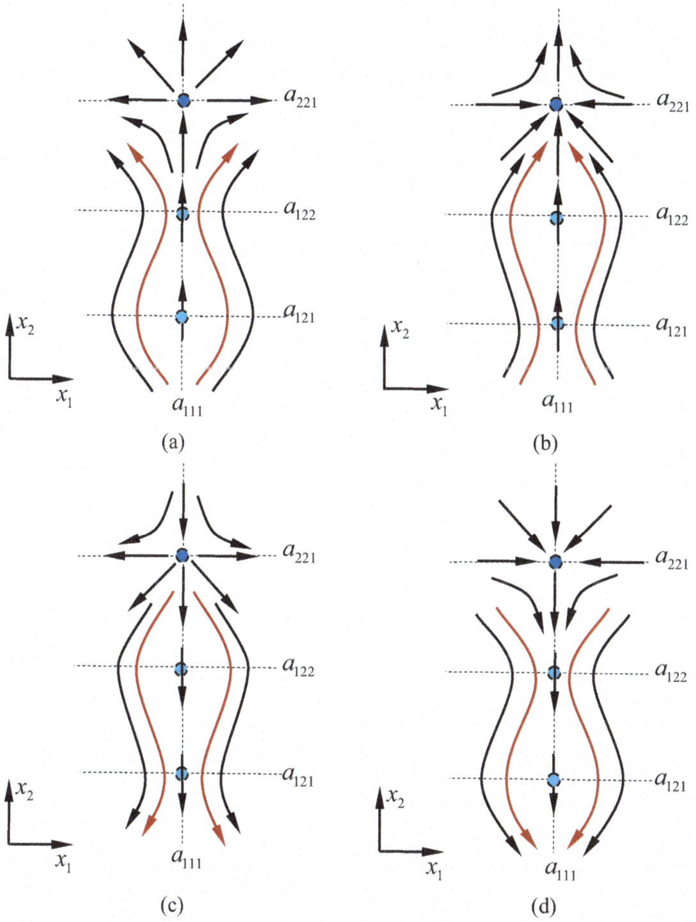

Fig. 2.9 Phase portraits ($a_{122} < a_{221}$) for two-dimensional systems on the x_1-direction with $x_1^* = a_{111}$ and on the x_2-direction with $x_2^* = a_{121}, a_{122}, a_{221}$. The equilibriums flows: (**a**) ($a_{110} > 0$, $a_{220} > 0$), (**b**) ($a_{110} < 0$, $a_{220} > 0$), (**c**) ($a_{110} > 0$, $a_{220} < 0$), (**d**) ($a_{110} < 0$, $a_{220} < 0$)

$$\begin{Bmatrix} (a_{111}, a_{221}) \\ (a_{111}, a_{122}) \\ (a_{111}, a_{121}) \end{Bmatrix} = \begin{Bmatrix} \underbrace{(\text{SO, US})}_{\text{upper-saddle source}} \\ \underbrace{(\text{DP : UP, pF})}_{\text{hyperbolic flow } (+)} \\ \underbrace{(\text{UP : DP, pF})}_{\text{hyperbolic-secant flow } (+)} \end{Bmatrix} \text{ for } a_{110} > 0 \text{ and } a_{220} > 0; \qquad (2.46)$$

$$\left\{\begin{array}{l}(a_{111},a_{221})\\(a_{111},a_{122})\\(a_{111},a_{121})\end{array}\right\}=\left\{\begin{array}{c}\underbrace{(\mathrm{SI,US})}_{\text{upper-saddle sink}}\\\underbrace{(\mathrm{UP:DP,pF})}_{\text{hyperbolic-secant flow }(+)}\\\underbrace{(\mathrm{DP:UP,pF})}_{\text{hyperbolic flow }(+)}\end{array}\right\}\text{ for }a_{110}<0\text{ and }a_{220}>0;\qquad(2.47)$$

$$\left\{\begin{array}{l}(a_{111},a_{221})\\(a_{111},a_{122})\\(a_{111},a_{121})\end{array}\right\}=\left\{\begin{array}{c}\underbrace{(\mathrm{SO,LS})}_{\text{lower-saddle source}}\\\underbrace{(\mathrm{UP:DP,nF})}_{\text{hyperbolic-secant flow }(-)}\\\underbrace{(\mathrm{DP:UP,nF})}_{\text{hyperbolic flow }(-)}\end{array}\right\}\text{ for }a_{110}>0\text{ and }a_{220}<0;\qquad(2.48)$$

$$\left\{\begin{array}{l}(a_{111},a_{221})\\(a_{111},a_{122})\\(a_{111},a_{121})\end{array}\right\}=\left\{\begin{array}{c}\underbrace{(\mathrm{SI,LS})}_{\text{lower-saddle sink}}\\\underbrace{(\mathrm{DP:UP,nF})}_{\text{hyperbolic flow }(-)}\\\underbrace{(\mathrm{UP:DP,nF})}_{\text{hyperbolic-secant flow }(-)}\end{array}\right\}\text{ for }a_{110}<0\text{ and }a_{220}<0.\qquad(2.49)$$

The hyperbolic and hyperbolic-secant flows with upper-saddle sink and source and lower-saddle sink and source are presented. The upper and lower, saddle-sink and saddle-source are the appearing bifurcations for saddle and sink and for saddle and source, respectively. All the equilibriums are on a line of $x_1^*=a_{111}$ to form a series of equilibriums. The saddle-sink and saddle-source are on the top of the series of equilibriums. The hyperbolic and hyperbolic-secant flows are connected.

(ii) For $a_{221}\in(a_{121},a_{122})$, the equilibriums of $x_1^*=a_{111}$ with $x_2^*=a_{121},a_{122},a_{221}$ have the following properties.

$$\left\{\begin{array}{l}(a_{111},a_{122})\\(a_{111},a_{221})\\(a_{111},a_{121})\end{array}\right\}=\left\{\begin{array}{c}\underbrace{(\mathrm{DP:UP,pF})}_{\text{hyperbolic flow }(+)}\\\underbrace{(\mathrm{SI,US})}_{\text{upper-saddle source}}\\\underbrace{(\mathrm{UP:DP,pF})}_{\text{hyperbolic-secant flow }(+)}\end{array}\right\}\text{ for }a_{110}>0\text{ and }a_{220}>0;\qquad(2.50)$$

$$\left\{\begin{array}{c}(a_{111},a_{122})\\(a_{111},a_{221})\\(a_{111},a_{121})\end{array}\right\}=\left\{\begin{array}{c}\underbrace{(\mathrm{UP}:\mathrm{DP},\mathrm{pF})}_{\text{hyperbolic-secant flow }(+)}\\\underbrace{(\mathrm{SO},\mathrm{US})}_{\text{upper-saddle source}}\\\underbrace{(\mathrm{DP}:\mathrm{UP},\mathrm{pF})}_{\text{hyperbolic flow }(+)}\end{array}\right\}\text{ for }a_{110}<0\text{ and }a_{220}>0;\tag{2.51}$$

$$\left\{\begin{array}{c}(a_{111},a_{122})\\(a_{111},a_{221})\\(a_{111},a_{121})\end{array}\right\}=\left\{\begin{array}{c}\underbrace{(\mathrm{UP}:\mathrm{DP},\mathrm{nF})}_{\text{hyperbolic-secant flow }(-)}\\\underbrace{(\mathrm{SI},\mathrm{LS})}_{\text{lower-saddle sink}}\\\underbrace{(\mathrm{DP}:\mathrm{UP},\mathrm{nF})}_{\text{hyperbolic flow }(-)}\end{array}\right\}\text{ for }a_{110}>0\text{ and }a_{220}<0;\tag{2.52}$$

$$\left\{\begin{array}{c}(a_{111},a_{122})\\(a_{111},a_{221})\\(a_{111},a_{121})\end{array}\right\}=\left\{\begin{array}{c}\underbrace{(\mathrm{DP}:\mathrm{UP},\mathrm{nF})}_{\text{hyperbolic flow }(-)}\\\underbrace{(\mathrm{SO},\mathrm{LS})}_{\text{lower-saddle source}}\\\underbrace{(\mathrm{UP}:\mathrm{DP},\mathrm{nF})}_{\text{hyperbolic-secant flow }(-)}\end{array}\right\}\text{ for }a_{110}<0\text{ and }a_{220}<0,\tag{2.53}$$

as shown in Fig. 2.10. The saddle-sources and sinks in the series of equilibriums lie between hyperbolic and hyperbolic-secant flows. The hyperbolic and hyperbolic-secant flows are separated by the saddle-source or saddle-sinks.

(iii) For $a_{221} < a_{121}$, the equilibriums of $x_1^* = a_{111}$ with $x_2^* = a_{121}, a_{122}, a_{221}$ have the following properties as in Eqs. (2.46)–(2.49), as shown in Fig. 2.11. The dynamical behaviors of equilibriums for $a_{221} < a_{121}$ are identical to the case of $a_{122} < a_{221}$. Just the locations of equilibriums are different. The saddle-sink and saddle-source are on the bottom of the series of equilibriums. The hyperbolic and hyperbolic-secant flows are connected.

2.3.2 Up-Down-to-Down-Up Upper-Saddles (Lower-Saddles)

Consider a dynamical system as

$$\begin{aligned}\dot{x}_1 &= a_{110}(x_1 - a_{111})(x_2 - a_{121})(x_2 - a_{122}),\\ \dot{x}_2 &= a_{220}(x_2 - a_{221})^2.\end{aligned}\tag{2.54}$$

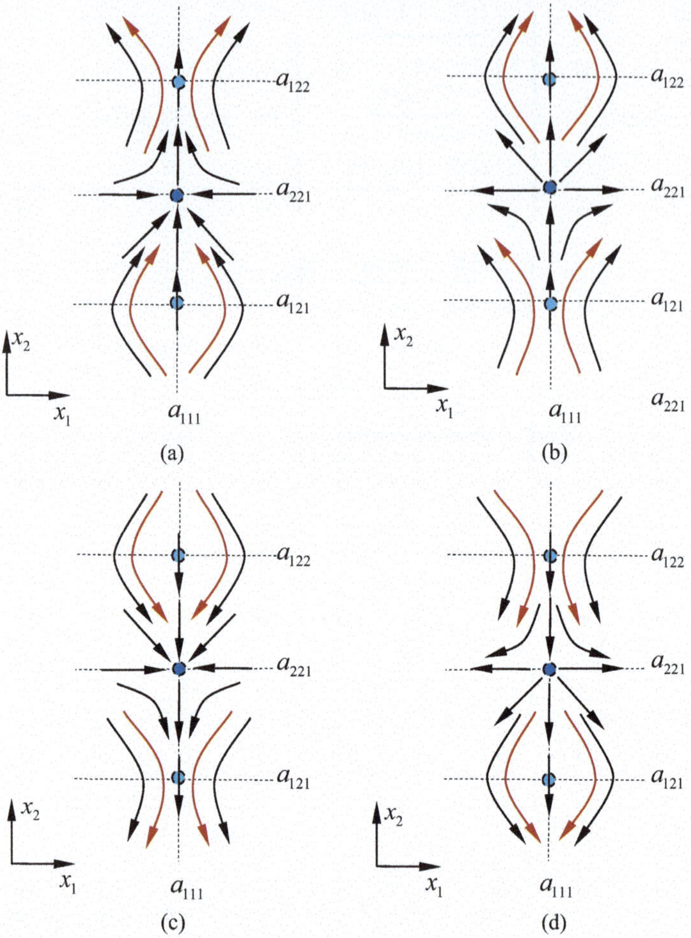

Fig. 2.10 Phase portraits ($a_{121} < a_{221}$) for two-dimensional systems on the x_1-direction with $x_1^* = a_{111}$ and on the x_2-direction with $x_2^* = a_{121}, a_{122}, a_{221}$. The equilibriums flows: **(a)** ($a_{110} > 0$, $a_{220} > 0$), **(b)** ($a_{110} < 0$, $a_{220} > 0$), **(c)** ($a_{110} > 0$, $a_{220} < 0$), **(d)** ($a_{110} < 0$, $a_{220} < 0$)

(i) For $a_{121} = a_{221}$, the corresponding first integral manifold is

$$\ln \frac{|x_1 - a_{111}|}{|x_{10} - a_{111}|} = \frac{a_{110}}{a_{220}} \left[(x_2 - x_{20}) + (a_{221} - a_{122}) \ln \frac{|x_2 - a_{221}|}{|x_{20} - a_{221}|} \right]. \tag{2.55}$$

The equilibriums of $x_2^* = a_{121}, a_{122}, a_{221}$ with $x_1^* = a_{111}$ for $a_{121} = a_{221}$ have the following properties.

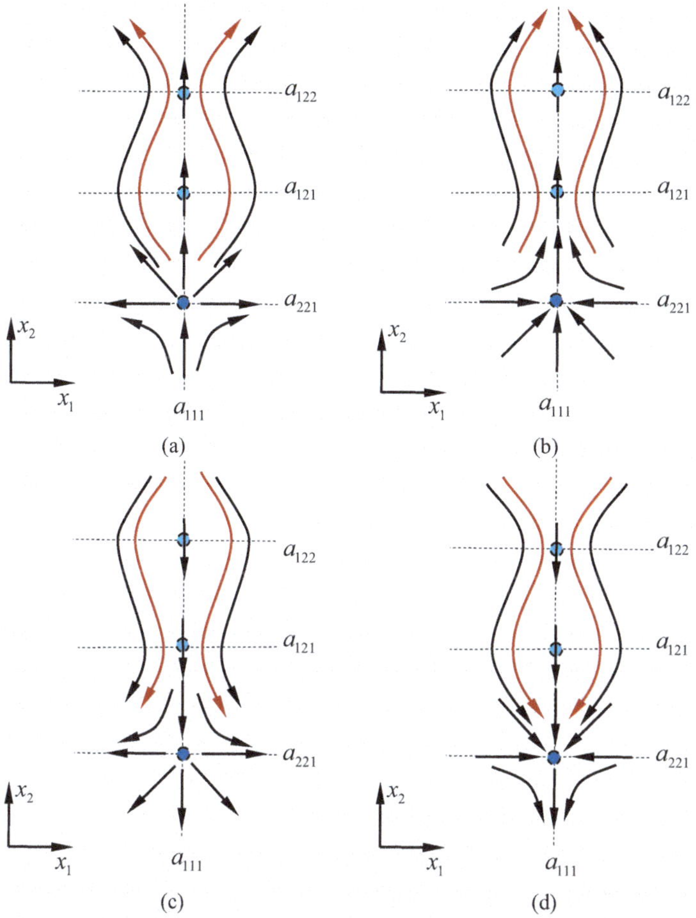

Fig. 2.11 Phase portraits ($a_{221} < a_{121}$) for two-dimensional systems on the x_1-direction with $x_1^* = a_{111}$ and on the x_2-direction with $x_2^* = a_{121}, a_{122}, a_{221}$. The equilibriums flows: (**a**) ($a_{110} > 0$, $a_{220} > 0$), (**b**) ($a_{110} < 0$, $a_{220} > 0$), (**c**) ($a_{110} > 0$, $a_{220} < 0$), (**d**) ($a_{110} < 0$, $a_{220} < 0$)

$$\left\{\begin{matrix}(a_{111}, a_{122})\\(a_{111}, a_{221})\end{matrix}\right\} = \left\{\begin{matrix}\underbrace{(\text{DP} : \text{UP}, \text{pF})}_{\text{hyperbolic flow } (+)}\\ \underbrace{(\text{DU} : \text{UD}, \text{US})}_{\text{down-up to up-down upper-saddle}}\end{matrix}\right\} \text{ for } a_{110} > 0 \text{ and } a_{220} > 0; \quad (2.56)$$

$$\left\{\begin{matrix}(a_{111}, a_{122})\\(a_{111}, a_{221})\end{matrix}\right\} = \left\{\begin{matrix}\underbrace{(\text{UP} : \text{DP}, \text{pF})}_{\text{hyperbolic-secant flow } (+)}\\ \underbrace{(\text{UD} : \text{DU} :, \text{US})}_{\text{up-down to down-up upper-saddle}}\end{matrix}\right\} \text{ for } a_{110} < 0 \text{ and } a_{220} > 0; \quad (2.57)$$

$$\left\{\begin{array}{l}(a_{111},a_{122})\\(a_{111},a_{221})\end{array}\right\}=\left\{\begin{array}{c}\underbrace{(\mathrm{UP}:\mathrm{DP},\mathrm{nF})}_{\text{hyperbolic-secant flow }(-)}\\\underbrace{(\mathrm{UD}:\mathrm{DU},\mathrm{LS})}_{\text{up-down to down-up lower-saddle}}\end{array}\right\}\text{ for } a_{110}>0 \text{ and } a_{220}<0; \quad (2.58)$$

$$\left\{\begin{array}{l}(a_{111},a_{122})\\(a_{111},a_{221})\end{array}\right\}=\left\{\begin{array}{c}\underbrace{(\mathrm{DP}:\mathrm{UP},\mathrm{nF})}_{\text{hyperbolic flow }(-)}\\\underbrace{(\mathrm{DU}:\mathrm{UD},\mathrm{LS})}_{\text{down-up to up-down lower-saddle}}\end{array}\right\}\text{ for } a_{110}<0 \text{ and } a_{220}<0. \quad (2.59)$$

The infinite-equilibriums of $x_2^*=a_{121}=a_{221}$ are

$$\begin{aligned}&(\bar{x}_1,a_{221})=\underbrace{(\mathrm{DU},\mathrm{US})}_{\text{down-up upper-saddle}}\quad \text{if } \bar{x}\in(-\infty,a_{111}),\\&(\bar{x}_1,a_{221})=\underbrace{(\mathrm{UD},\mathrm{US})}_{\text{up-down upper-saddle}}\quad \text{if } \bar{x}\in(a_{111},\infty),\\&\text{for } a_{110}>0 \text{ and } a_{220}>0;\end{aligned} \quad (2.60)$$

$$\begin{aligned}&(\bar{x}_1,a_{221})=\underbrace{(\mathrm{UD},\mathrm{US})}_{\text{up-down upper-saddle}}\quad \text{if } \bar{x}\in(-\infty,a_{111}),\\&(\bar{x}_1,a_{221})=\underbrace{(\mathrm{DU},\mathrm{US})}_{\text{down-up upper-saddle}}\quad \text{if } \bar{x}\in(a_{111},\infty),\\&\text{for } a_{110}<0 \text{ and } a_{220}>0;\end{aligned} \quad (2.61)$$

$$\begin{aligned}&(\bar{x}_1,a_{221})=\underbrace{(\mathrm{UD},\mathrm{LS})}_{\text{up-down lower-saddle}}\quad \text{if } \bar{x}\in(-\infty,a_{111}),\\&(\bar{x}_1,a_{221})=\underbrace{(\mathrm{DU},\mathrm{LS})}_{\text{down-up lower-saddle}}\quad \text{if } \bar{x}\in(a_{111},\infty),\\&\text{for } a_{110}>0 \text{ and } a_{220}<0;\end{aligned} \quad (2.62)$$

$$\begin{aligned}&(\bar{x}_1,a_{221})=\underbrace{(\mathrm{DU},\mathrm{LS})}_{\text{down-up lower-saddle}}\quad \text{if } \bar{x}\in(-\infty,a_{111}),\\&(\bar{x}_1,a_{221})=\underbrace{(\mathrm{UD},\mathrm{LS})}_{\text{up-down lower-saddle}}\quad \text{if } \bar{x}\in(a_{111},\infty),\\&\text{for } a_{110}<0 \text{ and } a_{220}<0.\end{aligned} \quad (2.63)$$

The above infinite-equilibriums are shown in Fig. 2.12. The hyperbolic-flow or hyperbolic-secant flows are on the upper sides of the up-down, down-up upper-saddle, and lower-saddle infinite-equilibriums.

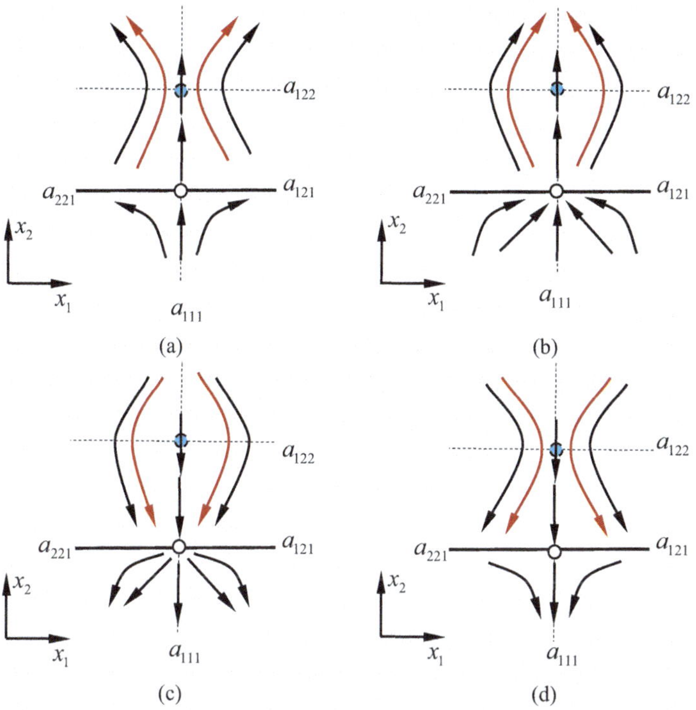

Fig. 2.12 Phase portraits ($a_{121} = a_{221}$) for two-dimensional systems on the x_1-direction with $x_1^* = a_{111}$ and on the x_2-direction with $x_2^* = a_{121}, a_{122}, a_{221}$. The equilibriums flows: **(a)** ($a_{110} > 0$, $a_{220} > 0$), **(b)** ($a_{110} < 0$, $a_{220} > 0$), **(c)** ($a_{110} > 0$, $a_{220} < 0$), **(d)** ($a_{110} < 0$, $a_{220} < 0$)

(ii) For $a_{221} = a_{122}$, the corresponding first integral manifold is

$$\ln \frac{|x_1 - a_{111}|}{|x_{10} - a_{111}|} = \frac{a_{110}}{a_{220}} \left[(x_2 - x_{20}) + (a_{221} - a_{121}) \ln \frac{|x_2 - a_{221}|}{|x_{20} - a_{221}|} \right]. \tag{2.64}$$

The equilibriums of $x_2^* = a_{121}, a_{122}, a_{221}$ with $x_1^* = a_{111}$ for$a_{122} = a_{221}$ are

$$\left\{ \begin{matrix} (a_{111}, a_{221}) \\ (a_{111}, a_{121}) \end{matrix} \right\} = \left\{ \begin{matrix} \underbrace{(\text{UD : DU, US})}_{\text{up-down to down-up upper-saddle}} \\ \underbrace{(\text{UP : DP, pF})}_{\text{hyperbolic-secant flow } (+)} \end{matrix} \right\} \text{ for } a_{110} > 0 \text{ and } a_{220} > 0; \tag{2.65}$$

$$\left\{\begin{array}{l}(a_{111},a_{221})\\(a_{111},a_{121})\end{array}\right\}=\left\{\begin{array}{c}\underbrace{(\text{DU}:\text{UD},\text{US})}_{\text{down-up to up-down upper-saddle}}\\\underbrace{(\text{DP}:\text{UP},\text{pF})}_{\text{hyperbolic flow }(+)}\end{array}\right\}\text{ for }a_{110}<0\text{ and }a_{220}>0;\quad(2.66)$$

$$\left\{\begin{array}{l}(a_{111},a_{221})\\(a_{111},a_{121})\end{array}\right\}=\left\{\begin{array}{c}\underbrace{(\text{DU}:\text{UD},\text{LS})}_{\text{down-up to up-down lower-saddle}}\\\underbrace{(\text{DP}:\text{UP},\text{nF})}_{\text{hyperbolic flow }(-)}\end{array}\right\}\text{ for }a_{110}>0\text{ and }a_{220}<0;\quad(2.67)$$

$$\left\{\begin{array}{l}(a_{111},a_{221})\\(a_{111},a_{121})\end{array}\right\}=\left\{\begin{array}{c}\underbrace{(\text{UD}:\text{DU},\text{LS})}_{\text{up-down to down-up lower-saddle}}\\\underbrace{(\text{UP}:\text{DP},\text{nF})}_{\text{hyperbolic-secant flow }(-)}\end{array}\right\}\text{ for }a_{110}<0\text{ and }a_{220}<0.\quad(2.68)$$

The infinite-equilibriums of $x_2^* = a_{122} = a_{221}$ are

$$\begin{aligned}(\bar{x}_1,a_{221})&=\underbrace{(\text{UD},\text{US})}_{\text{up-down upper-saddle}}\quad\text{if }\bar{x}\in(-\infty,a_{111}),\\(\bar{x}_1,a_{221})&=\underbrace{(\text{DU},\text{US})}_{\text{down-up upper-saddle}}\quad\text{if }\bar{x}\in(a_{111},\infty),\\&\text{for }a_{110}>0\text{ and }a_{220}>0;\end{aligned}\quad(2.69)$$

$$\begin{aligned}(\bar{x}_1,a_{221})&=\underbrace{(\text{DU},\text{US})}_{\text{down-up upper-saddle}}\quad\text{if }\bar{x}\in(-\infty,a_{111}),\\(\bar{x}_1,a_{221})&=\underbrace{(\text{UD},\text{US})}_{\text{up-down upper-saddle}}\quad\text{if }\bar{x}\in(a_{111},\infty),\\&\text{for }a_{110}<0\text{ and }a_{220}>0;\end{aligned}\quad(2.70)$$

$$\begin{aligned}(\bar{x}_1,a_{221})&=\underbrace{(\text{DU},\text{LS})}_{\text{down-up lower-saddle}}\quad\text{if }\bar{x}\in(-\infty,a_{111}),\\(\bar{x}_1,a_{221})&=\underbrace{(\text{UD},\text{LS})}_{\text{up-down lower-saddle}}\quad\text{if }\bar{x}\in(a_{111},\infty),\\&\text{for }a_{110}>0\text{ and }a_{220}<0;\end{aligned}\quad(2.71)$$

$$\begin{aligned}(\bar{x}_1,a_{221})&=\underbrace{(\text{UD},\text{LS})}_{\text{up-down lower-saddle}}\quad\text{if }\bar{x}\in(-\infty,a_{111}),\\(\bar{x}_1,a_{221})&=\underbrace{(\text{DU},\text{LS})}_{\text{down-up lower-saddle}}\quad\text{if }\bar{x}\in(a_{111},\infty),\\&\text{for }a_{110}<0\text{ and }a_{220}<0.\end{aligned}\quad(2.72)$$

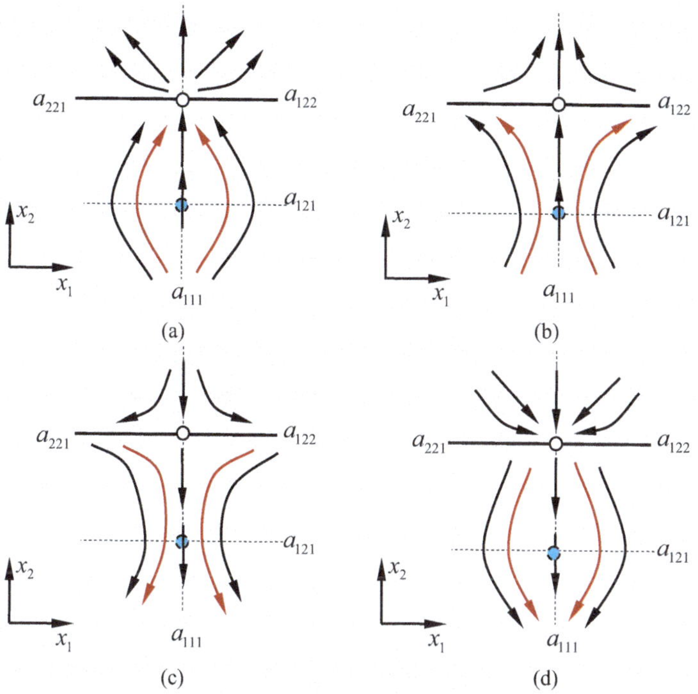

Fig. 2.13 Phase portraits ($a_{122} = a_{221}$) for two-dimensional systems on the x_1-direction with $x_1^* = a_{111}$ and on the x_2-direction with $x_2^* = a_{121}, a_{122}, a_{221}$. The equilibriums flows: (**a**) ($a_{110} > 0$, $a_{220} > 0$), (**b**) ($a_{110} < 0$, $a_{220} > 0$), (**c**) ($a_{110} > 0$, $a_{220} < 0$), (**d**) ($a_{110} < 0$, $a_{220} < 0$)

The corresponding infinite-equilibriums are sketched in Fig. 2.13. The hyperbolic-flow or hyperbolic-secant flows are on the lower sides of the up-down, down-up upper-saddle, and lower-saddle infinite-equilibriums. The infinite-equilibriums are for the switching bifurcations of the saddle-source (sink) with connected hyperbolic (hyperbolic-secant) flows.

2.4 Nonsingular Cubic Systems

In this section, the series of the connected hyperbolic and hyperbolic-secant flows with connected saddle and source (sink) are presented, and the switching bifurcations of two series are discussed through the inflection-source (sink) infinite-equilibriums.

2.4.1 Equilibrium Series with Connected Hyperbolic Flows

Consider a dynamical system as

$$\begin{aligned}\dot{x}_1 &= a_{110}(x_1 - a_{111})(x_2 - a_{121})(x_2 - a_{122}),\\ \dot{x}_2 &= a_{220}(x_2 - a_{221})(x_2 - a_{222}),\end{aligned} \tag{2.73}$$

and the corresponding first integral manifold is

$$\ln\frac{|x_1 - a_{111}|}{|x_{10} - a_{111}|} = \frac{a_{110}}{a_{220}}\Bigg[(x_2 - x_{20}) + \frac{(a_{221} - a_{121})(a_{221} - a_{122})}{a_{221} - a_{222}}\ln\frac{|x_2 - a_{221}|}{|x_{20} - a_{221}|} + \frac{(a_{222} - a_{121})(a_{222} - a_{121})}{a_{222} - a_{221}}\ln\frac{|x_2 - a_{222}|}{|x_{20} - a_{222}|}\Bigg]. \tag{2.74}$$

(i) For the case of $a_{122} < a_{221}$, the equilibrium series of $x_1^* = a_{111}$ with $x_2^* = a_{121}, a_{122}, a_{221}, a_{222}$ have the following properties, as shown in Fig. 2.14.

$$\left\{\begin{matrix}(a_{111}, a_{122}) & (a_{111}, a_{222})\\ (a_{111}, a_{121}) & (a_{111}, a_{221})\end{matrix}\right\} = \left\{\begin{matrix}\underbrace{(\mathrm{DP}:\mathrm{UP},\mathrm{pF})}_{\text{hyperbolic flow }(+)} & \underbrace{(\mathrm{SO},\mathrm{SO})}_{\text{source}}\\ \underbrace{(\mathrm{UP}:\mathrm{DP},\mathrm{pF})}_{\text{hyperbolic-secant flow }(+)} & \underbrace{(\mathrm{SO},\mathrm{SI})}_{\text{saddle}}\end{matrix}\right\} \tag{2.75}$$

for $a_{110} > 0$ and $a_{220} > 0$;

$$\left\{\begin{matrix}(a_{111}, a_{122}) & (a_{111}, a_{222})\\ (a_{111}, a_{121}) & (a_{111}, a_{221})\end{matrix}\right\} = \left\{\begin{matrix}\underbrace{(\mathrm{UP}:\mathrm{DP},\mathrm{pF})}_{\text{hyperbolic-secant flow }(+)} & \underbrace{(\mathrm{SI},\mathrm{SO})}_{\text{saddle}}\\ \underbrace{(\mathrm{DP}:\mathrm{UP},\mathrm{pF})}_{\text{hyperbolic flow }(+)} & \underbrace{(\mathrm{SI},\mathrm{SI})}_{\text{sink}}\end{matrix}\right\} \tag{2.76}$$

for $a_{110} < 0$ and $a_{220} > 0$;

$$\left\{\begin{matrix}(a_{111}, a_{122}) & (a_{111}, a_{222})\\ (a_{111}, a_{121}) & (a_{111}, a_{221})\end{matrix}\right\} = \left\{\begin{matrix}\underbrace{(\mathrm{UP}:\mathrm{DP},\mathrm{nF})}_{\text{hyperbolic-secant flow }(-)} & \underbrace{(\mathrm{SO},\mathrm{SI})}_{\text{saddle}}\\ \underbrace{(\mathrm{DP}:\mathrm{UP},\mathrm{nF})}_{\text{hyperbolic flow }(-)} & \underbrace{(\mathrm{SO},\mathrm{SO})}_{\text{source}}\end{matrix}\right\} \tag{2.77}$$

for $a_{110} > 0$ and $a_{220} < 0$;

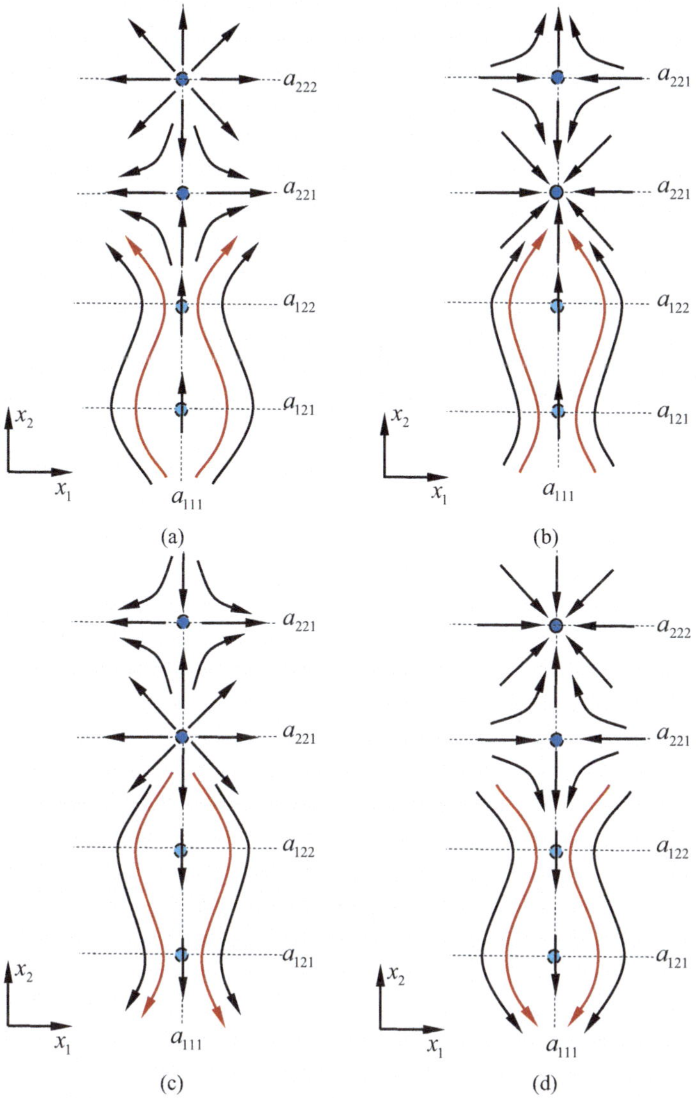

Fig. 2.14 Phase portraits ($a_{122} < a_{221}$) for two-dimensional systems on the x_1-direction with $x_1^* = a_{111}$ and on the x_2-direction with $x_2^* = a_{121}, a_{122}, a_{221}, a_{222}$. The equilibriums flows: (**a**) ($a_{110} > 0$, $a_{220} > 0$), (**b**) ($a_{110} < 0$, $a_{220} > 0$), (**c**) ($a_{110} > 0$, $a_{220} < 0$), (**d**) ($a_{110} < 0$, $a_{220} < 0$)

$$\left\{\begin{matrix} (a_{111}, a_{122}) & (a_{111}, a_{222}) \\ (a_{111}, a_{121}) & (a_{111}, a_{221}) \end{matrix}\right\} = \left\{\begin{matrix} \underbrace{(\text{DP}:\text{UP},\text{nF})}_{\text{hyperbolic flow }(-)} & \underbrace{(\text{SI},\text{SI})}_{\text{sink}} \\ \underbrace{(\text{UP}:\text{DP},\text{nF})}_{\text{hyperbolic-secant flow }(-)} & \underbrace{(\text{SI},\text{SO})}_{\text{saddle}} \end{matrix}\right\} \tag{2.78}$$

for $a_{110} < 0$ and $a_{220} < 0$.

The connected saddle and source or sink with connected hyperbolic and hyperbolic-secant flows exist on the series of equilibriums. The hyperbolic and hyperbolic-secant flows with directrix are connected together on the bottom. The saddle and source or sink on the top are from the appearing bifurcation of saddle-saddle or saddle-sink.

(ii) For $a_{221} \in (a_{121}, a_{122})$ and $a_{122} < a_{222}$, the equilibriums of $x_1^* = a_{111}$ with $x_2^* = a_{121}, a_{221}, a_{222}$ have the following properties, as shown in Fig. 2.15.

$$\left\{\begin{matrix} (a_{111}, a_{221}) & (a_{111}, a_{222}) \\ (a_{111}, a_{121}) & (a_{111}, a_{122}) \end{matrix}\right\} = \left\{\begin{matrix} \underbrace{(\text{SI},\text{SI})}_{\text{sink}} & \underbrace{(\text{SO},\text{SO})}_{\text{source}} \\ \underbrace{(\text{UP}:\text{DP},\text{pF})}_{\text{hyperbolic-secant flow }(+)} & \underbrace{(\text{UP}:\text{DP},\text{nF})}_{\text{hyperbolic-secant flow }(-)} \end{matrix}\right\}$$

for $a_{110} > 0$ and $a_{220} > 0$;

(2.79)

$$\left\{\begin{matrix} (a_{111}, a_{221}) & (a_{111}, a_{222}) \\ (a_{111}, a_{121}) & (a_{111}, a_{122}) \end{matrix}\right\} = \left\{\begin{matrix} \underbrace{(\text{SO},\text{SI})}_{\text{saddle}} & \underbrace{(\text{SI},\text{SO})}_{\text{saddle}} \\ \underbrace{(\text{DP}:\text{UP},\text{pF})}_{\text{hyperbolic flow }(+)} & \underbrace{(\text{DP}:\text{UP},\text{nF})}_{\text{hyperbolic flow }(-)} \end{matrix}\right\} \tag{2.80}$$

for $a_{110} < 0$ and $a_{220} > 0$;

$$\left\{\begin{matrix} (a_{111}, a_{221}) & (a_{111}, a_{222}) \\ (a_{111}, a_{121}) & (a_{111}, a_{122}) \end{matrix}\right\} = \left\{\begin{matrix} \underbrace{(\text{SI},\text{SO})}_{\text{saddle}} & \underbrace{(\text{SO},\text{SI})}_{\text{saddle}} \\ \underbrace{(\text{DP}:\text{UP},\text{nF})}_{\text{hyperbolic flow }(-)} & \underbrace{(\text{DP}:\text{UP},\text{pF})}_{\text{hyperbolic flow }(+)} \end{matrix}\right\} \tag{2.81}$$

for $a_{110} > 0$ and $a_{220} < 0$;

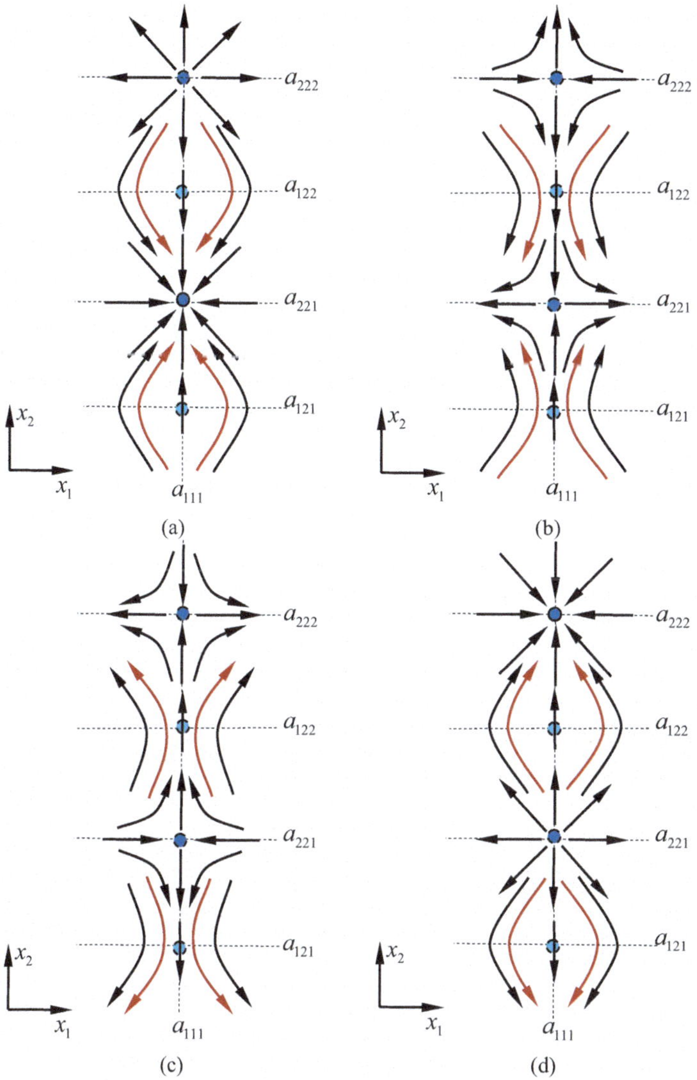

Fig. 2.15 Phase portraits ($a_{221} \in (a_{121}, a_{122})$ and $a_{122} < a_{222}$) for two-dimensional systems on the x_1-direction with $x_1^* = a_{111}$ and on the x_2-direction with $x_2^* = a_{121}, a_{122}, a_{221}, a_{222}$. The equilibriums flows: **(a)** ($a_{110} > 0$, $a_{220} > 0$), **(b)** ($a_{110} < 0$, $a_{220} > 0$), **(c)** ($a_{110} > 0$, $a_{220} < 0$), **(d)** ($a_{110} < 0$, $a_{220} < 0$)

$$\left\{\begin{matrix} (a_{111},a_{221}) & (a_{111},a_{222}) \\ (a_{111},a_{121}) & (a_{111},a_{122}) \end{matrix}\right\} = \left\{\begin{matrix} \underbrace{(\text{SO},\text{SO})}_{\text{source}} & \underbrace{(\text{SI},\text{SI})}_{\text{sink}} \\ \underbrace{(\text{UP}:\text{DP},\text{nF})}_{\text{hyperbolic-secant flow }(-)} & \underbrace{(\text{UP}:\text{DP},\text{pF})}_{\text{hyperbolic-secant flow }(+)} \end{matrix}\right\}$$
for $a_{110} < 0$ and $a_{220} < 0$.
(2.82)

The saddle and source or sink with hyperbolic and hyperbolic-secant flows exist on the series of equilibriums. The hyperbolic and hyperbolic-secant flows with directrix are separated through saddles or source and sink. The source and sink are separated by a hyperbolic-secant flow. The (SI,SO)-saddle and (SO,SI)-saddle are separated by a hyperbolic flow.

(iii) For a_{121}, $a_{122} \in (a_{221}, a_{222})$, the equilibriums of $x_1^* = a_{111}$ with $x_2^* = a_{121}, a_{122}$, a_{221}, a_{222} have the following properties, as shown in Fig. 2.16.

$$\left\{\begin{matrix} (a_{111},a_{121}) & (a_{111},a_{222}) \\ (a_{111},a_{221}) & (a_{111},a_{122}) \end{matrix}\right\} = \left\{\begin{matrix} \underbrace{(\text{DP}:\text{UP},\text{nF})}_{\text{hyperbolic flow }(-)} & \underbrace{(\text{SO},\text{SO})}_{\text{source}} \\ \underbrace{(\text{SO},\text{SI})}_{\text{saddle}} & \underbrace{(\text{UP}:\text{DP},\text{nF})}_{\text{hyperbolic-secant flow }(-)} \end{matrix}\right\} \quad (2.83)$$
for $a_{110} > 0$ and $a_{220} > 0$;

$$\left\{\begin{matrix} (a_{111},a_{121}) & (a_{111},a_{222}) \\ (a_{111},a_{221}) & (a_{111},a_{122}) \end{matrix}\right\} = \left\{\begin{matrix} \underbrace{(\text{UP}:\text{DP},\text{nF})}_{\text{hyperbolic-secant flow }(-)} & \underbrace{(\text{SI},\text{SO})}_{\text{saddle}} \\ \underbrace{(\text{SI},\text{SI})}_{\text{sink}} & \underbrace{(\text{DP}:\text{UP},\text{nF})}_{\text{hyperbolic flow }(-)} \end{matrix}\right\} \quad (2.84)$$
for $a_{110} < 0$ and $a_{220} > 0$;

$$\left\{\begin{matrix} (a_{111},a_{121}) & (a_{111},a_{222}) \\ (a_{111},a_{221}) & (a_{111},a_{122}) \end{matrix}\right\} = \left\{\begin{matrix} \underbrace{(\text{UP}:\text{DP},\text{pF})}_{\text{hyperbolic-secant flow }(+)} & \underbrace{(\text{SO},\text{SI})}_{\text{saddle}} \\ \underbrace{(\text{SO},\text{SO})}_{\text{source}} & \underbrace{(\text{DP}:\text{UP},\text{pF})}_{\text{hyperbolic flow }(+)} \end{matrix}\right\} \quad (2.85)$$
for $a_{110} > 0$ and $a_{220} < 0$;

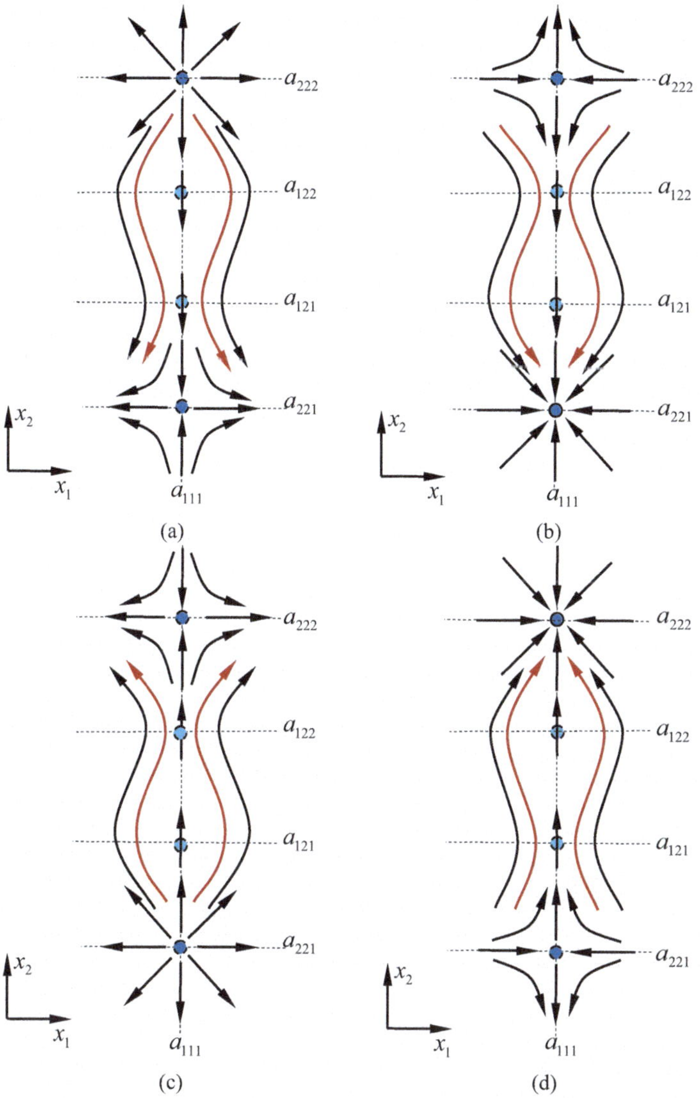

Fig. 2.16 Phase portraits ($a_{121}, a_{122} \in (a_{221}, a_{222})$) for two-dimensional systems on the x_1-direction with $x_1^* = a_{111}$ and on the x_2-direction with $x_2^* = a_{121}, a_{122}, a_{221}, a_{222}$. The equilibriums flows: (**a**) ($a_{110} > 0$, $a_{220} > 0$), (**b**) ($a_{110} < 0$, $a_{220} > 0$), (**c**) ($a_{110} > 0$, $a_{220} < 0$), (**d**) ($a_{110} < 0$, $a_{220} < 0$)

$$\left\{\begin{matrix}(a_{111},a_{121}) & (a_{111},a_{222})\\(a_{111},a_{221}) & (a_{111},a_{122})\end{matrix}\right\}=\left\{\begin{matrix}\underbrace{(\text{DP}:\text{UP},\text{pF})}_{\text{hyperbolic flow }(+)} & \underbrace{(\text{SI},\text{SI})}_{\text{sink}}\\ \underbrace{(\text{SI},\text{SO})}_{\text{saddle}} & \underbrace{(\text{UP}:\text{DP},\text{pF})}_{\text{hyperbolic-secant flow }(+)}\end{matrix}\right\} \tag{2.86}$$

for $a_{110}<0$ and $a_{220}<0$.

The saddle and source or sink with hyperbolic and hyperbolic-secant flows exist on the series of equilibriums. The hyperbolic and hyperbolic-secant flows with directrix are connected together. Source and saddle or sink and saddle are separated by the hyperbolic and hyperbolic-secant flows.

(iv) For $a_{121}\in(a_{221},a_{222})$ and $a_{222}<a_{122}$, the equilibriums of $x_1^*=a_{111}$ with $x_2^*=a_{121},a_{221},a_{222}$ have the following properties, as shown in Fig. 2.17.

$$\left\{\begin{matrix}(a_{111},a_{121}) & (a_{111},a_{122})\\(a_{111},a_{221}) & (a_{111},a_{222})\end{matrix}\right\}=\left\{\begin{matrix}\underbrace{(\text{DP}:\text{UP},\text{nF})}_{\text{hyperbolic flow }(-)} & \underbrace{(\text{DP}:\text{UP},\text{pF})}_{\text{hyperbolic flow }(+)}\\ \underbrace{(\text{SO},\text{SI})}_{\text{saddle}} & \underbrace{(\text{SI},\text{SO})}_{\text{saddle}}\end{matrix}\right\} \tag{2.87}$$

for $a_{110}>0$ and $a_{220}>0$;

$$\left\{\begin{matrix}(a_{111},a_{121}) & (a_{111},a_{122})\\(a_{111},a_{221}) & (a_{111},a_{222})\end{matrix}\right\}=\left\{\begin{matrix}\underbrace{(\text{UP}:\text{DP},\text{nF})}_{\text{hyperbolic-secant flow }(-)} & \underbrace{(\text{UP}:\text{DP},\text{pF})}_{\text{hyperbolic-secant flow }(+)}\\ \underbrace{(\text{SI},\text{SI})}_{\text{sink}} & \underbrace{(\text{SO},\text{SO})}_{\text{source}}\end{matrix}\right\}$$

for $a_{110}<0$ and $a_{220}>0$;

(2.88)

$$\left\{\begin{matrix}(a_{111},a_{121}) & (a_{111},a_{122})\\(a_{111},a_{221}) & (a_{111},a_{222})\end{matrix}\right\}=\left\{\begin{matrix}\underbrace{(\text{UP}:\text{DP},\text{pF})}_{\text{hyperbolic-secant flow }(+)} & \underbrace{(\text{DP}:\text{UP},\text{nF})}_{\text{hyperbolic-secant flow }(-)}\\ \underbrace{(\text{SO},\text{SO})}_{\text{source}} & \underbrace{(\text{SI},\text{SI})}_{\text{sink}}\end{matrix}\right\}$$

for $a_{110}>0$ and $a_{220}<0$;

(2.89)

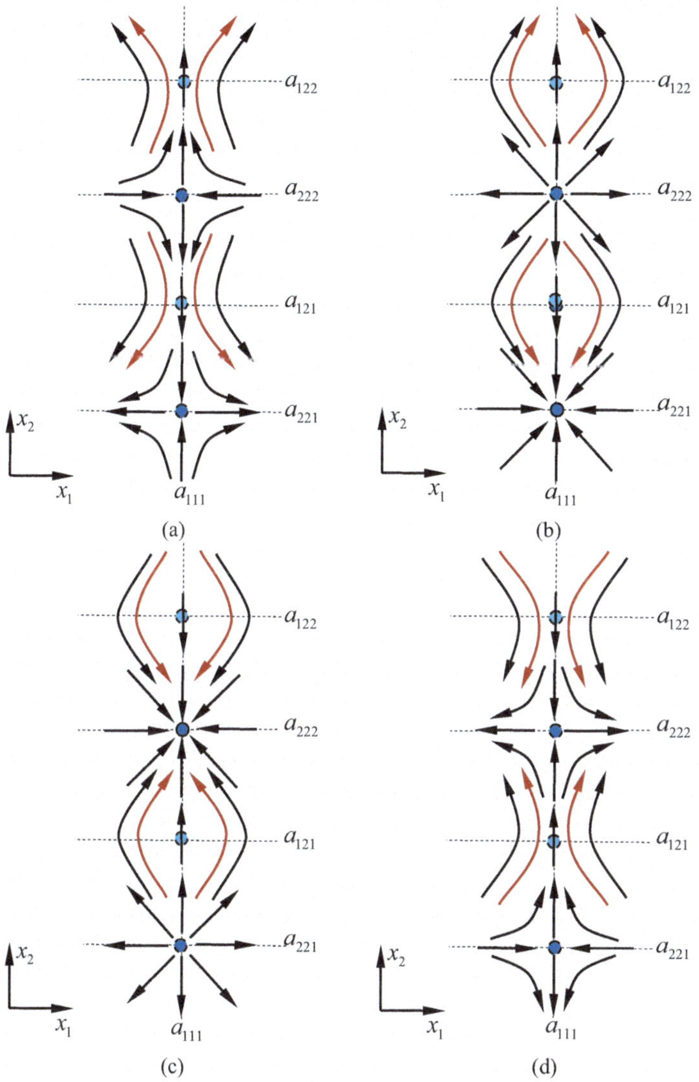

Fig. 2.17 Phase portraits ($a_{121} \in (a_{221}, a_{222})$ and $a_{222} < a_{122}$) for two-dimensional systems on the x_1-direction with $x_1^* = a_{111}$ and on the x_2-direction with $x_2^* = a_{121}, a_{122}, a_{221}, a_{222}$. The equilibriums flows: (**a**) ($a_{110} > 0$, $a_{220} > 0$), (**b**) ($a_{110} < 0$, $a_{220} > 0$), (**c**) ($a_{110} > 0$, $a_{220} < 0$), (**d**) ($a_{110} < 0$, $a_{220} < 0$)

$$\left\{\begin{matrix}(a_{111},a_{121}) & (a_{111},a_{122})\\ (a_{111},a_{221}) & (a_{111},a_{222})\end{matrix}\right\}=\left\{\begin{matrix}\underbrace{(\mathrm{DP}:\mathrm{UP},\mathrm{pF})}_{\text{hyperbolic flow }(+)} & \underbrace{(\mathrm{DP}:\mathrm{UP},\mathrm{nF})}_{\text{hyperbolic flow }(-)}\\ \underbrace{(\mathrm{SI},\mathrm{SO})}_{\text{saddle}} & \underbrace{(\mathrm{SO},\mathrm{SI})}_{\text{saddle}}\end{matrix}\right\} \tag{2.90}$$

for $a_{110}<0$ and $a_{220}<0$.

The saddle and source or sink with hyperbolic and hyperbolic-secant flows exist on the series of equilibriums. The hyperbolic and hyperbolic-secant flows with directrix are separated through saddles or source and sink. The source and sink are separated by a hyperbolic-secant flow. The (SI,SO)-saddle and (SO,SI)-saddle are separated by a hyperbolic flow.

(v) For the case $a_{222}<a_{121}$, the equilibrium series of $x_1^*=a_{111}$ with $x_2^*=a_{121},a_{122},a_{221},a_{222}$ have the properties as in Eqs. (2.75)–(2.78), as shown in Fig. 2.18. The saddle and source or sink with hyperbolic and hyperbolic-secant flows exist on the series of equilibriums. The hyperbolic and hyperbolic-secant flows with directrix are connected together on the top.

2.4.2 *Inflection-Source (Sink) Switching Bifurcations*

In this section, the inflection-source (sink) switching bifurcations for the hyperbolic (hyperbolic-secant flow) with connected saddle, source and sink.

A. Consider a dynamical system as

$$\begin{aligned}\dot{x}_1&=a_{110}(x_1-a_{111})(x_2-a_{121})(x_2-a_{122}),\\ \dot{x}_2&=a_{220}(x_2-a_{221})(x_2-a_{222}),\end{aligned} \tag{2.91}$$

and the corresponding first integral manifold for $a_{12s_1}=a_{22l_1}$ (s_1, s_2, l_1, $l_2\in\{1,2\}$, $s_1\neq s_2$, $l_1\neq l_2$) is

$$\ln\frac{|x_1-a_{111}|}{|x_{10}-a_{111}|}=\frac{a_{110}}{a_{220}}\left[(x_2-x_{20})+(a_{22l_2}-a_{12s_2})\ln\frac{|x_2-a_{22l_2}|}{|x_{20}-a_{22l_2}|}\right]. \tag{2.92}$$

(i) The equilibrium series of $x_2^*=a_{121},a_{122},a_{221},a_{222}$ with $x_1^*=a_{111}$ for $a_{221}=a_{121}$ are

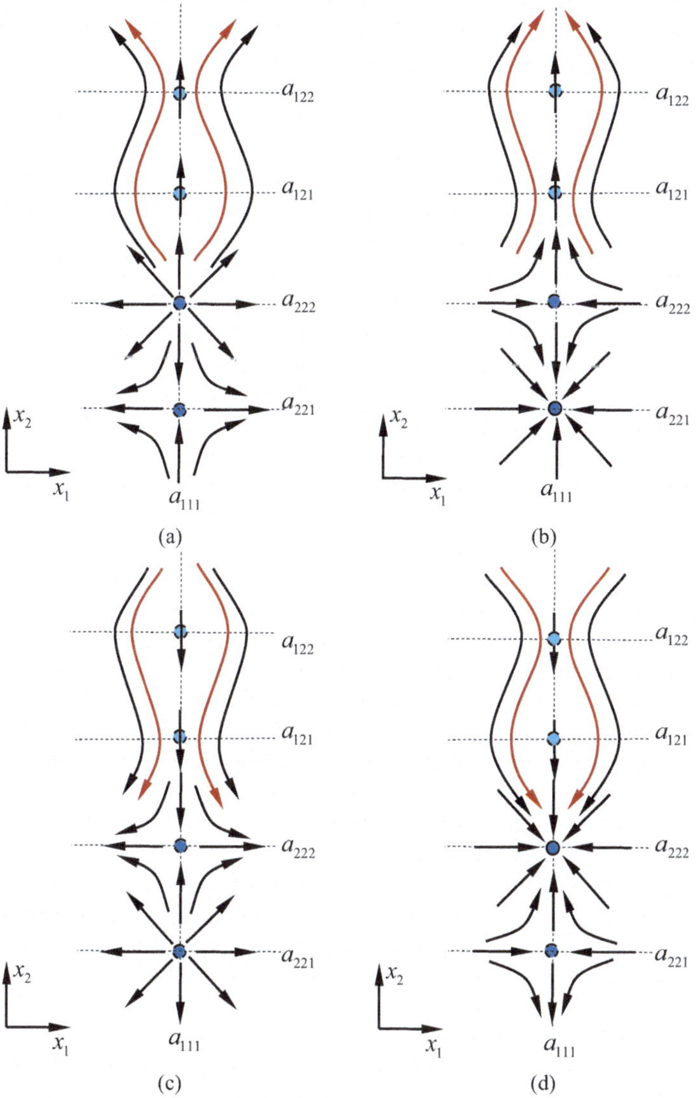

Fig. 2.18 Phase portraits ($a_{222} < a_{121}$) for two-dimensional systems on the x_1-direction with $x_1^* = a_{111}$ and on the x_2-direction with $x_2^* = a_{121}, a_{122}, a_{221}, a_{222}$. The equilibriums flows: (**a**) ($a_{110} > 0$, $a_{220} > 0$), (**b**) ($a_{110} < 0$, $a_{220} > 0$), (**c**) ($a_{110} > 0$, $a_{220} < 0$), (**d**) ($a_{110} < 0$, $a_{220} < 0$)

$$\left\{\begin{matrix}(a_{111},a_{222})\\(a_{111},a_{122})\\(a_{111},a_{121})\end{matrix}\right\}=\left\{\begin{matrix}\underbrace{(\text{SO},\text{SO})}_{\text{source}}\\\underbrace{(\text{UP}:\text{DP},\text{nF})}_{\text{hyperbolic-secant flow }(-)}\\\underbrace{(\text{DI}:\text{II},\text{SI})}_{\text{inflection sink}}\end{matrix}\right\}\text{ for }a_{110}>0,a_{220}>0, \tag{2.93}$$

$$\left\{\begin{matrix}(a_{111},a_{222})\\(a_{111},a_{122})\\(a_{111},a_{121})\end{matrix}\right\}=\left\{\begin{matrix}\underbrace{(\text{SI},\text{SO})}_{\text{saddle}}\\\underbrace{(\text{DP}:\text{UP},\text{nF})}_{\text{hyperbolic flow }(-)}\\\underbrace{(\text{II}:\text{DI},\text{SI})}_{\text{inflection sink}}\end{matrix}\right\}\text{ for }a_{110}<0,a_{220}>0, \tag{2.94}$$

$$\left\{\begin{matrix}(a_{111},a_{222})\\(a_{111},a_{122})\\(a_{111},a_{121})\end{matrix}\right\}=\left\{\begin{matrix}\underbrace{(\text{SO},\text{SI})}_{\text{saddle}}\\\underbrace{(\text{DP}:\text{UP},\text{pF})}_{\text{hyperbolic flow }(+)}\\\underbrace{(\text{II}:\text{DI},\text{SO})}_{\text{inflection source}}\end{matrix}\right\}\text{ for }a_{110}>0,a_{220}<0, \tag{2.95}$$

$$\left\{\begin{matrix}(a_{111},a_{222})\\(a_{111},a_{122})\\(a_{111},a_{121})\end{matrix}\right\}=\left\{\begin{matrix}\underbrace{(\text{SI},\text{SI})}_{\text{sink}}\\\underbrace{(\text{UP}:\text{DP},\text{pF})}_{\text{hyperbolic-secant flow }(+)}\\\underbrace{(\text{DI}:\text{II},\text{SO})}_{\text{inflection source}}\end{matrix}\right\}\text{ for }a_{110}<0,a_{220}<0. \tag{2.96}$$

The infinite-equilibrium of $x_2^* = a_{121} = a_{221}$ is summarized as follows.

$$\begin{aligned}&(\bar{x}_1,a_{221})=\underbrace{(\text{DI},\text{SI})}_{\text{decreasing-inflection sink}}\quad\text{if }\bar{x}_1\in(-\infty,a_{111}),\\&(\bar{x}_1,a_{221})=\underbrace{(\text{II},\text{SI})}_{\text{increasing-inflection sink}}\quad\text{if }\bar{x}_1\in(a_{111},\infty),\\&\text{for }a_{110}>0,a_{220}>0;\end{aligned} \tag{2.97}$$

$$\begin{aligned}&(\bar{x}_1,a_{221})=\underbrace{(\text{II},\text{SI})}_{\text{increasing-inflection sink}}\quad\text{if }\bar{x}_1\in(-\infty,a_{111}),\\&(\bar{x}_1,a_{221})=\underbrace{(\text{DI},\text{SI})}_{\text{decreasing-inflection sink}}\quad\text{if }\bar{x}_1\in(a_{111},\infty),\\&\text{for }a_{110}<0,a_{220}>0;\end{aligned} \tag{2.98}$$

$$\begin{aligned}
(\bar{x}_1, a_{221}) &= \underbrace{(\text{II, SO})}_{\text{increasing-inflection source}} \quad \text{if } \bar{x}_1 \in (-\infty, a_{111}),\\
(\bar{x}_1, a_{221}) &= \underbrace{(\text{DI, SO})}_{\text{decreasing-inflection source}} \quad \text{if } \bar{x}_1 \in (a_{111}, \infty),
\end{aligned} \tag{2.99}$$
for $a_{110} > 0, a_{220} < 0$;

$$\begin{aligned}
(\bar{x}_1, a_{221}) &= \underbrace{(\text{DI, SO})}_{\text{decreasing-inflection source}} \quad \text{if } \bar{x}_1 \in (-\infty, a_{111}),\\
(\bar{x}_1, a_{221}) &= \underbrace{(\text{II, SO})}_{\text{increasing-inflection source}} \quad \text{if } \bar{x}_1 \in (a_{111}, \infty),
\end{aligned} \tag{2.100}$$
for $a_{110} < 0, a_{220} < 0$.

The inflection-source and sink infinite-equilibriums of $x_2^* = a_{121} = a_{221}$ are the switching bifurcations of sink, source and saddles with hyperbolic flows for $a_{221} \in (a_{121}, a_{122})$ and $a_{221} < a_{121}$ with $a_{122} < a_{222}$, as shown in Fig. 2.19. In additions, the saddle and hyperbolic flows exist, or a hyperbolic-secant flow with sink or source exist. The infinite-equilibrium is on the bottom of the series of equilibriums.

(ii) The equilibrium series of $x_2^* = a_{121}, a_{122}, a_{221}, a_{222}$ with $x_1^* = a_{111}$ for $a_{221} = a_{122}$ are

$$\left\{\begin{array}{l} (a_{111}, a_{222}) \\ (a_{111}, a_{122}) \\ (a_{111}, a_{121}) \end{array}\right\} = \left\{\begin{array}{c} \underbrace{(\text{SO, SO})}_{\text{source}} \\ \underbrace{(\text{II : DI, SI})}_{\text{inflection sink}} \\ \underbrace{(\text{UP : DP, pF})}_{\text{hyperbolic-secant flow }(+)} \end{array}\right\} \text{ for } a_{110} > 0, a_{220} > 0, \tag{2.101}$$

$$\left\{\begin{array}{l} (a_{111}, a_{222}) \\ (a_{111}, a_{122}) \\ (a_{111}, a_{121}) \end{array}\right\} = \left\{\begin{array}{c} \underbrace{(\text{SI, SO})}_{\text{saddle}} \\ \underbrace{(\text{DI : II, SI})}_{\text{inflection sink}} \\ \underbrace{(\text{DP : UP, pF})}_{\text{hyperbolic flow }(+)} \end{array}\right\} \text{ for } a_{110} < 0, a_{220} > 0, \tag{2.102}$$

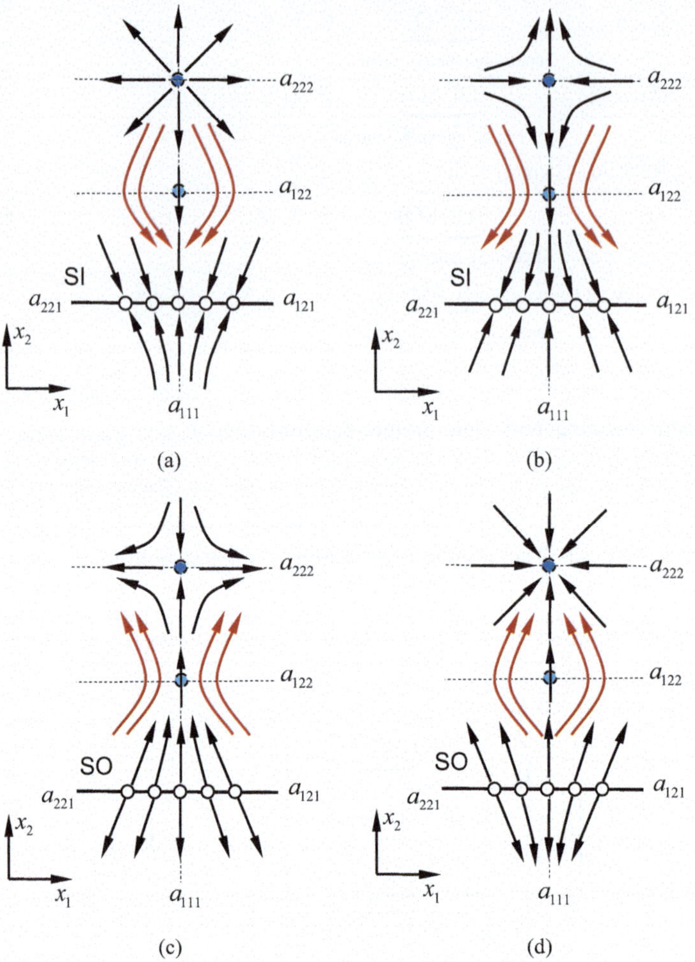

Fig. 2.19 Phase portraits ($a_{121} = a_{221}$) for two-dimensional systems on the x_1-direction with $x_1^* = a_{111}$ and on the x_2-direction with $x_2^* = a_{121}, a_{122}, a_{221}, a_{222}$. The equilibriums flows: (**a**) ($a_{110} > 0$, $a_{220} > 0$), (**b**) ($a_{110} < 0$, $a_{220} > 0$), (**c**) ($a_{110} > 0$, $a_{220} < 0$), (**d**) ($a_{110} < 0$, $a_{220} < 0$)

$$\begin{Bmatrix} (a_{111}, a_{222}) \\ (a_{111}, a_{122}) \\ (a_{111}, a_{121}) \end{Bmatrix} = \begin{Bmatrix} \underbrace{(\text{SO}, \text{SI})}_{\text{saddle}} \\ \underbrace{(\text{DI} : \text{II}, \text{SO})}_{\text{inflection source}} \\ \underbrace{(\text{DP} : \text{UP}, \text{nF})}_{\text{hyperbolic flow }(-)} \end{Bmatrix} \text{ for } a_{110} > 0, a_{220} < 0, \tag{2.103}$$

$$\left\{\begin{array}{c}(a_{111},a_{222})\\(a_{111},a_{122})\\(a_{111},a_{121})\end{array}\right\}=\left\{\begin{array}{c}\underbrace{\text{(SI, SI)}}_{\text{sink}}\\\underbrace{\text{(II : DI, SO)}}_{\text{inflection source}}\\\underbrace{\text{(UP : DP, nF)}}_{\text{hyperbolic-secant flow }(-)}\end{array}\right\}\text{ for }a_{110}<0,a_{220}<0. \tag{2.104}$$

The infinite-equilibrium of $x_2^*=a_{122}=a_{221}$ is summarized as follows.

$$\begin{array}{l}(\bar{x}_1,a_{221})=\underbrace{\text{(II, SI)}}_{\text{increasing-inflection sink}}\quad\text{if }\bar{x}_1\in(-\infty,a_{111}),\\(\bar{x}_1,a_{221})=\underbrace{\text{(DI, SI)}}_{\text{decreasing-inflection sink}}\quad\text{if }\bar{x}_1\in(a_{111},\infty),\\\text{for }a_{110}>0,a_{220}>0;\end{array} \tag{2.105}$$

$$\begin{array}{l}(\bar{x}_1,a_{221})=\underbrace{\text{(DI, SI)}}_{\text{decreasing-inflection sink}}\quad\text{if }\bar{x}_1\in(-\infty,a_{111}),\\(\bar{x}_1,a_{221})=\underbrace{\text{(II, SI)}}_{\text{increasing-inflection sink}}\quad\text{if }\bar{x}_1\in(a_{111},\infty),\\\text{for }a_{110}<0,a_{220}>0;\end{array} \tag{2.106}$$

$$\begin{array}{l}(\bar{x}_1,a_{221})=\underbrace{\text{(DI, SO)}}_{\text{decreasing-inflection source}}\quad\text{if }\bar{x}_1\in(-\infty,a_{111}),\\(\bar{x}_1,a_{221})=\underbrace{\text{(II, SO)}}_{\text{increasing-inflection source}}\quad\text{if }\bar{x}_1\in(a_{111},\infty),\\\text{for }a_{110}>0,a_{220}<0;\end{array} \tag{2.107}$$

$$\begin{array}{l}(\bar{x}_1,a_{221})=\underbrace{\text{(II, SO)}}_{\text{increasing-inflection source}}\quad\text{if }\bar{x}_1\in(-\infty,a_{111}),\\(\bar{x}_1,a_{221})=\underbrace{\text{(DI, SO)}}_{\text{decreasing-inflection source}}\quad\text{if }\bar{x}_1\in(a_{111},\infty),\\\text{for }a_{110}<0,a_{220}<0.\end{array} \tag{2.108}$$

The inflection-source and sink infinite-equilibriums of $x_2^*=a_{122}=a_{221}$ are the switching bifurcation of sink, source and saddles with hyperbolic flows for $a_{221}\in(a_{121},a_{122})$ and $a_{122}<a_{221}$, as shown in Fig. 2.20. In addition, the saddle and hyperbolic flows exist, or a hyperbolic-secant flow with sink or source exist. The infinite-equilibrium is in the middle between saddle and hyperbolic-flow or between the hyperbolic-secant flow and sink or source. In the series of equilibriums, the saddle, sink, and source are on the bottom, and the hyperbolic and hyperbolic-secant flows are on the top.

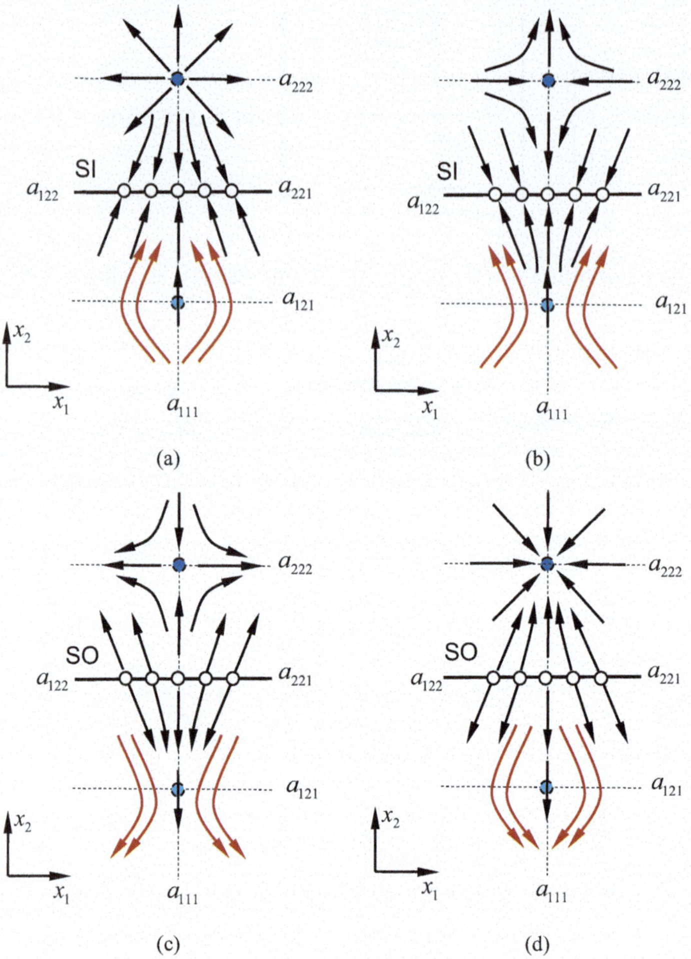

Fig. 2.20 Phase portraits ($a_{122} = a_{221}$) for two-dimensional systems on the x_1-direction with $x_1^* = a_{111}$ and on the x_2-direction with $x_2^* = a_{121}, a_{122}, a_{221}, a_{222}$. The equilibriums flows: (**a**) ($a_{110} > 0$, $a_{220} > 0$), (**b**) ($a_{110} < 0$, $a_{220} > 0$), (**c**) ($a_{110} > 0$, $a_{220} < 0$), (**d**) ($a_{110} < 0$, $a_{220} < 0$)

(iii) The equilibrium series of $x_2^* = a_{121}, a_{122}, a_{221}, a_{222}$ with $x_1^* = a_{111}$ for $a_{121} = a_{222}$ are

$$\left\{\begin{array}{c}(a_{111},a_{122})\\(a_{111},a_{222})\\(a_{111},a_{221})\end{array}\right\}=\left\{\begin{array}{c}\underbrace{(\mathrm{DP}:\mathrm{UP},\mathrm{pF})}_{\text{hyperbolic flow }(+)}\\\underbrace{(\mathrm{II}:\mathrm{DI},\mathrm{SO})}_{\text{inflection source}}\\\underbrace{(\mathrm{SO},\mathrm{SI})}_{\text{saddle}}\end{array}\right\}\text{ for }a_{110}>0,a_{220}>0, \tag{2.109}$$

$$\left\{\begin{array}{c}(a_{111},a_{122})\\(a_{111},a_{222})\\(a_{111},a_{221})\end{array}\right\}=\left\{\begin{array}{c}\underbrace{(\mathrm{UP}:\mathrm{DP},\mathrm{pF})}_{\text{hyperbolic-secant flow }(+)}\\\underbrace{(\mathrm{DI}:\mathrm{II},\mathrm{SO})}_{\text{inflection source}}\\\underbrace{(\mathrm{SI},\mathrm{SI})}_{\text{sink}}\end{array}\right\}\text{ for }a_{110}<0,a_{220}>0, \tag{2.110}$$

$$\left\{\begin{array}{c}(a_{111},a_{122})\\(a_{111},a_{222})\\(a_{111},a_{221})\end{array}\right\}=\left\{\begin{array}{c}\underbrace{(\mathrm{UP}:\mathrm{DP},\mathrm{nF})}_{\text{hyperbolic-secant flow }(-)}\\\underbrace{(\mathrm{DI}:\mathrm{II},\mathrm{SI})}_{\text{inflection sink}}\\\underbrace{(\mathrm{SO},\mathrm{SO})}_{\text{source}}\end{array}\right\}\text{ for }a_{110}>0,a_{220}<0, \tag{2.111}$$

$$\left\{\begin{array}{c}(a_{111},a_{122})\\(a_{111},a_{222})\\(a_{111},a_{221})\end{array}\right\}=\left\{\begin{array}{c}\underbrace{(\mathrm{DP}:\mathrm{UP},\mathrm{nF})}_{\text{hyperbolic flow }(-)}\\\underbrace{(\mathrm{II}:\mathrm{DI},\mathrm{SI})}_{\text{inflection sink}}\\\underbrace{(\mathrm{SI},\mathrm{SO})}_{\text{saddle}}\end{array}\right\}\text{for }a_{110}<0,a_{220}<0. \tag{2.112}$$

The infinite-equilibrium of $x_2^* = a_{121} = a_{222}$ is summarized as follows.

$$\begin{aligned}&(\bar{x}_1,a_{222})=\underbrace{(\mathrm{II},\mathrm{SO})}_{\text{increasing-inflection source}}\quad\text{if }\bar{x}_1\in(-\infty,a_{111}),\\&(\bar{x}_1,a_{222})=\underbrace{(\mathrm{DI},\mathrm{SO})}_{\text{decreasing-inflection source}}\quad\text{if }\bar{x}_1\in(a_{111},\infty),\\&\text{for }a_{110}>0,a_{220}>0;\end{aligned} \tag{2.113}$$

$$
\begin{aligned}
&(\bar{x}_1, a_{222}) = \underbrace{(\mathrm{DI}, \mathrm{SO})}_{\text{decreasing-inflection source}} \quad \text{if } \bar{x}_1 \in (-\infty, a_{111}),\\
&(\bar{x}_1, a_{222}) = \underbrace{(\mathrm{II}, \mathrm{SO})}_{\text{increasing-inflection source}} \quad \text{if } \bar{x}_1 \in (a_{111}, \infty),
\end{aligned} \tag{2.114}
$$

for $a_{110} < 0, a_{220} > 0$;

$$
\begin{aligned}
&(\bar{x}_1, a_{222}) = \underbrace{(\mathrm{DI}, \mathrm{SI})}_{\text{decreasing-inflection sink}} \quad \text{if } \bar{x}_1 \in (-\infty, a_{111}),\\
&(\bar{x}_1, a_{222}) = \underbrace{(\mathrm{II}, \mathrm{SI})}_{\text{increasing-inflection sink}} \quad \text{if } \bar{x}_1 \in (a_{111}, \infty),
\end{aligned} \tag{2.115}
$$

for $a_{110} > 0, a_{220} < 0$;

$$
\begin{aligned}
&(\bar{x}_1, a_{221}) = \underbrace{(\mathrm{II}, \mathrm{SI})}_{\text{increasing-inflection sink}} \quad \text{if } \bar{x}_1 \in (-\infty, a_{111}),\\
&(\bar{x}_1, a_{221}) = \underbrace{(\mathrm{DI}, \mathrm{SI})}_{\text{decreasing-inflection sink}} \quad \text{if } \bar{x}_1 \in (a_{111}, \infty),
\end{aligned} \tag{2.116}
$$

for $a_{110} < 0, a_{220} < 0$.

The infinite-equilibriums of $x_2^* = a_{122} = a_{222}$ are the switching bifurcations for the equilibriums for $a_{121} \in (a_{221}, a_{222})$ and $a_{121} < a_{221}$, as shown in Fig. 2.21. In additions, the saddle and hyperbolic flows exist, or a hyperbolic-secant flow with sink or source exist. The infinite-equilibrium is also in the middle between saddle and hyperbolic-flow or between the hyperbolic-secant flow and sink or source.

(iv) The equilibrium series of $x_2^* = a_{121}, a_{122}, a_{221}, a_{222}$ with $x_1^* = a_{111}$ for $a_{122} = a_{222}$ are

$$
\left\{\begin{array}{l} (a_{111}, a_{122}) \\ (a_{111}, a_{121}) \\ (a_{111}, a_{221}) \end{array}\right\} = \left\{\begin{array}{c} \underbrace{(\mathrm{DI}:\mathrm{II}, \mathrm{SO})}_{\text{inflection source}} \\ \underbrace{(\mathrm{DP}:\mathrm{UP}, \mathrm{nF})}_{\text{hyperbolic flow }(-)} \\ \underbrace{(\mathrm{SO}, \mathrm{SI})}_{\text{saddle}} \end{array}\right\} \text{ for } a_{110} > 0, a_{220} > 0, \tag{2.117}
$$

$$
\left\{\begin{array}{l} (a_{111}, a_{122}) \\ (a_{111}, a_{121}) \\ (a_{111}, a_{221}) \end{array}\right\} = \left\{\begin{array}{c} \underbrace{(\mathrm{II}:\mathrm{DI}, \mathrm{SO})}_{\text{inflection source}} \\ \underbrace{(\mathrm{UP}:\mathrm{DP}, \mathrm{nF})}_{\text{hyperbolic-secant flow }(-)} \\ \underbrace{(\mathrm{SI}, \mathrm{SI})}_{\text{sink}} \end{array}\right\} \text{ for } a_{110} < 0, a_{220} > 0, \tag{2.118}
$$

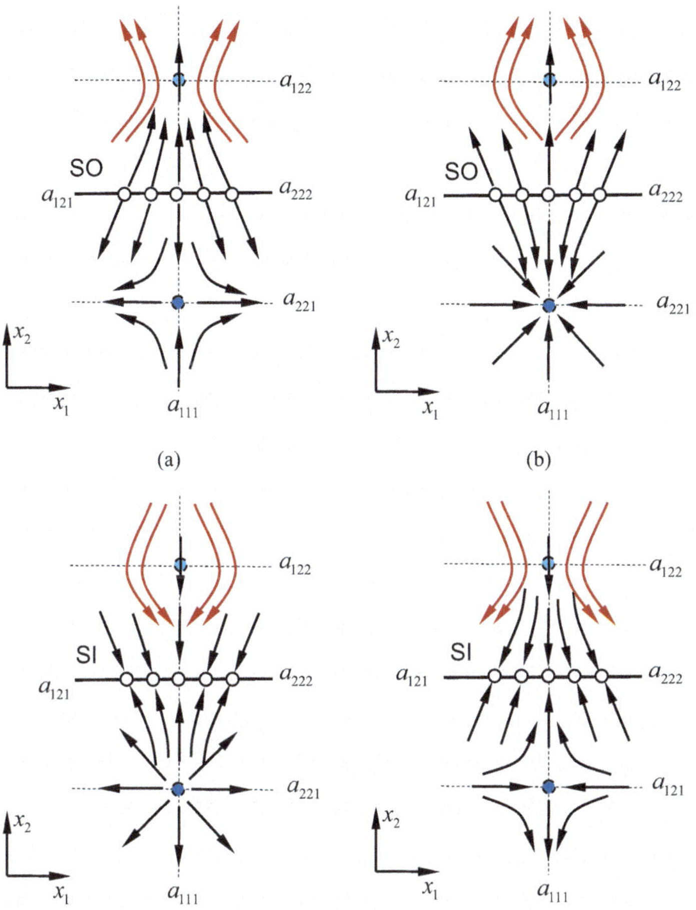

Fig. 2.21 Phase portraits ($a_{121} = a_{222}$) for two-dimensional systems on the x_1-direction with $x_1^* = a_{111}$ and on the x_2-direction with $x_2^* = a_{121}, a_{221}, a_{222}$. The equilibriums flows: **(a)** ($a_{110} > 0$, $a_{220} > 0$), **(b)** ($a_{110} < 0$, $a_{220} > 0$), **(c)** ($a_{110} > 0$, $a_{220} < 0$), **(d)** ($a_{110} < 0$, $a_{220} < 0$)

$$\begin{Bmatrix} (a_{111}, a_{122}) \\ (a_{111}, a_{121}) \\ (a_{111}, a_{221}) \end{Bmatrix} = \begin{Bmatrix} \underbrace{(\text{II} : \text{DI}, \text{SI})}_{\text{inflection sink}} \\ \underbrace{(\text{UP} : \text{DP}, \text{pF})}_{\text{hyperbolic-secant flow } (+)} \\ \underbrace{(\text{SO}, \text{SO})}_{\text{source}} \end{Bmatrix} \text{ for } a_{110} > 0, a_{220} < 0, \tag{2.119}$$

$$\left\{\begin{array}{c}(a_{111},a_{122})\\(a_{111},a_{222})\\(a_{111},a_{221})\end{array}\right\}=\left\{\begin{array}{c}\underbrace{(\text{DI}:\text{II},\text{SI})}_{\text{inflection sink}}\\\underbrace{(\text{DP}:\text{UP},\text{pF})}_{\text{hyperbolic flow }(+)}\\\underbrace{(\text{SI},\text{SO})}_{\text{saddle}}\end{array}\right\}\text{ for }a_{110}<0,a_{220}<0. \tag{2.120}$$

The infinite-equilibrium of $x_2^*=a_{122}=a_{222}$ is summarized as follows.

$$\begin{aligned}&(\bar{x}_1,a_{222})=\underbrace{(\text{DI},\text{SO})}_{\text{decreasing-inflection source}}\quad\text{if }\bar{x}_1\in(-\infty,a_{111}),\\&(\bar{x}_1,a_{222})=\underbrace{(\text{II},\text{SO})}_{\text{increasing-inflection source}}\quad\text{if }\bar{x}_1\in(a_{111},\infty),\\&\text{for }a_{110}>0,a_{220}>0;\end{aligned} \tag{2.121}$$

$$\begin{aligned}&(\bar{x}_1,a_{222})=\underbrace{(\text{II},\text{SO})}_{\text{increasing-inflection source}}\quad\text{if }\bar{x}_1\in(-\infty,a_{111}),\\&(\bar{x}_1,a_{222})=\underbrace{(\text{DI},\text{SO})}_{\text{decreasing-inflection source}}\quad\text{if }\bar{x}_1\in(a_{111},\infty),\\&\text{for }a_{110}<0,a_{220}>0.\end{aligned} \tag{2.122}$$

$$\begin{aligned}&(\bar{x}_1,a_{222})=\underbrace{(\text{II},\text{SI})}_{\text{increasing-inflection sink}}\quad\text{if }\bar{x}_1\in(-\infty,a_{111}),\\&(\bar{x}_1,a_{222})=\underbrace{(\text{DI},\text{SI})}_{\text{decreasing-inflection sink}}\quad\text{if }\bar{x}_1\in(a_{111},\infty),\\&\text{for }a_{110}>0,a_{220}<0;\end{aligned} \tag{2.123}$$

$$\begin{aligned}&(\bar{x}_1,a_{222})=\underbrace{(\text{DI},\text{SI})}_{\text{decreasing-inflection sink}}\quad\text{if }\bar{x}_1\in(-\infty,a_{111}),\\&(\bar{x}_1,a_{222})=\underbrace{(\text{II},\text{SI})}_{\text{increasing-inflection sink}}\quad\text{if }\bar{x}_1\in(a_{111},\infty),\\&\text{for }a_{110}<0,a_{220}<0.\end{aligned} \tag{2.124}$$

The inflection source and sink infinite-equilibriums of $x_2^*=a_{122}=a_{222}$ are for the switching bifurcations for sink, source and saddles for $a_{222}\in(a_{121},a_{122})$ and $a_{122}<a_{222}$, as presented in Fig. 2.22. In additions, the saddle and hyperbolic flows exist, or a hyperbolic-secant flow with sink or source exist. The infinite-equilibrium is on the top, and the saddle and hyperbolic-flow or the hyperbolic-secant flow and sink or source are on the bottom.

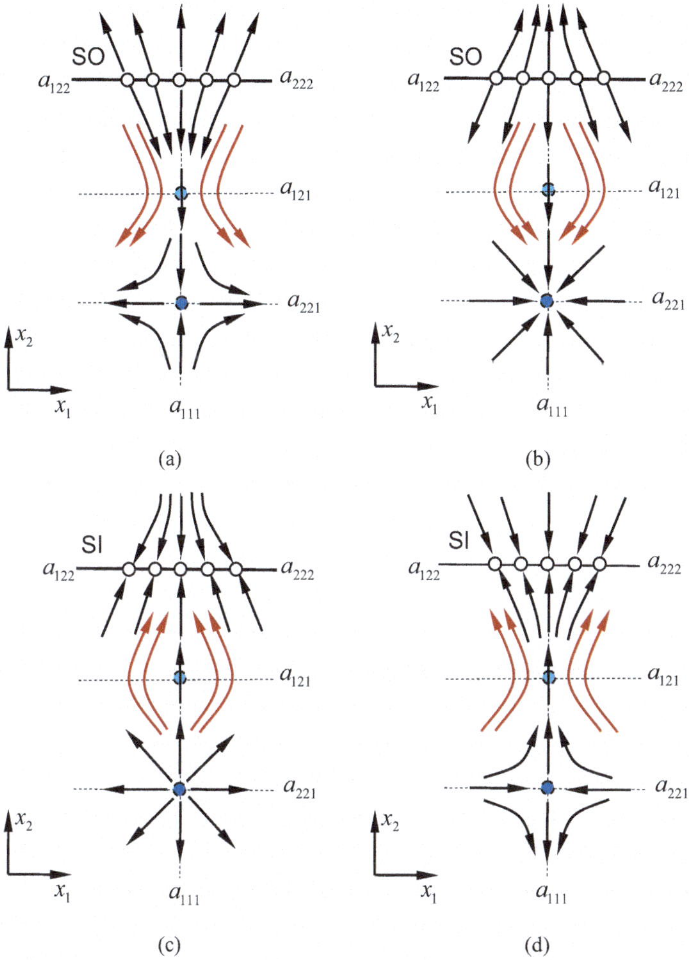

Fig. 2.22 Phase portraits ($a_{122} = a_{222}$) for two-dimensional systems on the x_1-direction with $x_1^* = a_{111}$ and on the x_2-direction with $x_2^* = a_{121}, a_{221}, a_{222}$. The equilibriums flows: (**a**) ($a_{110} > 0$, $a_{220} > 0$), (**b**) ($a_{110} < 0$, $a_{220} > 0$), (**c**) ($a_{110} > 0$, $a_{220} < 0$), (**d**) ($a_{110} < 0$, $a_{220} < 0$)

B. Consider a dynamical system as

$$\begin{aligned}\dot{x}_1 &= a_{110}(x_1 - a_{111})(x_2 - a_{121})(x_2 - a_{122}),\\ \dot{x}_2 &= a_{220}(x_2 - a_{221})(x_2 - a_{222}),\end{aligned} \tag{2.125}$$

and the corresponding first integral manifold for $a_{121} = a_{221}$ and $a_{122} = a_{222}$ is

$$\ln \frac{|x_1 - a_{111}|}{|x_{10} - a_{111}|} = \frac{a_{110}}{a_{220}}(x_2 - x_{20}). \tag{2.126}$$

The equilibrium series of $x_2^* = a_{121}, a_{122}, a_{221}, a_{222}$ with $x_1^* = a_{111}$ for $a_{121} = a_{221}$ and $a_{122} = a_{222}$ are

$$\left\{ \begin{matrix} (a_{111}, a_{122}) \\ (a_{111}, a_{121}) \end{matrix} \right\} = \left\{ \begin{matrix} (a_{111}, a_{222}) \\ (a_{111}, a_{221}) \end{matrix} \right\} = \left\{ \begin{matrix} \underbrace{(\text{DI}:\text{II},\text{SO})}_{\text{inflection source}} \\ \underbrace{(\text{DI}:\text{II},\text{SI})}_{\text{inflection sink}} \end{matrix} \right\} \text{for } a_{110} > 0, a_{220} > 0, \tag{2.127}$$

$$\left\{ \begin{matrix} (a_{111}, a_{122}) \\ (a_{111}, a_{121}) \end{matrix} \right\} = \left\{ \begin{matrix} (a_{111}, a_{222}) \\ (a_{111}, a_{221}) \end{matrix} \right\} = \left\{ \begin{matrix} \underbrace{(\text{II}:\text{DI},\text{SO})}_{\text{inflection source}} \\ \underbrace{(\text{II}:\text{DI},\text{SI})}_{\text{inflection sink}} \end{matrix} \right\} \text{for } a_{110} < 0, a_{220} > 0, \tag{2.128}$$

$$\left\{ \begin{matrix} (a_{111}, a_{122}) \\ (a_{111}, a_{121}) \end{matrix} \right\} = \left\{ \begin{matrix} (a_{111}, a_{222}) \\ (a_{111}, a_{221}) \end{matrix} \right\} = \left\{ \begin{matrix} \underbrace{(\text{II}:\text{DI},\text{SI})}_{\text{inflection sink}} \\ \underbrace{(\text{II}:\text{DI},\text{SO})}_{\text{inflection source}} \end{matrix} \right\} \text{for } a_{110} > 0, a_{220} < 0, \tag{2.129}$$

$$\left\{ \begin{matrix} (a_{111}, a_{122}) \\ (a_{111}, a_{121}) \end{matrix} \right\} = \left\{ \begin{matrix} (a_{111}, a_{222}) \\ (a_{111}, a_{221}) \end{matrix} \right\} = \left\{ \begin{matrix} \underbrace{(\text{DI}:\text{II},\text{SI})}_{\text{inflection sink}} \\ \underbrace{(\text{DI}:\text{II},\text{SO})}_{\text{inflection source}} \end{matrix} \right\} \text{for } a_{110} < 0, a_{220} < 0, \tag{2.130}$$

and the infinite-equilibrium of $x_2^* = a_{122} = a_{222}$ is summarized as follows.

$$\begin{aligned} (\bar{x}_1, a_{222}) &= \underbrace{(\text{DI},\text{SO})}_{\text{decreasing-inflection source}} \quad \text{if } \bar{x}_1 \in (-\infty, a_{111}), \\ (\bar{x}_1, a_{222}) &= \underbrace{(\text{II},\text{SO})}_{\text{increasing-inflection source}} \quad \text{if } \bar{x}_1 \in (a_{111}, \infty), \end{aligned} \tag{2.131}$$

for $a_{110} > 0, a_{220} > 0$;

$$(\bar{x}_1, a_{222}) = \underbrace{(\mathrm{II}, \mathrm{SO})}_{\text{increasing-inflection source}} \quad \text{if } \bar{x}_1 \in (-\infty, a_{111}),$$
$$(\bar{x}_1, a_{222}) = \underbrace{(\mathrm{DI}, \mathrm{SO})}_{\text{decreasing-inflection source}} \quad \text{if } \bar{x}_1 \in (a_{111}, \infty), \tag{2.132}$$
for $a_{110} < 0, a_{220} > 0$.

$$(\bar{x}_1, a_{222}) = \underbrace{(\mathrm{II}, \mathrm{SI})}_{\text{increasing-inflection sink}} \quad \text{if } \bar{x}_1 \in (-\infty, a_{111}),$$
$$(\bar{x}_1, a_{222}) = \underbrace{(\mathrm{DI}, \mathrm{SI})}_{\text{decreasing-inflection sink}} \quad \text{if } \bar{x}_1 \in (a_{111}, \infty), \tag{2.133}$$
for $a_{110} > 0, a_{220} < 0$;

$$(\bar{x}_1, a_{222}) = \underbrace{(\mathrm{DI}, \mathrm{SI})}_{\text{decreasing-inflection sink}} \quad \text{if } \bar{x}_1 \in (-\infty, a_{111}),$$
$$(\bar{x}_1, a_{222}) = \underbrace{(\mathrm{II}, \mathrm{SI})}_{\text{increasing-inflection sink}} \quad \text{if } \bar{x}_1 \in (a_{111}, \infty), \tag{2.134}$$
for $a_{110} < 0, a_{220} < 0$.

The infinite-equilibrium of $x_2^* = a_{121} = a_{221}$ is summarized as follows.

$$(\bar{x}_1, a_{221}) = \underbrace{(\mathrm{DI}, \mathrm{SI})}_{\text{decreasing-inflection sink}} \quad \text{if } \bar{x}_1 \in (-\infty, a_{111}),$$
$$(\bar{x}_1, a_{221}) = \underbrace{(\mathrm{II}, \mathrm{SI})}_{\text{increasing-inflection sink}} \quad \text{if } \bar{x}_1 \in (a_{111}, \infty), \tag{2.135}$$
for $a_{110} > 0, a_{220} > 0$,

$$(\bar{x}_1, a_{221}) = \underbrace{(\mathrm{II}, \mathrm{SI})}_{\text{increasing-inflection sink}} \quad \text{if } \bar{x}_1 \in (-\infty, a_{111}),$$
$$(\bar{x}_1, a_{221}) = \underbrace{(\mathrm{DI}, \mathrm{SI})}_{\text{decreasing-inflection sink}} \quad \text{if } \bar{x}_1 \in (a_{111}, \infty), \tag{2.136}$$
for $a_{110} < 0, a_{220} > 0$;

$$(\bar{x}_1, a_{221}) = \underbrace{(\mathrm{II}, \mathrm{SO})}_{\text{increasing-inflection source}} \quad \text{if } \bar{x}_1 \in (-\infty, a_{111}),$$
$$(\bar{x}_1, a_{221}) = \underbrace{(\mathrm{DI}, \mathrm{SO})}_{\text{decreasing-inflection source}} \quad \text{if } \bar{x}_1 \in (a_{111}, \infty), \tag{2.137}$$
for $a_{110} < 0, a_{220} > 0$,

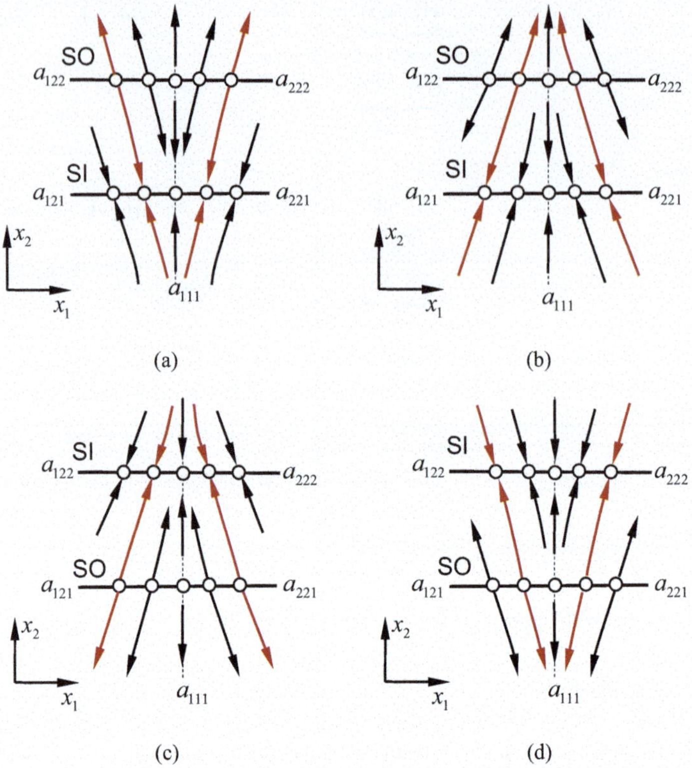

Fig. 2.23 Phase portraits ($a_{121} = a_{221}$ and $a_{122} = a_{222}$) for two-dimensional systems on the x_1-direction with $x_1^* = a_{111}$ and on the x_2-direction with $x_2^* = a_{121}, a_{221}, a_{222}$. The infinite-equilibriums: (**a**) ($a_{110} > 0$, $a_{220} > 0$), (**b**) ($a_{110} < 0$, $a_{220} > 0$), (**c**) ($a_{110} > 0$, $a_{220} < 0$), (**d**) ($a_{110} < 0$, $a_{220} < 0$)

$$
\begin{aligned}
(\bar{x}_1, a_{221}) &= \underbrace{(\mathrm{DI, SO})}_{\text{decreasing-inflection source}} \quad \text{if } \bar{x}_1 \in (-\infty, a_{111}), \\
(\bar{x}_1, a_{221}) &= \underbrace{(\mathrm{II, SO})}_{\text{increasing-inflection source}} \quad \text{if } \bar{x}_1 \in (a_{111}, \infty),
\end{aligned} \tag{2.138}
$$

for $a_{110} < 0, a_{220} < 0$.

The double inflection-sink and source infinite-equilibriums of $x_2^* = a_{121} = a_{221}$ and $x_2^* = a_{122} = a_{222}$ are presented in Fig. 2.23.

Chapter 3
Double-Saddles and Switching Bifurcations

In this chapter, the appearing and switching bifurcations in a crossing-linear and self-quadratic product system with a self-quadratic vector field are discussed, and the corresponding network of equilibriums are presented. The double-saddles with hyperbolic-to-hyperbolic flows are presented first for the appearing bifurcations. The up-down and down-up saddles are the switching bifurcations for the hyperbolic-to-hyperbolic-secant flow with connected double-saddles. The series of hyperbolic-to-hyperbolic-secant flows with connected saddle-source and saddle-sink are presented. The inflection-source (sink) on the infinite-equilibrium is for the switching bifurcations of hyperbolic-to-hyperbolic-secant flows with connected saddle-source (sink). The network of the paralleled hyperbolic and hyperbolic-secant flows with paralleled saddle-source and saddle-sink. The up-down-to-down-up and down-up-to-up-down upper-saddle (lower-saddle) are presented for the switching bifurcation of a hyperbolic (hyperbolic-secant) flow with saddle-source (sink). The networks of paralleled hyperbolic and hyperbolic-secant flows with connected saddle, sink, and source are presented. The connected inflection-sources (sinks) are presented for the paralleled hyperbolic and hyperbolic-secant flows with saddle and source (sink).

3.1 Double Self-Singular Cubic Systems

The two self-singularities in the cubic system generate the double-saddles for the appearing bifurcation of a network of saddles, source, and sink, and also for the appearing and switching bifurcations of saddle-source and saddle-sink. The up-down and down-up upper-saddle and lower-saddle infinite-equilibriums are the switching bifurcations of the double-saddles with hyperbolic-to-hyperbolic-secant flows.

A. C. J. Luo, *Two-dimensional Product-Cubic Systems, Vol. VII*,
https://doi.org/10.1007/978-3-031-48483-4_3

3.1.1 Double-Saddles with Hyperbolic-to-Hyperbolic-Secant Flows

In this section, double-saddles with connected hyperbolic-to-hyperbolic-secant flows are presented. The double-saddle bifurcations are for appearing bifurcations of a network of saddles, sink, and source. The appearing bifurcation of paralleled hyperbolic and hyperbolic-secant flows is presented by the hyperbolic-to-hyperbolic-secant flows.

Consider a dynamical system as

$$\begin{aligned}\dot{x}_1 &= a_{110}(x_1 - a_{111})^2(x_2 - a_{121}),\\ \dot{x}_2 &= a_{220}(x_2 - a_{221})^2,\end{aligned} \tag{3.1}$$

and the corresponding first integral manifold is

$$-\left(\frac{1}{x_1 - a_{111}} - \frac{1}{x_{10} - a_{111}}\right) = \frac{a_{110}}{a_{220}}\left[\ln\frac{|x_2 - a_{221}|}{|x_{20} - a_{221}|} - (a_{221} - a_{121})\left(\frac{1}{x_2 - a_{221}} - \frac{1}{x_{20} - a_{221}}\right)\right]. \tag{3.2}$$

(i) The equilibriums of $x_1^* = a_{111}, a_{211}$ and $x_2^* = a_{121}, a_{221}$ for $a_{121} < a_{221}$ are

$$\left\{\begin{matrix}(a_{111}, a_{221})\\(a_{111}, a_{121})\end{matrix}\right\} = \left\{\begin{matrix}\underbrace{(\text{US}, \text{US})}_{\text{double-saddle}}\\ \underbrace{(\text{UP}:\text{UP}, \text{pF})}_{\text{HS to HB flow }(+)}\end{matrix}\right\} \text{ for } a_{110} > 0 \text{ and } a_{220} > 0; \tag{3.3}$$

$$\left\{\begin{matrix}(a_{111}, a_{221})\\(a_{111}, a_{121})\end{matrix}\right\} = \left\{\begin{matrix}\underbrace{(\text{LS}, \text{US})}_{\text{double-saddle}}\\ \underbrace{(\text{DP}:\text{DP}, \text{pF})}_{\text{HB to HS flow }(+)}\end{matrix}\right\} \text{ for } a_{110} < 0 \text{ and } a_{220} > 0; \tag{3.4}$$

$$\left\{\begin{matrix}(a_{111}, a_{221})\\(a_{111}, a_{121})\end{matrix}\right\} = \left\{\begin{matrix}\underbrace{(\text{US}, \text{LS})}_{\text{double-saddle}}\\ \underbrace{(\text{DP}:\text{DP}, \text{nF})}_{\text{HB to HS flow }(-)}\end{matrix}\right\} \text{ for } a_{110} > 0 \text{ and } a_{220} < 0; \tag{3.5}$$

$$\left\{\begin{matrix}(a_{111}, a_{221})\\(a_{111}, a_{121})\end{matrix}\right\} = \left\{\begin{matrix}\underbrace{(\text{LS}, \text{LS})}_{\text{double-saddle}}\\ \underbrace{(\text{UP}:\text{UP}, \text{nF})}_{\text{HS to HB flow }(-)}\end{matrix}\right\} \text{ for } a_{110} < 0 \text{ and } a_{220} < 0; \tag{3.6}$$

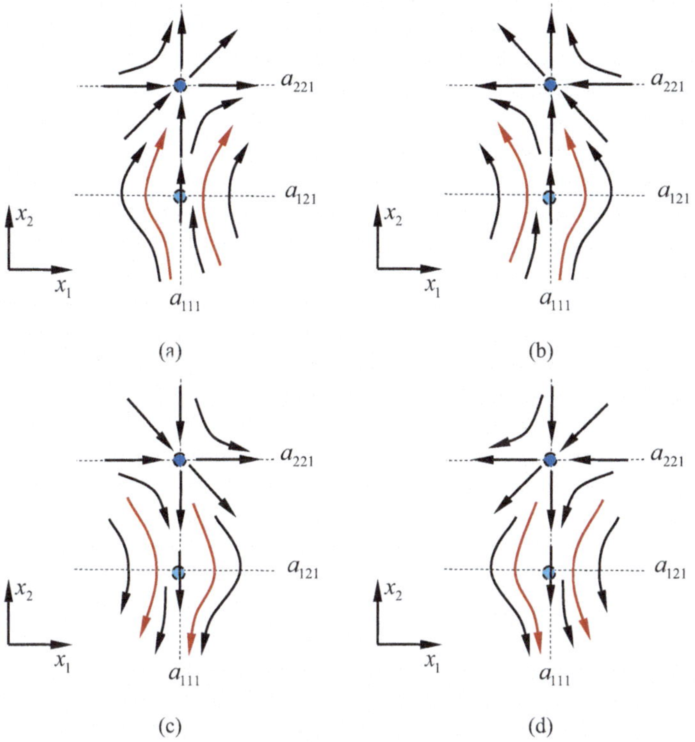

Fig. 3.1 Phase portraits ($a_{121} < a_{221}$) for two-dimensional systems on the x_1-direction with $x_1^* = a_{111}$ and on the x_2-direction with $x_2^* = a_{121}, a_{221}$. (**a**) ($a_{110} > 0$, $a_{220} > 0$), (**b**) ($a_{110} < 0$, $a_{220} > 0$), (**c**) ($a_{110} > 0$, $a_{220} < 0$), (**d**) ($a_{110} < 0$, $a_{220} < 0$)

as shown in Fig. 3.1. "HS to HB" and "HB to HS" are for the hyperbolic-secant to hyperbolic asymptotic flow and for the hyperbolic to hyperbolic-secant asymptotic flow, respectively. The double-saddle bifurcations of the upper-saddle and lower-saddle are presented for saddle-sink and saddle-source states and also for the appearing bifurcation of saddles, sink, and source. The hyperbolic-secant to hyperbolic flow bifurcations are the appearing bifurcations of the hyperbolic and hyperbolic-secant flows, which are also presented.

(ii) The equilibriums of $x_1^* = a_{111}$ and $x_2^* = a_{121}, a_{221}$ for $a_{221} < a_{121}$ are

$$\left\{ \begin{array}{c} (a_{111}, a_{121}) \\ (a_{111}, a_{221}) \end{array} \right\} = \left\{ \begin{array}{c} \underbrace{(\text{UP} : \text{UP}, \text{pF})}_{\text{HS to HB flow } (+)} \\ \underbrace{(\text{LS}, \text{US})}_{\text{double-saddle}} \end{array} \right\} \text{ for } a_{110} > 0 \text{ and } a_{220} > 0; \tag{3.7}$$

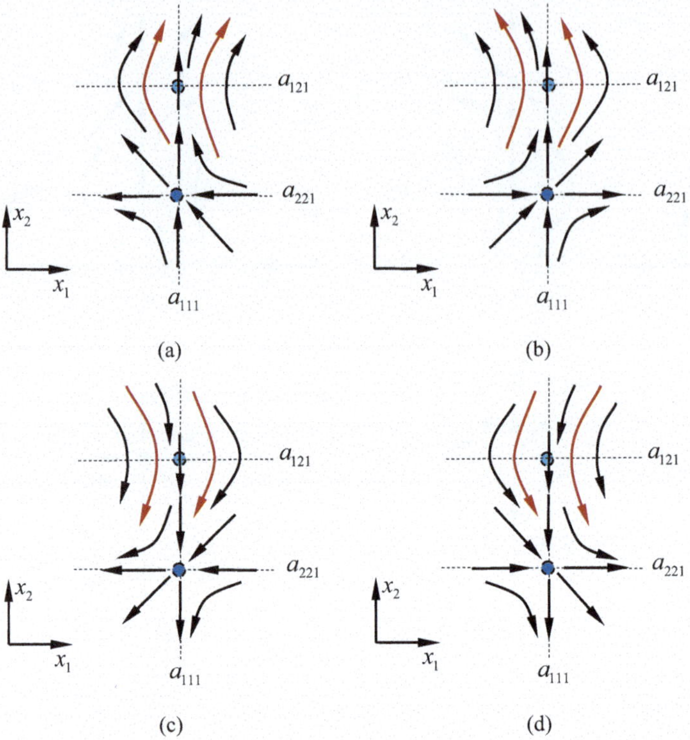

Fig. 3.2 Phase portraits ($a_{221} < a_{121}$) for two-dimensional systems on the x_1-direction with $x_1^* = a_{111}$ and on the x_2-direction with $x_2^* = a_{121}, a_{221}$. **(a)** ($a_{110} > 0$, $a_{220} > 0$), **(b)** ($a_{110} < 0$, $a_{220} > 0$), **(c)** ($a_{110} > 0$, $a_{220} < 0$), **(d)** ($a_{110} < 0$, $a_{220} < 0$)

$$\left\{ \begin{array}{l} (a_{111}, a_{121}) \\ (a_{111}, a_{221}) \end{array} \right\} = \left\{ \begin{array}{c} \underbrace{(\mathrm{DP} : \mathrm{DP}, \mathrm{pF})}_{\text{HB to HS flow }(+)} \\ \underbrace{(\mathrm{US}, \mathrm{US})}_{\text{double-saddle}} \end{array} \right\} \text{ for } a_{110} < 0 \text{ and } a_{220} > 0; \tag{3.8}$$

$$\left\{ \begin{array}{l} (a_{111}, a_{121}) \\ (a_{111}, a_{221}) \end{array} \right\} = \left\{ \begin{array}{c} \underbrace{(\mathrm{DP} : \mathrm{DP}, \mathrm{nF})}_{\text{HB to HS flow }(-)} \\ \underbrace{(\mathrm{LS}, \mathrm{LS})}_{\text{double-saddle}} \end{array} \right\} \text{ for } a_{110} > 0 \text{ and } a_{220} < 0; \tag{3.9}$$

$$\left\{ \begin{array}{l} (a_{111}, a_{121}) \\ (a_{111}, a_{221}) \end{array} \right\} = \left\{ \begin{array}{c} \underbrace{(\mathrm{UP} : \mathrm{UP}, \mathrm{nF})}_{\text{HS to HB flow }(-)} \\ \underbrace{(\mathrm{US}, \mathrm{LS})}_{\text{double-saddle}} \end{array} \right\} \text{ for } a_{110} < 0 \text{ and } a_{220} < 0; \tag{3.10}$$

The double-saddle bifurcations of the upper-saddle and lower-saddle for saddle-sink and saddle-source are presented in Fig. 3.2 for saddle-sink and saddle-source states and for the appearing bifurcation of saddles, sink, and source. The hyperbolic-secant to hyperbolic flow bifurcations are also presented for the appearing bifurcations of the hyperbolic and hyperbolic-secant flows.

3.1.2 Up-Down and Down-Up Upper-Saddles (Lower-Saddles)

In this section, up-down and down-up upper-saddles (lower-saddles) are for the switching bifurcations of the double-saddle with connected hyperbolic-to-hyperbolic-secant flows.

Consider a dynamical system for $a_{121}=a_{221}$ with up-down and down-up upper-saddle and lower-saddle infinite-equilibriums as

$$\begin{aligned}\dot{x}_1 &= a_{110}(x_1-a_{111})^2(x_2-a_{121}),\\ \dot{x}_2 &= a_{220}(x_2-a_{221})^2,\end{aligned} \tag{3.11}$$

and the corresponding first integral manifold ($a_{121}=a_{221}$) is

$$-(\frac{1}{x_1-a_{111}}-\frac{1}{x_{10}-a_{111}})=\frac{a_{110}}{a_{220}}\ln\frac{|\,x_2-a_{221}\,|}{|\,x_{20}-a_{221}\,|}. \tag{3.12}$$

The equilibriums of $x_1^*=a_{111}$ and $x_2^*=a_{121},a_{221}$ for $a_{221}=a_{121}$ are

$$\begin{aligned}(a_{211},a_{221}) &= \underbrace{(\text{DU}:\text{DU},\text{US})}_{\text{down-up upper-saddle}}\text{ for } a_{110}>0 \text{ and } a_{220}>0;\\ (a_{211},a_{221}) &= \underbrace{(\text{UD}:\text{UD},\text{US})}_{\text{up-down upper-saddle}}\text{ for } a_{110}<0 \text{ and } a_{220}>0;\\ (a_{211},a_{221}) &= \underbrace{(\text{UD}:\text{UD},\text{LS})}_{\text{up-down lower-saddle}}\text{ for } a_{110}>0 \text{ and } a_{220}<0;\\ (a_{211},a_{221}) &= \underbrace{(\text{DU}:\text{DU},\text{LS})}_{\text{down-up lower-saddle}}\text{ for } a_{110}<0 \text{ and } a_{220}<0;\end{aligned} \tag{3.13}$$

and the up-down and down-up upper-saddle and lower-saddle infinite-equilibrium of $x_2^*=a_{121}=a_{221}$ are

$$\begin{aligned}(\bar{x}_1,a_{221}) &= \underbrace{(\text{DU},\text{US})}_{\text{down-up upper-saddle}}\quad \text{if } \bar{x}_1\in(-\infty,a_{111}),\\ (\bar{x}_1,a_{221}) &= \underbrace{(\text{DU},\text{US})}_{\text{down-up upper-saddle}}\quad \text{if } \bar{x}_1\in(a_{111},\infty),\\ &\text{for } a_{110}>0 \text{ and } a_{220}>0;\end{aligned} \tag{3.14}$$

$$\begin{aligned}
&(\bar{x}_1, a_{221}) = \underbrace{(\text{UD}, \text{US})}_{\text{up-down upper-saddle}} \quad \text{if } \bar{x}_1 \in (-\infty, a_{111}), \\
&(\bar{x}_1, a_{221}) = \underbrace{(\text{UD}, \text{US})}_{\text{up-down upper-saddle}} \quad \text{if } \bar{x}_1 \in (a_{111}, \infty), \\
&\text{for } a_{110} < 0 \text{ and } a_{220} > 0;
\end{aligned} \tag{3.15}$$

$$\begin{aligned}
&(\bar{x}_1, a_{221}) = \underbrace{(\text{UD}, \text{LS})}_{\text{up-down lower-saddle}} \quad \text{if } \bar{x}_1 \in (-\infty, a_{111}), \\
&(\bar{x}_1, a_{221}) = \underbrace{(\text{UD}, \text{LS})}_{\text{up-down lower-saddle}} \quad \text{if } \bar{x}_1 \in (a_{111}, \infty), \\
&\text{for } a_{110} > 0 \text{ and } a_{220} < 0;
\end{aligned} \tag{3.16}$$

$$\begin{aligned}
&(\bar{x}_1, a_{221}) = \underbrace{(\text{DU}, \text{LS})}_{\text{down-up lower-saddle}} \quad \text{if } \bar{x}_1 \in (-\infty, a_{111}), \\
&(\bar{x}_1, a_{221}) = \underbrace{(\text{DU}, \text{LS})}_{\text{down-up lower-saddle}} \quad \text{if } \bar{x}_1 \in (a_{111}, \infty), \\
&\text{for } a_{110} < 0 \text{ and } a_{220} < 0;
\end{aligned} \tag{3.17}$$

as presented in Fig. 3.3a–d for the switching bifurcations of the double-saddles and hyperbolic-to-hyperbolic-secant flows.

3.2 Self-Singularity on the Product-Cubic Vector Fields

In this section, the two connected saddle-sink and saddle-source with hyperbolic singular flows will be presented, and the inflection-source and sink infinite-equilibriums are the switching bifurcations for the saddle-source (sink) with the hyperbolic singular flows.

3.2.1 Saddle-Nodes with Hyperbolic to Hyperbolic-Secant Flows

In this section, series of connected saddle-source and saddle-sink with a connected hyperbolic-to-hyperbolic-secant flow will be presented.

Consider a dynamical system as

$$\begin{aligned}
\dot{x}_1 &= a_{110}(x_1 - a_{111})^2(x_2 - a_{121}), \\
\dot{x}_2 &= a_{220}(x_2 - a_{221})(x_2 - a_{222}),
\end{aligned} \tag{3.18}$$

and the corresponding first integral manifold is

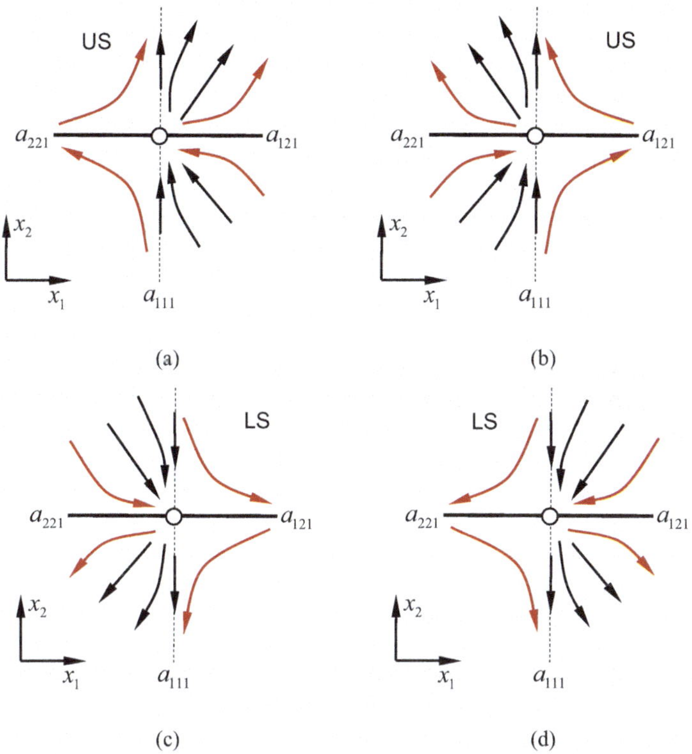

Fig. 3.3 Phase portraits of infinite-equilibrium ($a_{221} = a_{121}$) for two-dimensional systems on the x_1-direction with $x_1^* = a_{111}$ and on the x_2-direction with $x_2^* = a_{121} = a_{221}$. The infinite-equilibriums: (**a**) ($a_{110} > 0$, $a_{220} > 0$), (**b**) ($a_{110} < 0$, $a_{220} > 0$), (**c**) ($a_{110} > 0$, $a_{220} < 0$), (**d**) ($a_{110} < 0$, $a_{220} < 0$)

$$\begin{aligned} -\left(\frac{1}{x_1 - a_{111}} - \frac{1}{x_{10} - a_{111}}\right) = \frac{a_{110}}{a_{220}}\Big(&\frac{a_{221} - a_{121}}{a_{221} - a_{222}} \ln \frac{|\,x_2 - a_{221}\,|}{|\,x_{20} - a_{221}\,|} \\ &+ \frac{a_{222} - a_{121}}{a_{222} - a_{221}} \ln \frac{|\,x_2 - a_{222}\,|}{|\,x_{20} - a_{222}\,|}\Big). \end{aligned} \tag{3.19}$$

(i) The equilibriums of $x_1^* = a_{111}$ and $x_2^* = a_{121}, a_{221}, a_{222}$ for $a_{121} < a_{221}$ are

$$\left\{\begin{array}{c} (a_{111}, a_{222}) \\ (a_{111}, a_{221}) \\ (a_{111}, a_{121}) \end{array}\right\} = \left\{\begin{array}{c} \underbrace{(\text{US}, \text{SO})}_{\text{upper-saddle source}} \\ \underbrace{(\text{US}, \text{SI})}_{\text{upper-saddle sink}} \\ \underbrace{(\text{UP} : \text{UP}, \text{pF})}_{\text{HS to HB flow } (+)} \end{array}\right\} \text{ for } a_{110} > 0 \text{ and } a_{220} > 0; \tag{3.20}$$

$$\left\{\begin{array}{l}(a_{111}, a_{222})\\(a_{111}, a_{221})\\(a_{111}, a_{121})\end{array}\right\} = \left\{\begin{array}{c}\underbrace{(\text{LS}, \text{SO})}_{\text{lower-saddle source}}\\\underbrace{(\text{LS}, \text{SI})}_{\text{lower-saddle sink}}\\\underbrace{(\text{DP}:\text{DP}, \text{pF})}_{\text{HB to HS flow }(+)}\end{array}\right\} \text{ for } a_{110} < 0 \text{ and } a_{220} > 0; \tag{3.21}$$

$$\left\{\begin{array}{l}(a_{111}, a_{222})\\(a_{111}, a_{221})\\(a_{111}, a_{121})\end{array}\right\} = \left\{\begin{array}{c}\underbrace{(\text{US}, \text{SI})}_{\text{upper-saddle sink}}\\\underbrace{(\text{US}, \text{SO})}_{\text{upper-saddle source}}\\\underbrace{(\text{DP}:\text{DP}, \text{nF})}_{\text{HB to HS flow }(-)}\end{array}\right\} \text{ for } a_{110} > 0 \text{ and } a_{220} < 0; \tag{3.22}$$

$$\left\{\begin{array}{l}(a_{111}, a_{222})\\(a_{111}, a_{221})\\(a_{111}, a_{121})\end{array}\right\} = \left\{\begin{array}{c}\underbrace{(\text{LS}, \text{SI})}_{\text{lower-saddle sink}}\\\underbrace{(\text{LS}, \text{SO})}_{\text{lower-saddle source}}\\\underbrace{(\text{UP}:\text{UP}, \text{nF})}_{\text{HS to HB flow }(-)}\end{array}\right\} \text{ for } a_{110} < 0 \text{ and } a_{220} < 0; \tag{3.23}$$

as shown in Fig. 3.4. After a double-saddle appearing bifurcation occurs, the saddle-sink and saddle-source will exist. The upper-saddle-sink and upper-saddle source, lower-saddle-sink, and lower-saddle-source are presented. The saddle-sink and saddle-source bifurcations are the appearing bifurcation of saddle and sink and for saddle and source. The hyperbolic-secant and hyperbolic flow bifurcations for hyperbolic and hyperbolic-secant flows are also presented. The saddle-sink and saddle-source in the series of equilibriums are above the hyperbolic-to-hyperbolic-secant flows.

(ii) The equilibriums of $x_1^* = a_{111}$ and $x_2^* = a_{121}, a_{221}, a_{222}$ for $a_{121} \in (a_{221}, a_{222})$ are

$$\left\{\begin{array}{l}(a_{111}, a_{222})\\(a_{111}, a_{121})\\(a_{111}, a_{221})\end{array}\right\} = \left\{\begin{array}{c}\underbrace{(\text{US}, \text{SO})}_{\text{upper-saddle source}}\\\underbrace{(\text{DP}:\text{DP}, \text{nF})}_{\text{HB to HS flow }(-)}\\\underbrace{(\text{LS}, \text{SI})}_{\text{lower-saddle sink}}\end{array}\right\} \text{ for } a_{110} > 0 \text{ and } a_{220} > 0; \tag{3.24}$$

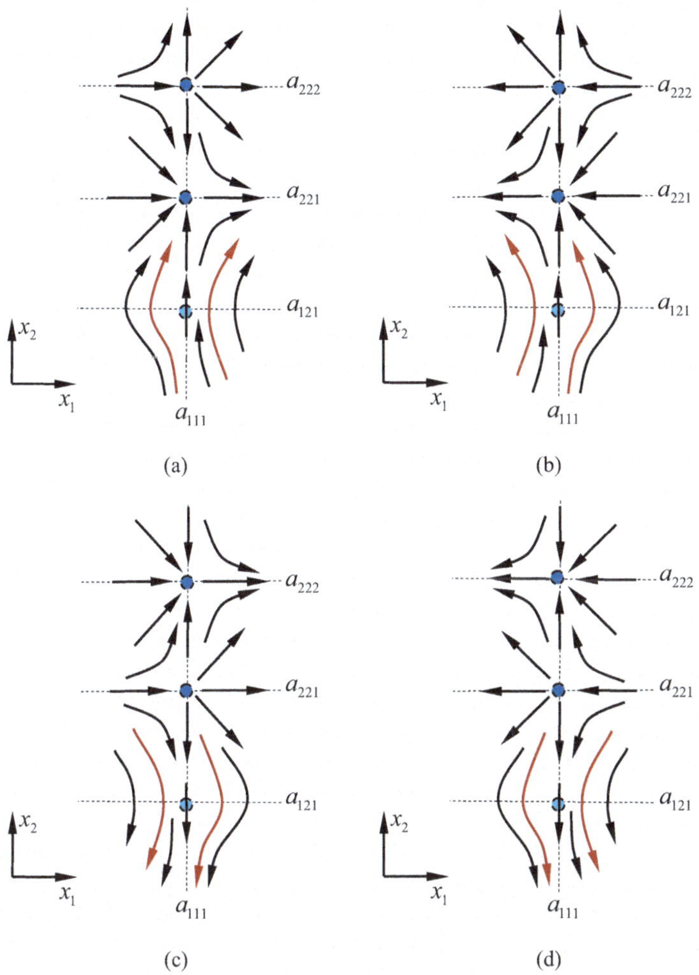

Fig. 3.4 Phase portraits ($a_{121} < a_{221}$) for two-dimensional systems on the x_1-direction with $x_1^* = a_{111}$ and on the x_2-direction with $x_2^* = a_{121}, a_{221}, a_{222}$. The saddle-nodes with hyperbolic to hyperbolic-secant flows: **(a)** ($a_{110} > 0, a_{220} > 0$), **(b)** ($a_{110} < 0, a_{220} > 0$), **(c)** ($a_{110} > 0, a_{220} < 0$), **(d)** ($a_{110} < 0, a_{220} < 0$)

$$\left\{\begin{array}{c}(a_{111}, a_{222}) \\ (a_{111}, a_{121}) \\ (a_{111}, a_{221})\end{array}\right\} = \left\{\begin{array}{c}\underbrace{(\text{LS}, \text{SO})}_{\text{lower-saddle source}} \\ \underbrace{(\text{UP}:\text{UP}, \text{nF})}_{\text{HS to HB flow }(-)} \\ \underbrace{(\text{US}, \text{SI})}_{\text{upper-saddle sink}}\end{array}\right\} \text{ for } a_{110} < 0 \text{ and } a_{220} > 0; \tag{3.25}$$

$$\left\{\begin{array}{c}(a_{111},a_{222})\\(a_{111},a_{121})\\(a_{111},a_{221})\end{array}\right\}=\left\{\begin{array}{c}\underbrace{(\mathrm{US},\mathrm{SI})}_{\text{upper-saddle sink}}\\\underbrace{(\mathrm{UP}:\mathrm{UP},\mathrm{pF})}_{\text{HS to HB flow }(+)}\\\underbrace{(\mathrm{LS},\mathrm{SO})}_{\text{lower-saddle source}}\end{array}\right\}\text{ for }a_{110}>0\text{ and }a_{220}<0;\tag{3.26}$$

$$\left\{\begin{array}{c}(a_{111},a_{222})\\(a_{111},a_{221})\\(a_{111},a_{121})\end{array}\right\}=\left\{\begin{array}{c}\underbrace{(\mathrm{LS},\mathrm{SI})}_{\text{lower-saddle sink}}\\\underbrace{(\mathrm{DP}:\mathrm{DP},\mathrm{pF})}_{\text{HB to HS flow }(+)}\\\underbrace{(\mathrm{US},\mathrm{SO})}_{\text{upper-saddle source}}\end{array}\right\}\text{ for }a_{110}<0\text{ and }a_{220}<0;\tag{3.27}$$

as shown in Fig. 3.5. The saddle-sink and saddle-source are presented, which are the appearing bifurcations for saddle and sink and for saddle. The hyperbolic-secant and hyperbolic flow bifurcations for hyperbolic and hyperbolic-secant flows are also presented. The hyperbolic-to-hyperbolic-secant flows lies between the saddle-sink and saddle-source in the series of equilibriums.

(iii) The equilibriums of $x_1^*=a_{111}$ and $x_2^*=a_{121},a_{221},a_{222}$ for $a_{222}<a_{121}$ are

$$\left\{\begin{array}{c}(a_{111},a_{222})\\(a_{111},a_{221})\\(a_{111},a_{121})\end{array}\right\}=\left\{\begin{array}{c}\underbrace{(\mathrm{UP}:\mathrm{UP},\mathrm{pF})}_{\text{HS to HB flow }(+)}\\\underbrace{(\mathrm{LS},\mathrm{SO})}_{\text{lower-saddle source}}\\\underbrace{(\mathrm{LS},\mathrm{SI})}_{\text{lower-saddle sink}}\end{array}\right\}\text{ for }a_{110}>0\text{ and }a_{220}>0;\tag{3.28}$$

$$\left\{\begin{array}{c}(a_{111},a_{121})\\(a_{111},a_{222})\\(a_{111},a_{221})\end{array}\right\}=\left\{\begin{array}{c}\underbrace{(\mathrm{DP}:\mathrm{DP},\mathrm{pF})}_{\text{HB to HS flow }(+)}\\\underbrace{(\mathrm{US},\mathrm{SO})}_{\text{upper-saddle source}}\\\underbrace{(\mathrm{US},\mathrm{SI})}_{\text{upper-saddle sink}}\end{array}\right\}\text{ for }a_{110}<0\text{ and }a_{220}>0;\tag{3.29}$$

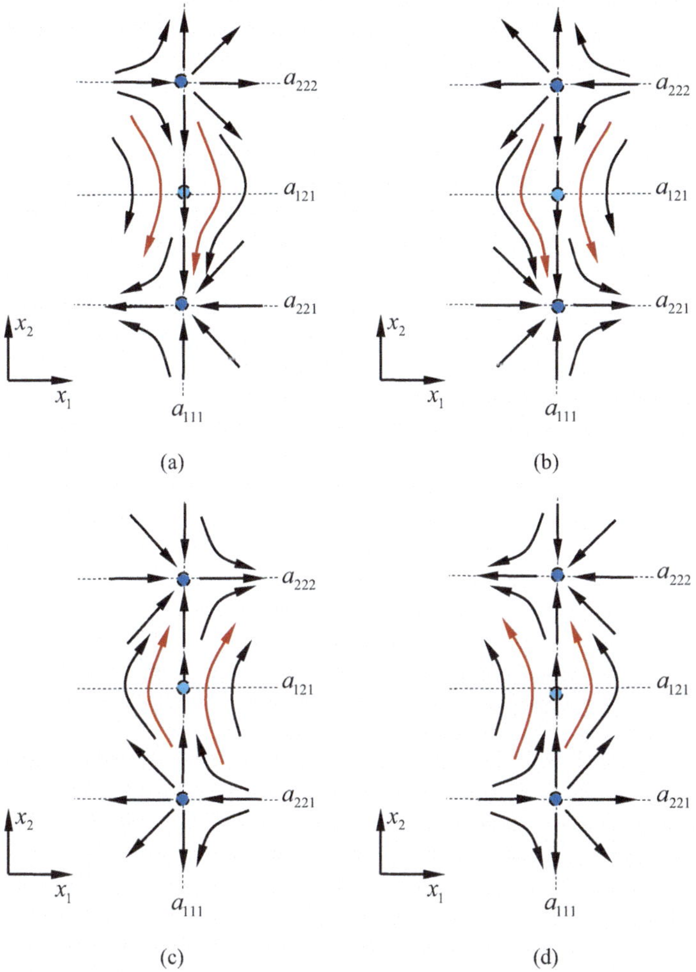

Fig. 3.5 Phase portraits ($a_{121} \in (a_{221}, a_{222})$) for two-dimensional systems on the x_1-direction with $x_1^* = a_{111}$ and on the x_2-direction with $x_2^* = a_{121}, a_{221}, a_{222}$. The saddle-nodes with hyperbolic to hyperbolic-secant flows: **(a)** ($a_{110} > 0$, $a_{220} > 0$), **(b)** ($a_{110} < 0$, $a_{220} > 0$), **(c)** ($a_{110} > 0$, $a_{220} < 0$), **(d)** ($a_{110} < 0$, $a_{220} < 0$)

$$\begin{Bmatrix} (a_{111}, a_{121}) \\ (a_{111}, a_{222}) \\ (a_{111}, a_{221}) \end{Bmatrix} = \begin{Bmatrix} \underbrace{(\mathrm{DP}:\mathrm{DP}, \mathrm{nF})}_{\text{HB to HS flow }(-)} \\ \underbrace{(\mathrm{LS}, \mathrm{SI})}_{\text{lower-saddle sink}} \\ \underbrace{(\mathrm{LS}, \mathrm{SI})}_{\text{lower-saddle source}} \end{Bmatrix} \text{ for } a_{110} > 0 \text{ and } a_{220} < 0; \tag{3.30}$$

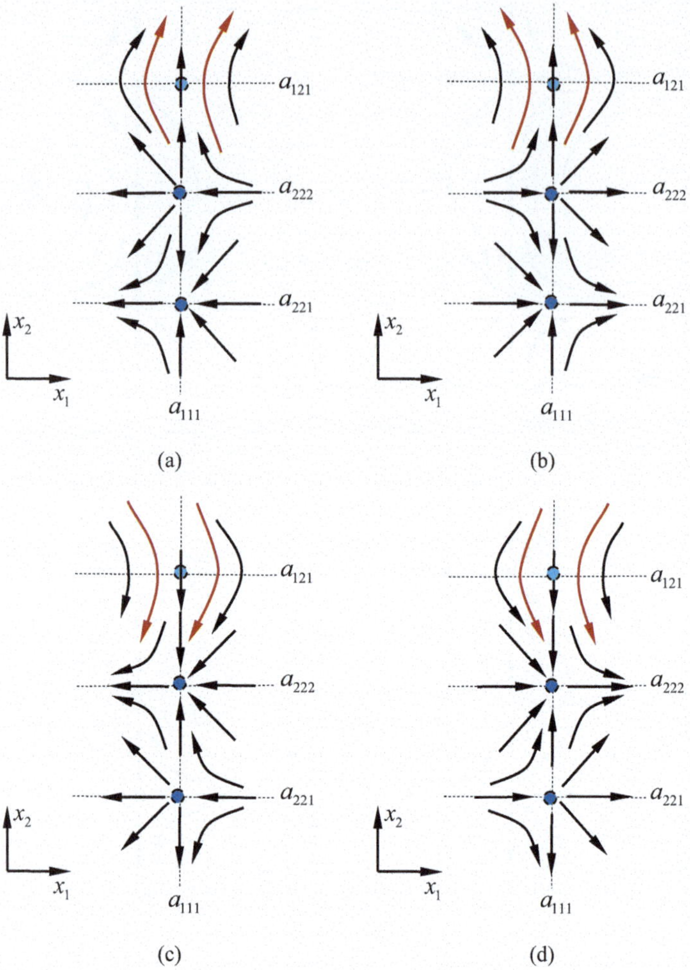

Fig. 3.6 Phase portraits ($a_{222} < a_{121}$) for two-dimensional systems on the x_1-direction with $x_1^* = a_{111}$ and on the x_2-direction with $x_2^* = a_{121}, a_{221}, a_{222}$. The saddle-nodes with hyperbolic to hyperbolic-secant flows: **(a)** ($a_{110} > 0, a_{220} > 0$), **(b)** ($a_{110} < 0, a_{220} > 0$), **(c)** ($a_{110} > 0, a_{220} < 0$), **(d)** ($a_{110} < 0, a_{220} < 0$)

$$\left\{\begin{array}{c}(a_{111}, a_{121})\\(a_{111}, a_{222})\\(a_{111}, a_{221})\end{array}\right\} = \left\{\begin{array}{c}\underbrace{(\text{UP}:\text{UP},\text{nF})}_{\text{HS to HB flow }(-)}\\\underbrace{(\text{US},\text{SI})}_{\text{upper-saddle sink}}\\\underbrace{(\text{US},\text{SO})}_{\text{upper-saddle source}}\end{array}\right\} \text{ for } a_{110} < 0 \text{ and } a_{220} < 0; \tag{3.31}$$

as shown in Fig. 3.6. The saddle-sink and saddle-source bifurcations for saddle and sink and for saddle and source are presented. The hyperbolic-secant and hyperbolic flow bifurcations for hyperbolic and hyperbolic-secant flows are also sketched. The saddle-sink and saddle-source in the series of equilibriums are below the hyperbolic-to-hyperbolic-secant flows.

3.2.2 *Inflection-Source (Sink) Infinite-Equilibriums*

In this section, the inflection-source (sink) is for the switching bifurcation of a connected hyperbolic-to-hyperbolic-secant flows with a connected saddle-source (sink).

For the above two cases, the corresponding switching bifurcations will be discussed through the inflection source and sink infinite-equilibriums.

(i) Consider a dynamical system for $a_{221} = a_{121}$ as

$$\begin{aligned} \dot{x}_1 &= a_{110}(x_1 - a_{111})^2(x_2 - a_{121}), \\ \dot{x}_2 &= a_{220}(x_2 - a_{221})(x_2 - a_{222}), \end{aligned} \tag{3.32}$$

and the corresponding first integral manifold for $a_{221} = a_{121}$ is

$$-(\frac{1}{x_1 - a_{111}} - \frac{1}{x_{10} - a_{111}}) = \frac{a_{110}}{a_{220}} \ln \frac{|\, x_2 - a_{222} \,|}{|\, x_{20} - a_{222} \,|}. \tag{3.33}$$

The infinite-equilibriums of $x_2^* = a_{221} = a_{121}$ are

$$\begin{Bmatrix} (a_{111}, a_{222}) \\ (a_{111}, a_{221}) \end{Bmatrix} = \begin{Bmatrix} \underbrace{(\text{US, SO})}_{\text{upper-saddle source}} \\ \underbrace{(\text{DI : DI, SI})}_{\text{decreasing-inflection sink}} \end{Bmatrix} \text{ for } a_{110} > 0, a_{220} > 0, \tag{3.34}$$

$$\begin{Bmatrix} (a_{111}, a_{222}) \\ (a_{111}, a_{221}) \end{Bmatrix} = \begin{Bmatrix} \underbrace{(\text{LS, SO})}_{\text{lower-saddle source}} \\ \underbrace{(\text{II : II, SI})}_{\text{increasing-inflection sink}} \end{Bmatrix} \text{ for } a_{110} < 0, a_{220} > 0, \tag{3.35}$$

$$\begin{Bmatrix} (a_{111}, a_{222}) \\ (a_{111}, a_{221}) \end{Bmatrix} = \begin{Bmatrix} \underbrace{(\text{US, SI})}_{\text{upper-saddle sink}} \\ \underbrace{(\text{II : II, SO})}_{\text{increasing-inflection source}} \end{Bmatrix} \text{ for } a_{110} > 0, a_{220} < 0, \tag{3.36}$$

$$\left\{\begin{array}{l}(a_{111},a_{222})\\(a_{111},a_{221})\end{array}\right\}=\left\{\begin{array}{c}\underbrace{(\text{LS, SI})}_{\text{lower-saddle sink}}\\\underbrace{(\text{DI : DI, SO})}_{\text{decreasing-inflection source}}\end{array}\right\}\text{ for } a_{110}<0, a_{220}<0. \tag{3.37}$$

and the infinite-equilibrium of $x_2^*=a_{121}=a_{221}$ is summarized as follows.

$$\begin{aligned}&(\bar{x}_1,a_{221})=\underbrace{(\text{DI, SI})}_{\text{decreasing-inflection sink}} \quad \text{if } \bar{x}_1\in(-\infty,a_{111}),\\&(\bar{x}_1,a_{221})=\underbrace{(\text{DI, SI})}_{\text{decreasing-inflection sink}} \quad \text{if } \bar{x}_1\in(a_{111},\infty),\\&\text{for } a_{110}>0, a_{220}>0;\end{aligned} \tag{3.38}$$

$$\begin{aligned}&(\bar{x}_1,a_{221})=\underbrace{(\text{II, SI})}_{\text{decreasing-inflection sink}} \quad \text{if } \bar{x}_1\in(-\infty,a_{111}),\\&(\bar{x}_1,a_{221})=\underbrace{(\text{II, SI})}_{\text{decreasing-inflection sink}} \quad \text{if } \bar{x}_1\in(a_{111},\infty),\\&\text{for } a_{110}<0, a_{220}>0;\end{aligned} \tag{3.39}$$

$$\begin{aligned}&(\bar{x}_1,a_{221})=\underbrace{(\text{II, SO})}_{\text{increasing-inflection source}} \quad \text{if } \bar{x}_1\in(-\infty,a_{111}),\\&(\bar{x}_1,a_{221})=\underbrace{(\text{II, SO})}_{\text{increasing-inflection source}} \quad \text{if } \bar{x}_1\in(a_{111},\infty),\\&\text{for } a_{110}>0, a_{220}<0;\end{aligned} \tag{3.40}$$

$$\begin{aligned}&(\bar{x}_1,a_{221})=\underbrace{(\text{DI, SO})}_{\text{decreasing-inflection source}} \quad \text{if } \bar{x}_1\in(-\infty,a_{111}),\\&(\bar{x}_1,a_{221})=\underbrace{(\text{DI, SO})}_{\text{decreasing-inflection source}} \quad \text{if } \bar{x}_1\in(a_{111},\infty),\\&\text{for } a_{110}<0, a_{220}<0.\end{aligned} \tag{3.41}$$

The inflection-sink and source infinite-equilibriums of $x_2^*=a_{121}=a_{221}$ are the switching bifurcation of saddle-sink and saddle-source with hyperbolic-to-hyperbolic-secant flows for $a_{121}<a_{221}$ and $a_{221}<a_{121}<a_{222}$, as shown in Fig. 3.7. The inflection infinite-equilibrium is below the saddle-sink or saddle-source.

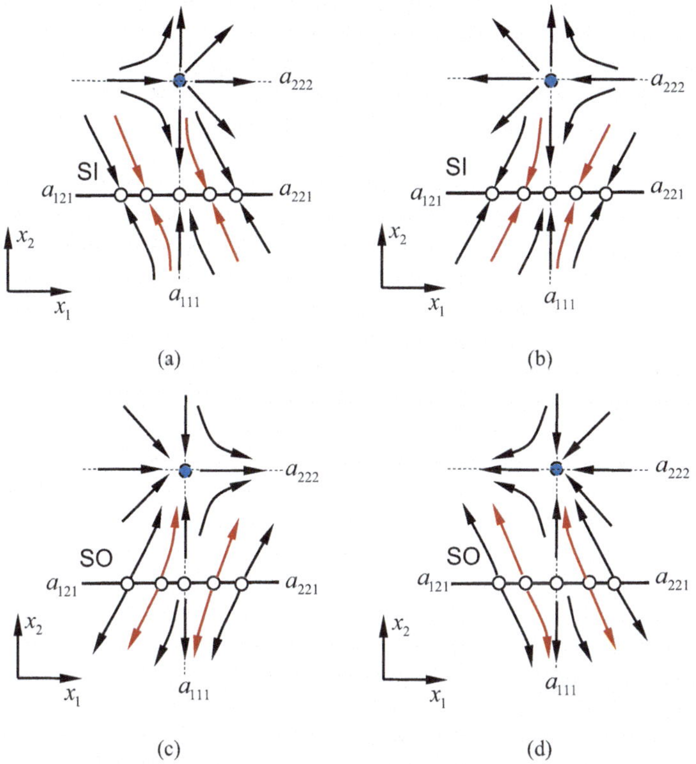

Fig. 3.7 Phase portraits ($a_{121} = a_{221}$) for two-dimensional systems on the x_1-direction with $x_1^* = a_{111}$ and on the x_2-direction with $x_2^* = a_{121}, a_{221}, a_{222}$. The inflection sink and source infinite-equilibriums with saddle-source and saddle-sink: **(a)** ($a_{110} > 0$, $a_{220} > 0$), **(b)** ($a_{110} < 0$, $a_{220} > 0$), **(c)** ($a_{110} > 0$, $a_{220} < 0$), **(d)** ($a_{110} < 0$, $a_{220} < 0$)

(ii) Consider a dynamical system for $a_{222} = a_{121}$ as

$$\begin{aligned}\dot{x}_1 &= a_{110}(x_1 - a_{111})^2(x_2 - a_{121}),\\ \dot{x}_2 &= a_{220}(x_2 - a_{221})(x_2 - a_{222}),\end{aligned} \tag{3.42}$$

and the corresponding first integral manifold for $a_{221} = a_{121}$ is

$$-\left(\frac{1}{x_1 - a_{111}} - \frac{1}{x_{10} - a_{111}}\right) = \frac{a_{110}}{a_{220}} \ln \frac{|\, x_2 - a_{221} \,|}{|\, x_{20} - a_{221} \,|}. \tag{3.43}$$

The infinite-equilibriums of $x_2^* = a_{222} = a_{121}$ are

$$\left\{\begin{array}{l}(a_{111}, a_{222})\\(a_{111}, a_{221})\end{array}\right\} = \left\{\begin{array}{c}\underbrace{(\text{II}:\text{II}, \text{SO})}_{\text{incerasing-inflection source}}\\ \underbrace{(\text{LS}, \text{SI})}_{\text{lower-saddle sink}}\end{array}\right\} \text{ for } a_{110} > 0, a_{220} > 0, \tag{3.44}$$

$$\left\{\begin{array}{l}(a_{111}, a_{222})\\(a_{111}, a_{221})\end{array}\right\} = \left\{\begin{array}{c}\underbrace{(\text{DI}:\text{DI}, \text{SO})}_{\text{decerasing-inflection source}}\\ \underbrace{(\text{US}, \text{SI})}_{\text{upper-saddle sink}}\end{array}\right\} \text{ for } a_{110} < 0, a_{220} > 0, \tag{3.45}$$

$$\left\{\begin{array}{l}(a_{111}, a_{222})\\(a_{111}, a_{221})\end{array}\right\} = \left\{\begin{array}{c}\underbrace{(\text{DI}:\text{DI}, \text{SI})}_{\text{decerasing-inflection sink}}\\ \underbrace{(\text{LS}, \text{SO})}_{\text{lower-saddle source}}\end{array}\right\} \text{ for } a_{110} > 0, a_{220} < 0, \tag{3.46}$$

$$\left\{\begin{array}{l}(a_{111}, a_{222})\\(a_{111}, a_{221})\end{array}\right\} = \left\{\begin{array}{c}\underbrace{(\text{II}:\text{II}, \text{SI})}_{\text{incerasing-inflection sink}}\\ \underbrace{(\text{US}, \text{SO})}_{\text{upper-saddle source}}\end{array}\right\} \text{ for } a_{110} < 0, a_{220} < 0, \tag{3.47}$$

and the infinite-equilibrium of $x_2^* = a_{121} = a_{222}$ is summarized as follows.

$$\begin{aligned}
&(\bar{x}_1, a_{222}) = \underbrace{(\text{II}, \text{SO})}_{\text{increasing-inflection source}} \quad \text{if } \bar{x}_1 \in (-\infty, a_{111}),\\
&(\bar{x}_1, a_{222}) = \underbrace{(\text{II}, \text{SO})}_{\text{increasing-inflection source}} \quad \text{if } \bar{x}_1 \in (a_{111}, \infty),\\
&\text{for } a_{110} > 0, a_{220} > 0;
\end{aligned} \tag{3.48}$$

$$\begin{aligned}
&(\bar{x}_1, a_{222}) = \underbrace{(\text{DI}, \text{SO})}_{\text{decreasing-inflection source}} \quad \text{if } \bar{x}_1 \in (-\infty, a_{111}),\\
&(\bar{x}_1, a_{222}) = \underbrace{(\text{DI}, \text{SO})}_{\text{decreasing-inflection source}} \quad \text{if } \bar{x}_1 \in (a_{111}, \infty),\\
&\text{for } a_{110} < 0, a_{220} > 0;
\end{aligned} \tag{3.49}$$

$$\begin{aligned}
&(\bar{x}_1, a_{222}) = \underbrace{(\text{DI}, \text{SI})}_{\text{decreasing-inflection sink}} \quad \text{if } \bar{x}_1 \in (-\infty, a_{111}),\\
&(\bar{x}_1, a_{222}) = \underbrace{(\text{DI}, \text{SI})}_{\text{decreasing-inflection sink}} \quad \text{if } \bar{x}_1 \in (a_{111}, \infty),\\
&\text{for } a_{110} > 0, a_{220} < 0;
\end{aligned} \tag{3.50}$$

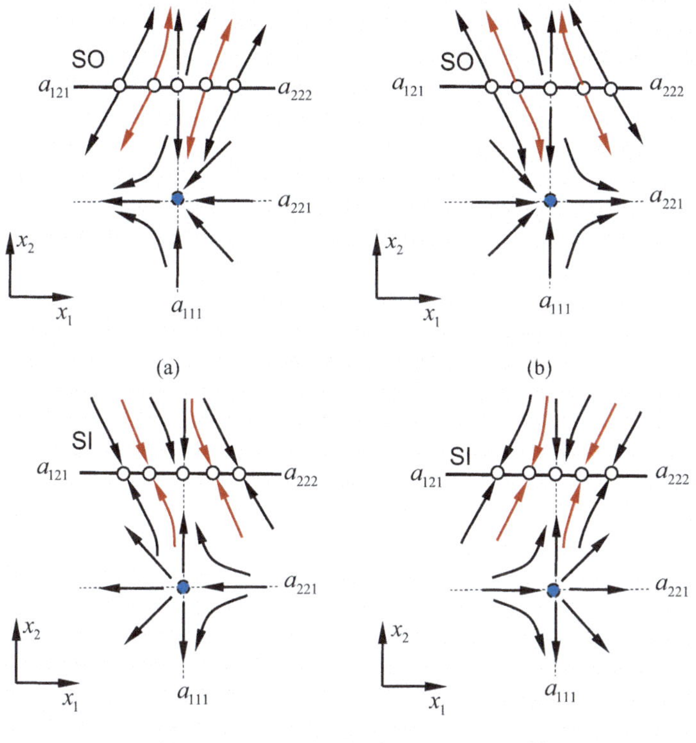

Fig. 3.8 Phase portraits ($a_{121} = a_{222}$) for two-dimensional systems on the x_1-direction with $x_1^* = a_{111}$ and on the x_2-direction with $x_2^* = a_{121}, a_{221}, a_{222}$. The inflection sink and source infinite-equilibriums with saddle-source and saddle-sink: **(a)** ($a_{110} > 0$, $a_{220} > 0$), **(b)** ($a_{110} < 0$, $a_{220} > 0$), **(c)** ($a_{110} > 0$, $a_{220} < 0$), **(d)** ($a_{110} < 0$, $a_{220} < 0$)

$$\begin{aligned}
(\bar{x}_1, a_{222}) &= \underbrace{(\text{II, SI})}_{\text{increasing-inflection sink}} \quad \text{if } \bar{x}_1 \in (-\infty, a_{111}),\\
(\bar{x}_1, a_{222}) &= \underbrace{(\text{II, SI})}_{\text{increasing-inflection sink}} \quad \text{if } \bar{x}_1 \in (a_{111}, \infty),\\
&\text{for } a_{110} < 0, a_{220} < 0.
\end{aligned} \tag{3.51}$$

The inflection-source and sink infinite-equilibriums of $x_2^* = a_{121} = a_{222}$ are for the switching bifurcation for of saddle-sink and saddle-source with hyperbolic-to-hyperbolic-secant flows for $a_{121} \in (a_{221}, a_{222})$ and $a_{121} > a_{222}$, as presented in Fig. 3.8. The inflection infinite-equilibrium is above the saddle-sink or saddle-source in the series of equilibriums.

3.3 Single Self-Singularity in the Self-Quadratic Vector Field

The paralleled saddle-sink and saddle-source with the paralleled hyperbolic flows will be discussed. The corresponding networks of paralleled saddle-sources (sinks) with the paralleled hyperbolic and hyperbolic-secant flows are presented, and the switching bifurcations of the corresponding networks are given through the up-down and down-up saddle infinite-equilibriums.

3.3.1 Saddle-Nodes with Paralleled Hyperbolic Flows

In this section, networks of paralleled hyperbolic and hyperbolic-secant flows with saddle-source and saddle-sinks will be presented.

Consider a dynamical system as

$$\begin{aligned}\dot{x}_1 &= a_{110}(x_1 - a_{111})(x_1 - a_{112})(x_2 - a_{121}),\\ \dot{x}_2 &= a_{220}(x_2 - a_{221})^2,\end{aligned} \tag{3.52}$$

and the corresponding first integral manifold is

$$\begin{aligned}&\frac{1}{a_{112} - a_{111}}\left(\ln \frac{|x_1 - a_{112}|}{|x_{10} - a_{112}|} - \ln \frac{|x_1 - a_{111}|}{|x_{10} - a_{111}|}\right)\\ &= \frac{a_{110}}{a_{220}}\left[\ln \frac{|x_2 - a_{221}|}{|x_{20} - a_{221}|} - (a_{221} - a_{121})\left(\frac{1}{x_2 - a_{221}} - \frac{1}{x_{20} - a_{221}}\right)\right].\end{aligned} \tag{3.53}$$

(i) For the case of $a_{121} < a_{221}$, the saddle-nodes and hyperbolic flows of $x_1^* = a_{111}, a_{112}$ with $x_2^* = a_{121}, a_{221}$ have the following properties, as shown in Fig. 3.9.

$$\begin{Bmatrix} (a_{111}, a_{221}) & (a_{112}, a_{221}) \\ (a_{111}, a_{121}) & (a_{112}, a_{121}) \end{Bmatrix} = \begin{Bmatrix} \underbrace{(\text{SI}, \text{US})}_{\text{upper-saddle sink}} & \underbrace{(\text{SO}, \text{US})}_{\text{upper-saddle source}} \\ \underbrace{(\text{UP}:\text{DP}, \text{pF})}_{\text{hyperbolic-secant flow }(+)} & \underbrace{(\text{DP}:\text{UP}, \text{pF})}_{\text{hyperbolic flow }(+)} \end{Bmatrix} \tag{3.54}$$

for $a_{110} > 0$ and $a_{220} > 0$;

$$\left\{\begin{array}{cc}(a_{111},a_{221}) & (a_{112},a_{221})\\(a_{111},a_{121}) & (a_{112},a_{121})\end{array}\right\}=\left\{\begin{array}{cc}\underbrace{(\text{SO},\text{US})}_{\text{upper-saddle source}} & \underbrace{(\text{SI},\text{US})}_{\text{upper-saddle sink}}\\ \underbrace{(\text{DP}:\text{UP},\text{pF})}_{\text{hyperbolic flow }(+)} & \underbrace{(\text{UP}:\text{DP},\text{pF})}_{\text{hyperbolic-secant flow }(+)}\end{array}\right\} \tag{3.55}$$

for $a_{110} < 0$ and $a_{220} > 0$;

$$\left\{\begin{array}{cc}(a_{111},a_{221}) & (a_{112},a_{221})\\(a_{111},a_{121}) & (a_{112},a_{121})\end{array}\right\}=\left\{\begin{array}{cc}\underbrace{(\text{SI},\text{LS})}_{\text{lower-saddle sink}} & \underbrace{(\text{SO},\text{LS})}_{\text{lower-saddle source}}\\ \underbrace{(\text{DP}:\text{UP},\text{nF})}_{\text{hyperbolic flow }(-)} & \underbrace{(\text{UP}:\text{DP},\text{nF})}_{\text{hyperbolic-secant flow }(-)}\end{array}\right\} \tag{3.56}$$

for $a_{110} > 0$ and $a_{220} < 0$;

$$\left\{\begin{array}{cc}(a_{111},a_{221}) & (a_{112},a_{221})\\(a_{111},a_{121}) & (a_{112},a_{121})\end{array}\right\}=\left\{\begin{array}{cc}\underbrace{(\text{SO},\text{LS})}_{\text{lower-saddle source}} & \underbrace{(\text{SI},\text{LS})}_{\text{lower-saddle sink}}\\ \underbrace{(\text{UP}:\text{DP},\text{nF})}_{\text{hyperbolic-secant flow }(-)} & \underbrace{(\text{DP}:\text{UP},\text{nF})}_{\text{hyperbolic flow }(-)}\end{array}\right\} \tag{3.57}$$

for $a_{110} < 0$ and $a_{220} < 0$.

The saddle-sink and saddle-source with hyperbolic and hyperbolic-secant flows are presented in Fig. 3.9. The saddle-sink and saddle-source are the appearing bifurcation of saddle and sink and the appearing bifurcation of saddle and source, respectively. The saddle-sink and saddle-source are above the hyperbolic and hyperbolic-secant flows.

(ii) For $a_{221} < a_{121}$, the equilibriums of $x_1^* = a_{111}, a_{112}$ with $x_2^* = a_{121}, a_{221}$ have the following properties.

$$\left\{\begin{array}{cc}(a_{111},a_{121}) & (a_{112},a_{121})\\(a_{111},a_{221}) & (a_{112},a_{221})\end{array}\right\}=\left\{\begin{array}{cc}\underbrace{(\text{UP}:\text{DP},\text{pF})}_{\text{hyperbolic-secant flow }(+)} & \underbrace{(\text{DP}:\text{UP},\text{pF})}_{\text{hyperbolic flow }(+)}\\ \underbrace{(\text{SO},\text{US})}_{\text{upper-saddle source}} & \underbrace{(\text{SI},\text{US})}_{\text{upper-saddle sink}}\end{array}\right\} \tag{3.58}$$

for $a_{110} > 0$ and $a_{220} > 0$;

$$\left\{\begin{array}{cc}(a_{111},a_{121}) & (a_{112},a_{121})\\(a_{111},a_{221}) & (a_{112},a_{221})\end{array}\right\}=\left\{\begin{array}{cc}\underbrace{(\text{DP}:\text{UP},\text{pF})}_{\text{hyperbolic flow }(+)} & \underbrace{(\text{UP}:\text{DP},\text{pF})}_{\text{hyperbolic-secant flow }(+)}\\ \underbrace{(\text{SI},\text{US})}_{\text{upper-saddle sink}} & \underbrace{(\text{SO},\text{US})}_{\text{upper-saddle source}}\end{array}\right\} \tag{3.59}$$

for $a_{110} < 0$ and $a_{220} > 0$;

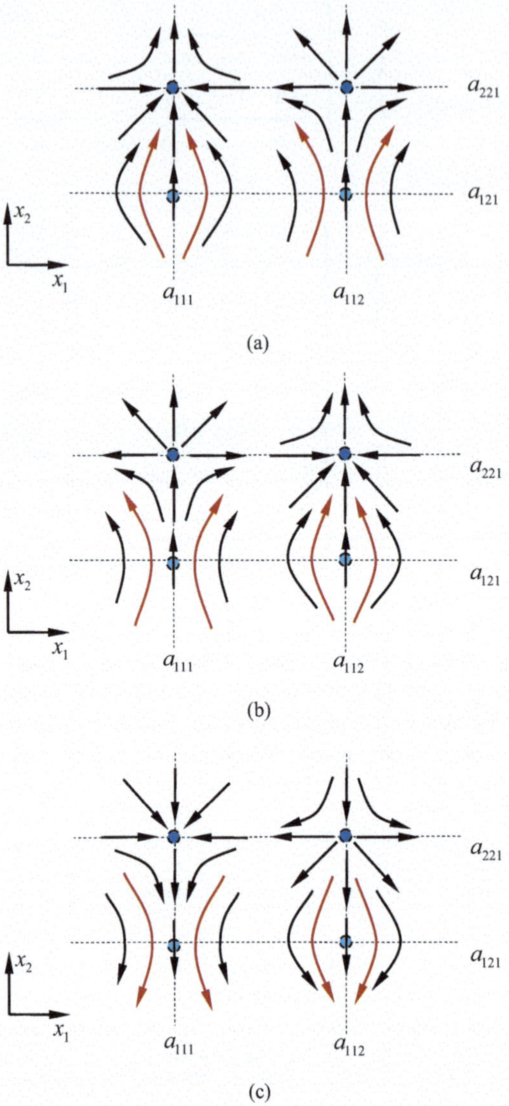

Fig. 3.9 Phase portraits ($a_{121} < a_{221}$) for two-dimensional systems on the x_1-direction with $x_1^* = a_{111}$ and on the x_2-direction with $x_2^* = a_{121}, a_{221}, a_{222}$. The saddle-nodes and hyperbolic flows: (**a**) ($a_{110} > 0$, $a_{220} > 0$), (**b**) ($a_{110} < 0$, $a_{220} > 0$), (**c**) ($a_{110} > 0$, $a_{220} < 0$), (**d**) ($a_{110} < 0$, $a_{220} < 0$)

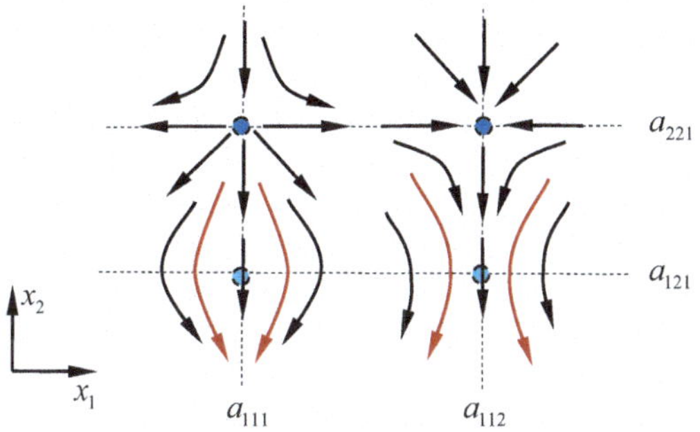

(d)

Fig. 3.9 (continued)

$$\left\{\begin{matrix}(a_{111},a_{121}) & (a_{112},a_{121})\\ (a_{111},a_{221}) & (a_{112},a_{221})\end{matrix}\right\} = \left\{\begin{matrix}\underbrace{(\mathrm{DP}:\mathrm{UP},\mathrm{nF})}_{\text{hyperbolic flow }(-)} & \underbrace{(\mathrm{UP}:\mathrm{DP},\mathrm{nF})}_{\text{hyperbolic-secant flow }(-)}\\ \underbrace{(\mathrm{SO},\mathrm{LS})}_{\text{lower-saddle source}} & \underbrace{(\mathrm{SI},\mathrm{LS})}_{\text{lower-saddle sink}}\end{matrix}\right\} \quad (3.60)$$

for $a_{110} > 0$ and $a_{220} < 0$;

$$\left\{\begin{matrix}(a_{111},a_{121}) & (a_{112},a_{121})\\ (a_{111},a_{221}) & (a_{112},a_{221})\end{matrix}\right\} = \left\{\begin{matrix}\underbrace{(\mathrm{UP}:\mathrm{DP},\mathrm{nF})}_{\text{hyperbolic-secant flow }(-)} & \underbrace{(\mathrm{DP}:\mathrm{UP},\mathrm{nF})}_{\text{hyperbolic flow }(-)}\\ \underbrace{(\mathrm{SI},\mathrm{LS})}_{\text{lower-saddle sink}} & \underbrace{(\mathrm{SO},\mathrm{LS})}_{\text{lower-saddle source}}\end{matrix}\right\} \quad (3.61)$$

for $a_{110} < 0$ and $a_{220} < 0$.

The paralleled saddle-sink and saddle-source with paralleled hyperbolic and hyperbolic-secant flows are presented in Fig. 3.10. The saddle-sink and saddle-source are the appearing bifurcation of saddle and sink and the appearing bifurcation of saddle and source, respectively. The saddle-sink and saddle-source are below the hyperbolic and hyperbolic-secant flows.

3.3.2 *Up-Down-to-Down-Up Upper-Saddles (Lower-Saddles)*

In this section, the up-down and down-up upper-saddle (lower-saddle) will be presented for the switching bifurcations of paralleled hyperbolic and hyperbolic-secant flows with corresponding connected saddle-source and saddle-sink.

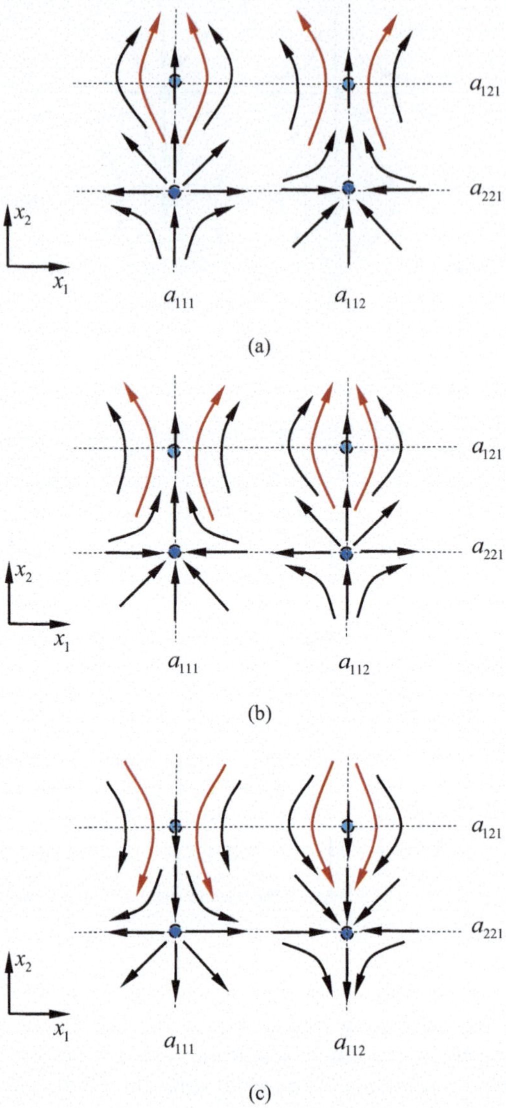

Fig. 3.10 Phase portraits ($a_{121} < a_{221}$) for two-dimensional systems on the x_1-direction with $x_1^* = a_{111}$ and on the x_2-direction with $x_2^* = a_{121}, a_{221}, a_{222}$. The equilibriums flows: (**a**) ($a_{110} > 0$, $a_{220} > 0$), (**b**) ($a_{110} < 0$, $a_{220} > 0$), (**c**) ($a_{110} > 0$, $a_{220} < 0$), (**d**) ($a_{110} < 0$, $a_{220} < 0$)

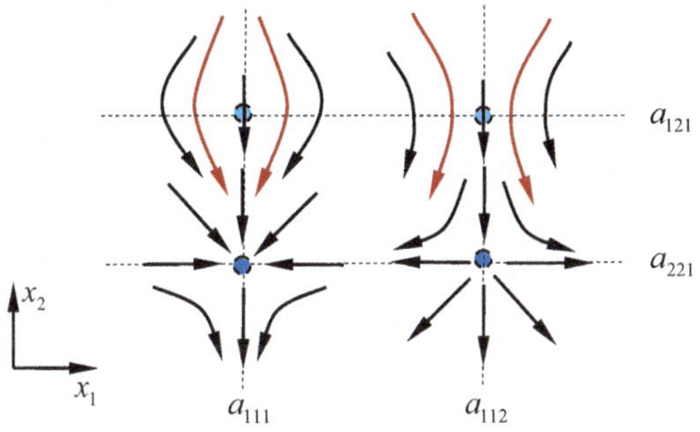

(d)

Fig. 3.10 (continued)

Consider a dynamical system with up-down and down-up upper-saddle and lower-saddle infinite-equilibrium as

$$\begin{aligned}\dot{x}_1 &= a_{110}(x_1 - a_{111})(x_1 - a_{112})(x_2 - a_{121}),\\ \dot{x}_2 &= a_{220}(x_2 - a_{221})^2.\end{aligned} \tag{3.62}$$

For $a_{121} = a_{221}$, the corresponding first integral manifold is

$$\frac{1}{a_{112} - a_{111}}\left(\ln\frac{|x_1 - a_{112}|}{|x_{10} - a_{112}|} - \ln\frac{|x_1 - a_{111}|}{|x_{10} - a_{111}|}\right) = \frac{a_{110}}{a_{220}}\ln\frac{|x_2 - a_{221}|}{|x_{20} - a_{221}|}. \tag{3.63}$$

The infinite-equilibriums of $x_2^* = a_{121} = a_{221}$ with $x_1^* = a_{111}, a_{112}$ have the following properties, as shown in Fig. 3.11.

$$\left\{\begin{matrix}(a_{112}, a_{221})\\(a_{111}, a_{221})\end{matrix}\right\} = \left\{\begin{matrix}\underbrace{(\text{UD}:\text{DU},\text{US})}_{\text{up-down-to-down-up upper-saddle}}\\ \underbrace{(\text{DU}:\text{UD},\text{US})}_{\text{down-up-to-up-down upper-saddle}}\end{matrix}\right\} \text{ for } a_{110} > 0, a_{220} > 0, \tag{3.64}$$

$$\left\{\begin{matrix}(a_{112}, a_{221})\\(a_{111}, a_{221})\end{matrix}\right\} = \left\{\begin{matrix}\underbrace{(\text{DU}:\text{UD},\text{US})}_{\text{down-up-to-up-down upper-saddle}}\\ \underbrace{(\text{UD}:\text{DU},\text{US})}_{\text{up-down-to-down-up upper-saddle}}\end{matrix}\right\} \text{ for } a_{110} < 0, a_{220} > 0, \tag{3.65}$$

$$\left\{\begin{matrix}(a_{112}, a_{221})\\(a_{111}, a_{221})\end{matrix}\right\} = \left\{\begin{matrix}\underbrace{(\text{DU}:\text{UD},\text{LS})}_{\text{down-up-to-up-down lower-saddle}}\\ \underbrace{(\text{UD}:\text{DU},\text{LS})}_{\text{up-down-to-down-up lower-saddle}}\end{matrix}\right\} \text{ for } a_{110} > 0, a_{220} < 0, \tag{3.66}$$

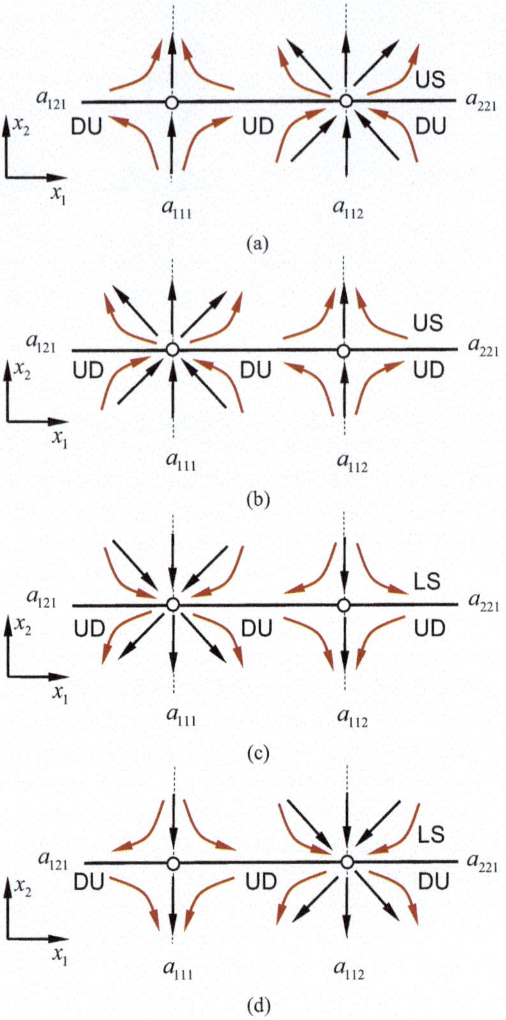

Fig. 3.11 Phase portraits ($a_{121} = a_{221}$) for two-dimensional systems on the x_1-direction with $x_1^* = a_{111}, a_{112}$ and on the x_2-direction with $x_2^* = a_{121}, a_{221}$. The up-down and down-up, upper-saddle and lower-saddle infinite-equilibriums: **(a)** ($a_{110} > 0$, $a_{220} > 0$), **(b)** ($a_{110} < 0$, $a_{220} > 0$), **(c)** ($a_{110} > 0$, $a_{220} < 0$), **(d)** ($a_{110} < 0$, $a_{220} < 0$)

$$\left\{ \begin{matrix} (a_{112}, a_{221}) \\ (a_{111}, a_{221}) \end{matrix} \right\} = \left\{ \begin{matrix} \underbrace{(\text{UD} : \text{DU}, \text{LS})}_{\text{up-down-to-down-up lower-saddle}} \\ \underbrace{(\text{DU} : \text{UD}, \text{LS})}_{\text{down-down-to-up-down lower-saddle}} \end{matrix} \right\} \text{ for } a_{110} < 0, a_{220} < 0, \quad (3.67)$$

and the infinite-equilibrium of $x_2^* = a_{121} = a_{222}$ is summarized as follows.

$$
\begin{aligned}
&(\bar{x}_1, a_{221}) = \underbrace{(\text{DU}, \text{US})}_{\text{down-up upper-saddle}} \quad \text{if } \bar{x}_1 \in (-\infty, a_{111}),\\
&(\bar{x}_1, a_{221}) = \underbrace{(\text{UD}, \text{US})}_{\text{up-down upper-saddle}} \quad \text{if } \bar{x}_1 \in (a_{111}, a_{112}),\\
&(\bar{x}_1, a_{221}) = \underbrace{(\text{DU}, \text{US})}_{\text{down-up upper-saddle}} \quad \text{if } \bar{x}_1 \in (a_{112}, \infty),\\
&\text{for } a_{110} > 0, a_{220} > 0;
\end{aligned}
\tag{3.68}
$$

$$
\begin{aligned}
&(\bar{x}_1, a_{221}) = \underbrace{(\text{UD}, \text{US})}_{\text{up-down upper-saddle}} \quad \text{if } \bar{x}_1 \in (-\infty, a_{111}),\\
&(\bar{x}_1, a_{221}) = \underbrace{(\text{DU}, \text{US})}_{\text{down-up upper-saddle}} \quad \text{if } \bar{x}_1 \in (a_{111}, a_{112}),\\
&(\bar{x}_1, a_{221}) = \underbrace{(\text{UD}, \text{US})}_{\text{up-down upper-saddle}} \quad \text{if } \bar{x}_1 \in (a_{112}, \infty),\\
&\text{for } a_{110} < 0, a_{220} > 0;
\end{aligned}
\tag{3.69}
$$

$$
\begin{aligned}
&(\bar{x}_1, a_{221}) = \underbrace{(\text{UD}, \text{LS})}_{\text{up-down lower-saddle}} \quad \text{if } \bar{x}_1 \in (-\infty, a_{111}),\\
&(\bar{x}_1, a_{221}) = \underbrace{(\text{DU}, \text{LS})}_{\text{down-up lower-saddle}} \quad \text{if } \bar{x}_1 \in (a_{111}, a_{112}),\\
&(\bar{x}_1, a_{221}) = \underbrace{(\text{UD}, \text{LS})}_{\text{up-down lower-saddle}} \quad \text{if } \bar{x}_1 \in (a_{112}, \infty),\\
&\text{for } a_{110} > 0, a_{220} < 0;
\end{aligned}
\tag{3.70}
$$

$$
\begin{aligned}
&(\bar{x}_1, a_{221}) = \underbrace{(\text{DU}, \text{LS})}_{\text{down-up lower-saddle}} \quad \text{if } \bar{x}_1 \in (-\infty, a_{111}),\\
&(\bar{x}_1, a_{221}) = \underbrace{(\text{UD}, \text{LS})}_{\text{up-down lower-saddle}} \quad \text{if } \bar{x}_1 \in (a_{111}, a_{112}),\\
&(\bar{x}_1, a_{221}) = \underbrace{(\text{DU}, \text{LS})}_{\text{down-up lower-saddle}} \quad \text{if } \bar{x}_1 \in (a_{112}, \infty),\\
&\text{for } a_{110} < 0, a_{220} < 0.
\end{aligned}
\tag{3.71}
$$

The down-up and up-down infinite-equilibriums of $x_2^* = a_{121} = a_{221}$ are up-down and down-up upper-saddle and lower-saddle, as presented in Fig. 3.11. The infinite-equilibrium is for the switching bifurcation of equilibriums for $a_{121} < a_{221}$ and $a_{121} > a_{221}$. The two equilibriums on the infinite-equilibrium are up-down-to-down-up and down-up-to-up-down, upper-saddle and lower-saddle.

3.4 Nonsingular Cubic Systems

In this section, the network of saddles, source, and sink with paralleled hyperbolic and hyperbolic-secant flows will be presented, and the corresponding switching bifurcations will be discussed through the inflection-source and sink infinite-equilibriums.

3.4.1 Equilibrium Networks with Paralleled Hyperbolic Flows

Consider a dynamical system as

$$\begin{aligned}\dot{x}_1 &= a_{110}(x_1 - a_{111})(x_1 - a_{112})(x_2 - a_{121}),\\ \dot{x}_2 &= a_{220}(x_2 - a_{221})(x_2 - a_{222}),\end{aligned} \tag{3.72}$$

and the corresponding first integral manifold is

$$\begin{aligned}&\frac{1}{a_{112} - a_{111}}(\ln\frac{|x_1 - a_{112}|}{|x_{10} - a_{112}|} - \ln\frac{|x_1 - a_{111}|}{|x_{10} - a_{111}|})\\ &= \frac{a_{110}}{a_{220}}(\frac{a_{221} - a_{121}}{a_{221} - a_{222}}\ln\frac{|x_2 - a_{221}|}{|x_{20} - a_{221}|} + \frac{a_{222} - a_{121}}{a_{222} - a_{221}}\ln\frac{|x_2 - a_{222}|}{|x_{20} - a_{222}|}).\end{aligned} \tag{3.73}$$

(i) The simple equilibrium network for $a_{211} < a_{111}$ and $a_{121} < a_{221}$ has the following properties.

$$\left\{\begin{matrix}(a_{111}, a_{222}) & (a_{112}, a_{222})\\ (a_{111}, a_{221}) & (a_{112}, a_{221})\\ (a_{111}, a_{121}) & (a_{112}, a_{121})\end{matrix}\right\} = \left\{\begin{matrix}\underbrace{(\text{SI, SO})}_{\text{saddle}} & \underbrace{(\text{SO, SO})}_{\text{source}}\\ \underbrace{(\text{SI, SI})}_{\text{sink}} & \underbrace{(\text{SO, SI})}_{\text{saddle}}\\ \underbrace{(\text{UP, DP, pF})}_{\text{hyperbolic-secant flow }(+)} & \underbrace{(\text{DP : UP, pF})}_{\text{hyperbolic flow }(+)}\end{matrix}\right\}$$

for $a_{110} > 0$ and $a_{220} > 0$;

$$\tag{3.74}$$

$$\left\{\begin{matrix}(a_{111},a_{222}) & (a_{112},a_{222})\\ (a_{111},a_{221}) & (a_{112},a_{221})\\ (a_{111},a_{121}) & (a_{112},a_{121})\end{matrix}\right\} = \left\{\begin{matrix}\underbrace{(\mathrm{SO},\mathrm{SO})}_{\text{source}} & \underbrace{(\mathrm{SI},\mathrm{SO})}_{\text{saddle}}\\ \underbrace{(\mathrm{SO},\mathrm{SI})}_{\text{saddle}} & \underbrace{(\mathrm{SI},\mathrm{SI})}_{\text{sink}}\\ \underbrace{(\mathrm{DP}:\mathrm{UP},\mathrm{pF})}_{\text{hyperbolic flow }(+)} & \underbrace{(\mathrm{UP},\mathrm{DP},\mathrm{pF})}_{\text{hyperbolic-secant flow }(+)}\end{matrix}\right\}$$

for $a_{110} < 0$ and $a_{220} > 0$;

(3.75)

$$\left\{\begin{matrix}(a_{111},a_{222}) & (a_{112},a_{222})\\ (a_{111},a_{221}) & (a_{112},a_{221})\\ (a_{111},a_{121}) & (a_{112},a_{121})\end{matrix}\right\} = \left\{\begin{matrix}\underbrace{(\mathrm{SI},\mathrm{SI})}_{\text{sink}} & \underbrace{(\mathrm{SO},\mathrm{SI})}_{\text{saddle}}\\ \underbrace{(\mathrm{SI},\mathrm{SO})}_{\text{saddle}} & \underbrace{(\mathrm{SO},\mathrm{SO})}_{\text{source}}\\ \underbrace{(\mathrm{DP}:\mathrm{UP},\mathrm{nF})}_{\text{hyperbolic flow }(-)} & \underbrace{(\mathrm{UP},\mathrm{DP},\mathrm{nF})}_{\text{hyperbolic-secant flow }(-)}\end{matrix}\right\} \tag{3.76}$$

for $a_{110} > 0$ and $a_{220} < 0$;

$$\left\{\begin{matrix}(a_{111},a_{222}) & (a_{112},a_{222})\\ (a_{111},a_{221}) & (a_{112},a_{221})\\ (a_{111},a_{121}) & (a_{112},a_{121})\end{matrix}\right\} = \left\{\begin{matrix}\underbrace{(\mathrm{SO},\mathrm{SI})}_{\text{saddle}} & \underbrace{(\mathrm{SI},\mathrm{SI})}_{\text{sink}}\\ \underbrace{(\mathrm{SO},\mathrm{SO})}_{\text{source}} & \underbrace{(\mathrm{SI},\mathrm{SO})}_{\text{saddle}}\\ \underbrace{(\mathrm{UP},\mathrm{DP},\mathrm{nF})}_{\text{hyperbolic-secant flow }(-)} & \underbrace{(\mathrm{DP}:\mathrm{UP},\mathrm{nF})}_{\text{hyperbolic flow }(-)}\end{matrix}\right\} \tag{3.77}$$

for $a_{110} < 0$ and $a_{220} < 0$;

as shown in Fig. 3.12. The directrix flows of $x_1 = a_{111}$, a_{112} are positive and negative. The networks of saddle, sink, and source with hyperbolic and hyperbolic-secant flows are presented. The saddles, sink, and sources in the network of equilibriums are above the hyperbolic and hyperbolic-secant flows.

(ii) The simple equilibrium networks for $a_{112} < a_{211}$ and $a_{121} < a_{221}$ have the following properties.

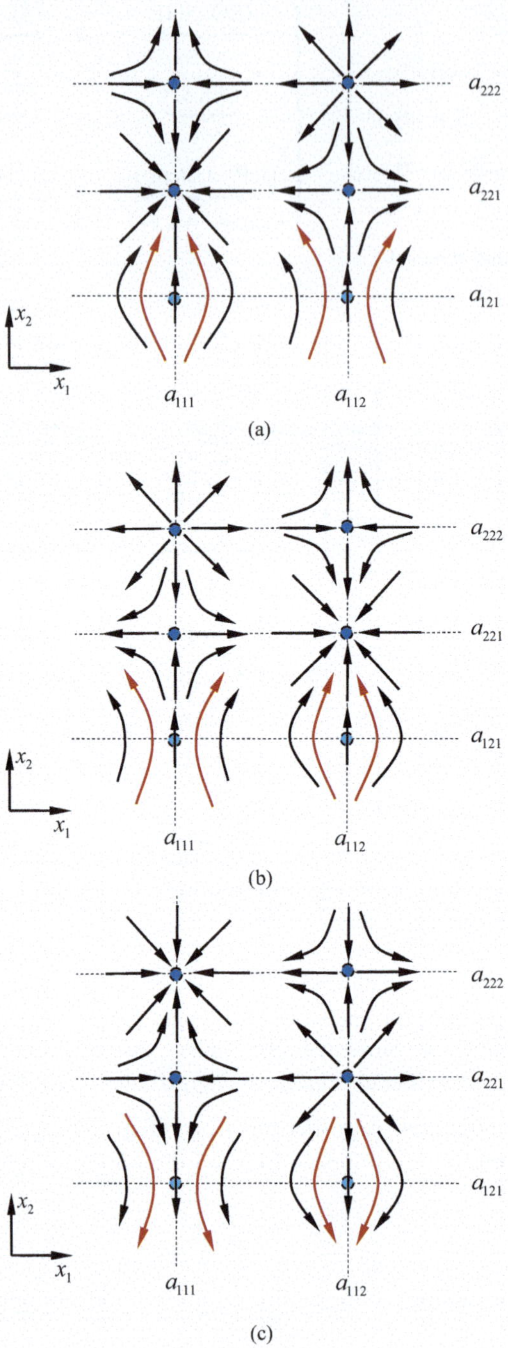

Fig. 3.12 Phase portraits ($a_{121} < a_{221}$) for two-dimensional systems on the x_1-direction with $x_1^* = a_{111}$ and on the x_2-direction with $x_2^* = a_{121}, a_{221}, a_{222}$. The simple equilibriums with hyperbolic flows: (**a**) ($a_{110} > 0$, $a_{220} > 0$), (**b**) ($a_{110} < 0$, $a_{220} > 0$), (**c**) ($a_{110} > 0$, $a_{220} < 0$), (**d**) ($a_{110} < 0$, $a_{220} < 0$)

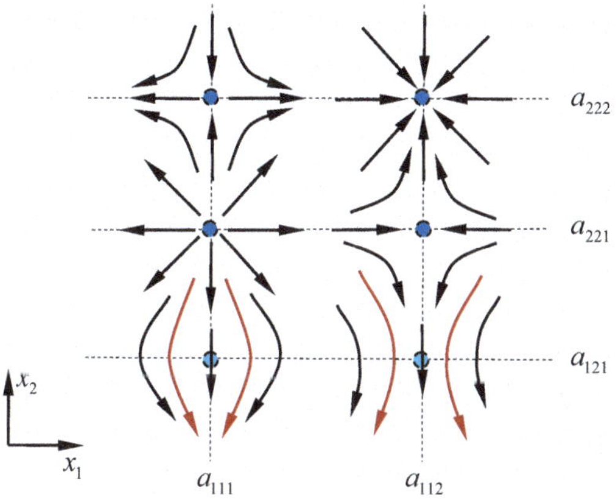

(d)

Fig. 3.12 (continued)

$$\left\{\begin{matrix}(a_{111},a_{121}) & (a_{211},a_{121})\\ (a_{111},a_{222}) & (a_{112},a_{222})\\ (a_{111},a_{221}) & (a_{111},a_{221})\end{matrix}\right\}=\left\{\begin{matrix}\underbrace{(\text{UP}:\text{DP},\text{pF})}_{\text{hyperbolic-secant flow }(+)} & \underbrace{(\text{DP}:\text{UP},\text{pF})}_{\text{hyperbolic flow }(+)}\\ \underbrace{(\text{SO},\text{SO})}_{\text{source}} & \underbrace{(\text{SI},\text{SO})}_{\text{saddle}}\\ \underbrace{(\text{SO},\text{SI})}_{\text{saddle}} & \underbrace{(\text{SI},\text{SI})}_{\text{sink}}\end{matrix}\right\}$$

for $a_{110}>0$ and $a_{220}>0$;

(3.78)

$$\left\{\begin{matrix}(a_{111},a_{121}) & (a_{211},a_{121})\\ (a_{111},a_{222}) & (a_{112},a_{222})\\ (a_{111},a_{221}) & (a_{111},a_{221})\end{matrix}\right\}=\left\{\begin{matrix}\underbrace{(\text{DP}:\text{UP},\text{pF})}_{\text{hyperbolic flow }(+)} & \underbrace{(\text{UP}:\text{DP},\text{pF})}_{\text{hyperbolic-secant flow }(+)}\\ \underbrace{(\text{SI},\text{SO})}_{\text{saddle}} & \underbrace{(\text{SO},\text{SO})}_{\text{source}}\\ \underbrace{(\text{SI},\text{SI})}_{\text{sink}} & \underbrace{(\text{SO},\text{SI})}_{\text{saddle}}\end{matrix}\right\}$$

for $a_{110}<0$ and $a_{220}>0$;

(3.79)

$$\left\{\begin{array}{cc}(a_{111},a_{121}) & (a_{211},a_{121})\\(a_{111},a_{222}) & (a_{112},a_{222})\\(a_{111},a_{221}) & (a_{111},a_{221})\end{array}\right\}=\left\{\begin{array}{cc}\underbrace{(\mathrm{DP}:\mathrm{UP},\mathrm{nF})}_{\text{hyperbolic flow }(-)} & \underbrace{(\mathrm{UP}:\mathrm{DP},\mathrm{nF})}_{\text{hyperbolic-secant flow }(-)}\\\underbrace{(\mathrm{SO},\mathrm{SI})}_{\text{saddle}} & \underbrace{(\mathrm{SI},\mathrm{SI})}_{\text{sink}}\\\underbrace{(\mathrm{SO},\mathrm{SO})}_{\text{source}} & \underbrace{(\mathrm{SI},\mathrm{SO})}_{\text{saddle}}\end{array}\right\} \tag{3.80}$$

for $a_{110} > 0$ and $a_{220} < 0$;

$$\left\{\begin{array}{cc}(a_{111},a_{121}) & (a_{211},a_{121})\\(a_{111},a_{222}) & (a_{112},a_{222})\\(a_{111},a_{221}) & (a_{111},a_{221})\end{array}\right\}=\left\{\begin{array}{cc}\underbrace{(\mathrm{UP}:\mathrm{DP},\mathrm{nF})}_{\text{hyperbolic-secant flow }(-)} & \underbrace{(\mathrm{DP}:\mathrm{UP},\mathrm{nF})}_{\text{hyperbolic flow }(-)}\\\underbrace{(\mathrm{SI},\mathrm{SI})}_{\text{sink}} & \underbrace{(\mathrm{SO},\mathrm{SI})}_{\text{saddle}}\\\underbrace{(\mathrm{SI},\mathrm{SO})}_{\text{saddle}} & \underbrace{(\mathrm{SO},\mathrm{SO})}_{\text{source}}\end{array}\right\} \tag{3.81}$$

for $a_{110} < 0$ and $a_{220} < 0$;

as shown in Fig. 3.13. The directrix flows of $x_1 = a_{111}$, a_{112} are positive and negative. The networks of saddle, sink, and source with hyperbolic and hyperbolic-secant flows are presented. The saddles, sink, and sources in the network of equilibriums are *below* the paralleled hyperbolic and hyperbolic-secant flows.

(iii) The simple equilibrium networks for $a_{211} \in (a_{111}, a_{112})$ and $a_{121} < a_{221}$ have the following properties.

$$\left\{\begin{array}{cc}(a_{111},a_{222}) & (a_{112},a_{222})\\(a_{111},a_{121}) & (a_{112},a_{121})\\(a_{111},a_{221}) & (a_{112},a_{221})\end{array}\right\}=\left\{\begin{array}{cc}\underbrace{(\mathrm{SI},\mathrm{SO})}_{\text{saddle}} & \underbrace{(\mathrm{SO},\mathrm{SO})}_{\text{source}}\\\underbrace{(\mathrm{DP}:\mathrm{UP},\mathrm{nF})}_{\text{hyperbolic flow }(-)} & \underbrace{(\mathrm{UP}:\mathrm{DP},\mathrm{nF})}_{\text{hyperbolic-secant flow }(-)}\\\underbrace{(\mathrm{SO},\mathrm{SI})}_{\text{saddle}} & \underbrace{(\mathrm{SI},\mathrm{SI})}_{\text{sink}}\end{array}\right\} \tag{3.82}$$

for $a_{110} > 0$ and $a_{220} > 0$;

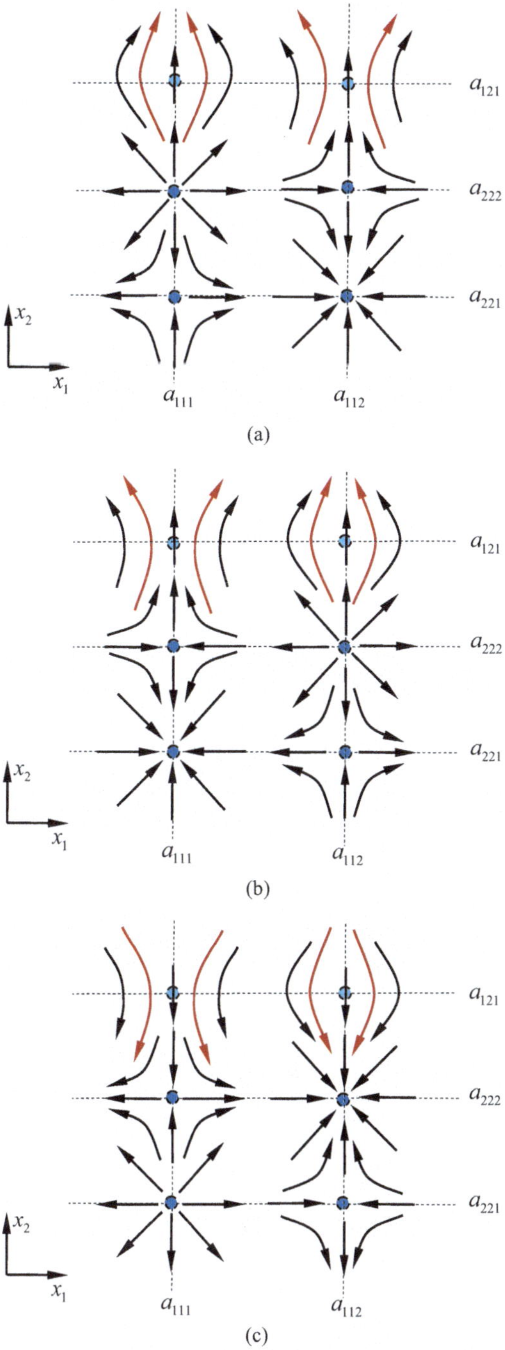

Fig. 3.13 Phase portraits ($a_{121} < a_{221}$) for two-dimensional systems on the x_1-direction with $x_1^* = a_{111}$ and on the x_2-direction with $x_2^* = a_{121}, a_{221}, a_{222}$. The simple equilibriums with hyperbolic flows: **(a)** ($a_{110} > 0, a_{220} > 0$), **(b)** ($a_{110} < 0, a_{220} > 0$), **(c)** ($a_{110} > 0, a_{220} < 0$), **(d)** ($a_{110} < 0, a_{220} < 0$)

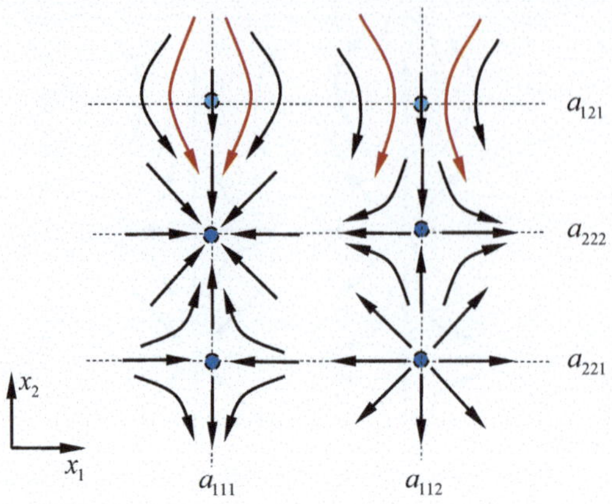

(d)

Fig. 3.13 (continued)

$$\left\{\begin{matrix} (a_{111},a_{222}) & (a_{112},a_{222}) \\ (a_{111},a_{121}) & (a_{112},a_{121}) \\ (a_{111},a_{221}) & (a_{112},a_{221}) \end{matrix}\right\} = \left\{\begin{matrix} \underbrace{(\mathrm{SO},\mathrm{SO})}_{\text{source}} & \underbrace{(\mathrm{SI},\mathrm{SO})}_{\text{saddle}} \\ \underbrace{(\mathrm{UP}:\mathrm{DP},\mathrm{nF})}_{\text{hyperbolic-secant flow }(-)} & \underbrace{(\mathrm{DP}:\mathrm{UP},\mathrm{nF})}_{\text{hyperbolic flow }(-)} \\ \underbrace{(\mathrm{SI},\mathrm{SI})}_{\text{sink}} & \underbrace{(\mathrm{SO},\mathrm{SI})}_{\text{saddle}} \end{matrix}\right\} \tag{3.83}$$

for $a_{110} < 0$ and $a_{220} > 0$;

$$\left\{\begin{matrix} (a_{111},a_{222}) & (a_{112},a_{222}) \\ (a_{111},a_{121}) & (a_{112},a_{121}) \\ (a_{111},a_{221}) & (a_{112},a_{221}) \end{matrix}\right\} = \left\{\begin{matrix} \underbrace{(\mathrm{SI},\mathrm{SI})}_{\text{sink}} & \underbrace{(\mathrm{SO},\mathrm{SI})}_{\text{saddle}} \\ \underbrace{(\mathrm{UP}:\mathrm{DP},\mathrm{pF})}_{\text{hyperbolic-secant flow }(+)} & \underbrace{(\mathrm{DP}:\mathrm{UP},\mathrm{pF})}_{\text{hyperbolic flow }(+)} \\ \underbrace{(\mathrm{SO},\mathrm{SO})}_{\text{source}} & \underbrace{(\mathrm{SI},\mathrm{SO})}_{\text{saddle}} \end{matrix}\right\}$$

for $a_{110} > 0$ and $a_{220} < 0$;

$$\tag{3.84}$$

$$\left\{\begin{matrix} (a_{111}, a_{222}) & (a_{112}, a_{222}) \\ (a_{111}, a_{121}) & (a_{112}, a_{121}) \\ (a_{111}, a_{221}) & (a_{112}, a_{221}) \end{matrix}\right\} = \left\{\begin{matrix} \underbrace{(\text{SO}, \text{SI})}_{\text{saddle}} & \underbrace{(\text{SI}, \text{SI})}_{\text{sink}} \\ \underbrace{(\text{DP} : \text{UP}, \text{pF})}_{\text{hyperbolic flow } (+)} & \underbrace{(\text{UP} : \text{DP}, \text{pF})}_{\text{hyperbolic-secant flow } (+)} \\ \underbrace{(\text{SI}, \text{SO})}_{\text{saddle}} & \underbrace{(\text{SO}, \text{SO})}_{\text{source}} \end{matrix}\right\}$$

for $a_{110} < 0$ and $a_{220} < 0$;

(3.85)

as shown in Fig. 3.14. The directrix flows of $x_1 = a_{111}, a_{112}$ are positive and negative. The networks of saddles, sink, and source with paralleled hyperbolic and hyperbolic-secant flows are presented. The saddles, sink, and sources in the network of equilibriums are *separated by* the hyperbolic and hyperbolic-secant flows. The saddle-to-saddle and source-to-sink are paralleled. The saddle-to-saddle is separated by a hyperbolic flow, and the source-to-sink is separated by a hyperbolic-secant flow.

3.4.2 Connected Inflection-Sources (Sinks)

In this section, discussed are the connected inflection-source and inflection-sinks on the infinite-equilibriums, which are for the switching bifurcations of the paralleled hyperbolic and hyperbolic-secant flows with corresponding connected saddle, source, and sink.

Consider a dynamical system for $a_{121} = a_{22s_1}$ $(s_1 = 1, 2)$ as

$$\begin{aligned} \dot{x}_1 &= a_{110}(x_1 - a_{111})(x_1 - a_{112})(x_2 - a_{121}), \\ \dot{x}_2 &= a_{220}(x_2 - a_{221})(x_2 - a_{222}), \end{aligned} \tag{3.86}$$

and the corresponding first integral manifold is

$$\frac{1}{a_{112} - a_{111}}\left(\ln \frac{|x_1 - a_{112}|}{|x_{10} - a_{112}|} - \ln \frac{|x_1 - a_{111}|}{|x_{10} - a_{111}|}\right) = \frac{a_{110}}{a_{220}} \ln \frac{|x_2 - a_{22s_2}|}{|x_{20} - a_{22s_2}|}. \tag{3.87}$$

(i) The equilibriums of $x_2^* = a_{121} = a_{222}$ have the following properties.

$$\left\{\begin{matrix} (a_{111}, a_{121}) & (a_{112}, a_{121}) \\ (a_{111}, a_{221}) & (a_{112}, a_{221}) \end{matrix}\right\} = \left\{\begin{matrix} \underbrace{(\text{II} : \text{DI}, \text{SO})}_{\text{inflection source}} & \underbrace{(\text{DI} : \text{II}, \text{SO})}_{\text{inflection source}} \\ \underbrace{(\text{SO}, \text{SI})}_{\text{saddle}} & \underbrace{(\text{SI}, \text{SI})}_{\text{sink}} \end{matrix}\right\} \tag{3.88}$$

for $a_{110} > 0, a_{220} > 0$,

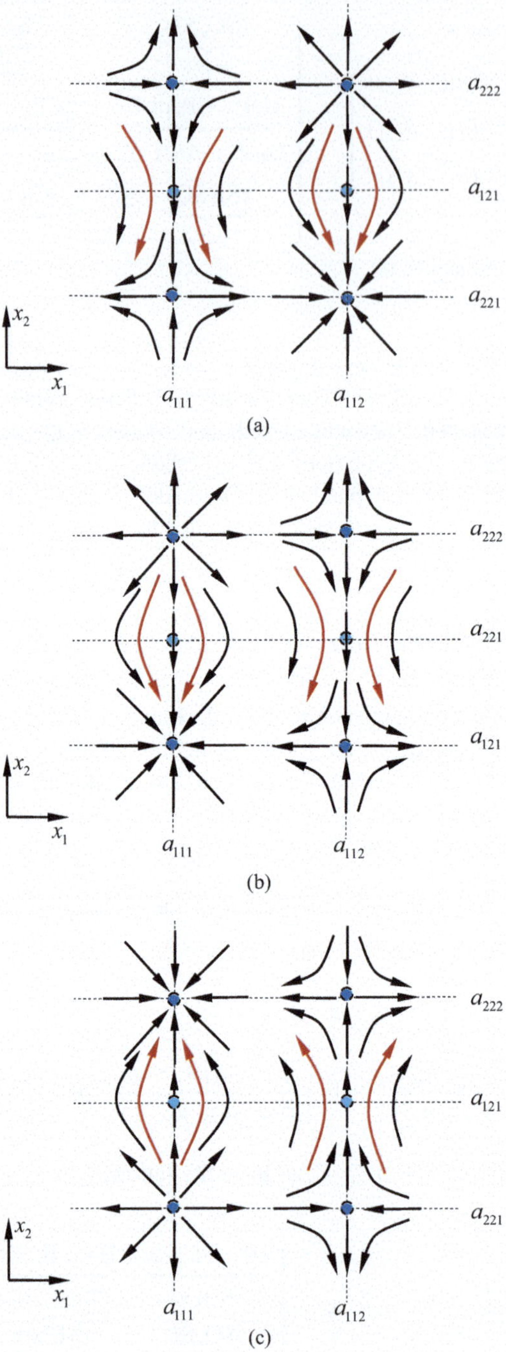

Fig. 3.14 Phase portraits ($a_{121} < a_{221}$) for two-dimensional systems on the x_1-direction with $x_1^* = a_{111}$ and on the x_2-direction with $x_2^* = a_{121}, a_{221}, a_{222}$. The simple equilibriums with hyperbolic flows: **(a)** ($a_{110} > 0, a_{220} > 0$), **(b)** ($a_{110} < 0, a_{220} > 0$), **(c)** ($a_{110} > 0, a_{220} < 0$), **(d)** ($a_{110} < 0, a_{220} < 0$)

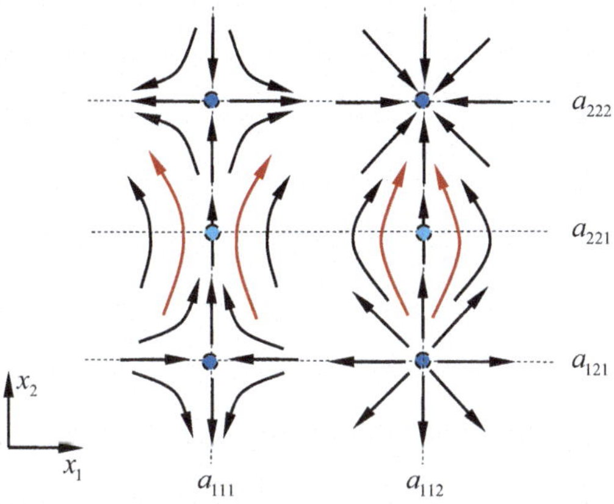

(d)

Fig. 3.14 (continued)

$$\left\{\begin{array}{cc}(a_{111},a_{121}) & (a_{112},a_{121})\\ (a_{111},a_{221}) & (a_{112},a_{221})\end{array}\right\} = \left\{\begin{array}{cc}\underbrace{(\mathrm{DI}:\mathrm{II},\mathrm{SO})}_{\text{inflection source}} & \underbrace{(\mathrm{II}:\mathrm{DI},\mathrm{SO})}_{\text{inflection source}}\\ \underbrace{(\mathrm{SI},\mathrm{SI})}_{\text{sink}} & \underbrace{(\mathrm{SO},\mathrm{SI})}_{\text{saddle}}\end{array}\right\} \tag{3.89}$$

for $a_{110} < 0, a_{220} > 0$,

$$\left\{\begin{array}{cc}(a_{111},a_{121}) & (a_{112},a_{121})\\ (a_{111},a_{221}) & (a_{112},a_{221})\end{array}\right\} = \left\{\begin{array}{cc}\underbrace{(\mathrm{DI}:\mathrm{II},\mathrm{SI})}_{\text{inflection sink}} & \underbrace{(\mathrm{II}:\mathrm{DI},\mathrm{SI})}_{\text{inflection sink}}\\ \underbrace{(\mathrm{SO},\mathrm{SO})}_{\text{source}} & \underbrace{(\mathrm{SI},\mathrm{SO})}_{\text{saddle}}\end{array}\right\} \tag{3.90}$$

for $a_{110} < 0, a_{220} > 0$, for $a_{110} > 0, a_{220} < 0$,

$$\left\{\begin{array}{cc}(a_{111},a_{121}) & (a_{112},a_{121})\\ (a_{111},a_{221}) & (a_{112},a_{221})\end{array}\right\} = \left\{\begin{array}{cc}\underbrace{(\mathrm{II}:\mathrm{DI},\mathrm{SI})}_{\text{inflection sink}} & \underbrace{(\mathrm{DI}:\mathrm{II},\mathrm{SI})}_{\text{inflection sink}}\\ \underbrace{(\mathrm{SI},\mathrm{SO})}_{\text{saddle}} & \underbrace{(\mathrm{SO},\mathrm{SO})}_{\text{source}}\end{array}\right\} \tag{3.91}$$

for $a_{110} < 0, a_{220} < 0$;

and the inflection-sink and source infinite-equilibrium is

$$\left.\begin{array}{ll}\underbrace{(\bar{x}_1,a_{222})}_{\text{infinite-equilibrium}}=\underbrace{(\text{II, SO})}_{\text{increasing-inflection source}} & \text{for } \bar{x}_1\in(-\infty,a_{111}),\\ \underbrace{(\bar{x}_1,a_{222})}_{\text{infinite-equilibrium}}=\underbrace{(\text{DI, SO})}_{\text{decreasing-inflection source}} & \text{for } \bar{x}_1\in(a_{111},a_{112}),\\ \underbrace{(\bar{x}_1,a_{222})}_{\text{infinite-equilibrium}}=\underbrace{(\text{II, SO})}_{\text{increasing-inflection source}} & \text{for } \bar{x}_1\in(a_{112},\infty)\\ \text{for } a_{110}>0, a_{220}>0, & \end{array}\right\} \tag{3.92}$$

$$\left.\begin{array}{ll}\underbrace{(\bar{x}_1,a_{222})}_{\text{infinite-equilibrium}}=\underbrace{(\text{DI, SO})}_{\text{decreasing-inflection source}} & \text{for } \bar{x}_1\in(-\infty,a_{111}),\\ \underbrace{(\bar{x}_1,a_{222})}_{\text{infinite-equilibrium}}=\underbrace{(\text{II, SO})}_{\text{increasing-inflection source}} & \text{for } \bar{x}_1\in(a_{111},a_{112}),\\ \underbrace{(\bar{x}_1,a_{222})}_{\text{infinite-equilibrium}}=\underbrace{(\text{DI, SO})}_{\text{decreasing-inflection source}} & \text{for } \bar{x}_1\in(a_{112},\infty)\\ \text{for } a_{110}>0, a_{220}>0, & \end{array}\right\} \tag{3.93}$$

$$\left.\begin{array}{ll}\underbrace{(\bar{x}_1,a_{222})}_{\text{infinite-equilibrium}}=\underbrace{(\text{DI, SI})}_{\text{decreasing-inflection sink}} & \text{for } \bar{x}_1\in(-\infty,a_{111}),\\ \underbrace{(\bar{x}_1,a_{222})}_{\text{infinite-equilibrium}}=\underbrace{(\text{II, SI})}_{\text{increasing-inflection sink}} & \text{for } \bar{x}_1\in(a_{111},a_{112}),\\ \underbrace{(\bar{x}_1,a_{222})}_{\text{infinite-equilibrium}}=\underbrace{(\text{DI, SI})}_{\text{decreasing-inflection sink}} & \text{for } \bar{x}_1\in(a_{112},\infty)\\ \text{for } a_{110}>0, a_{220}<0, & \end{array}\right\} \tag{3.94}$$

$$\left.\begin{array}{ll}\underbrace{(\bar{x}_1,a_{222})}_{\text{infinite-equilibrium}}=\underbrace{(\text{II, SI})}_{\text{increasing-inflection sink}} & \text{for } \bar{x}_1\in(-\infty,a_{111}),\\ \underbrace{(\bar{x}_1,a_{222})}_{\text{infinite-equilibrium}}=\underbrace{(\text{DI, SI})}_{\text{decreasing-inflection sink}} & \text{for } \bar{x}_1\in(a_{111},a_{112}),\\ \underbrace{(\bar{x}_1,a_{222})}_{\text{infinite-equilibrium}}=\underbrace{(\text{II, SI})}_{\text{increasing-inflection sink}} & \text{for } \bar{x}_1\in(a_{112},\infty)\\ \text{for } a_{110}<0, a_{220}<0, & \end{array}\right\} \tag{3.95}$$

as shown in Fig. 3.15. The inflection-source and sink infinite-equilibriums of $x_2^*=a_{121}=a_{222}$ are for the switching bifurcations, which are above the saddle and sink or source equilibriums.

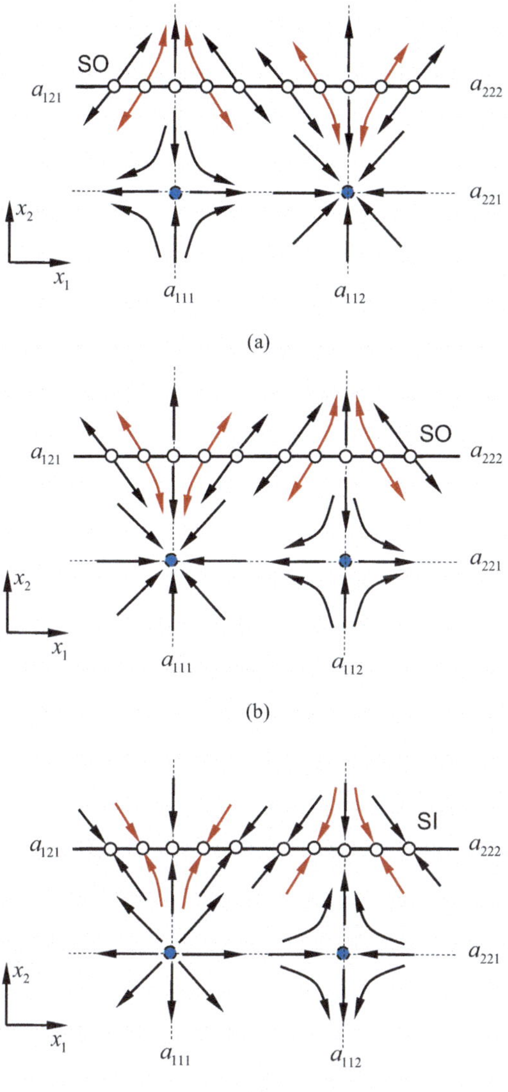

Fig. 3.15 Phase portraits ($a_{121} = a_{222}$) for two-dimensional systems on the x_1-direction with $x_1^* = a_{111}, a_{112}$ and on the x_2-direction with $x_2^* = a_{221}, a_{121} = a_{222}$. The inflection-sink and source infinite-equilibriums with sink, source, and saddles: (**a**) ($a_{110} > 0$, $a_{220} > 0$), (**b**) ($a_{110} < 0$, $a_{220} > 0$), (**c**) ($a_{110} > 0$, $a_{220} < 0$), (**d**) ($a_{110} < 0$, $a_{220} < 0$)

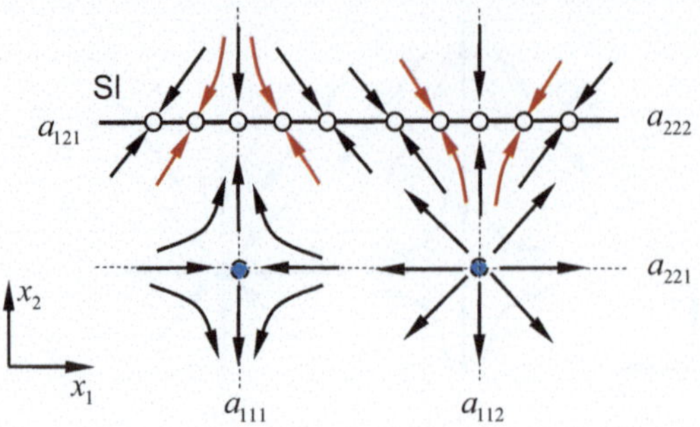

(d)

Fig. 3.15 (continued)

(ii) The equilibriums of $x_2^* = a_{121} = a_{221}$ have the following properties.

$$\begin{Bmatrix} (a_{111}, a_{222}) & (a_{112}, a_{222}) \\ (a_{111}, a_{221}) & (a_{112}, a_{221}) \end{Bmatrix} = \begin{Bmatrix} \underbrace{(\text{SI, SO})}_{\text{saddle}} & \underbrace{(\text{SO, SO})}_{\text{source}} \\ \underbrace{(\text{DI : II, SI})}_{\text{inflection sink}} & \underbrace{(\text{II : DI, SI})}_{\text{inflection sink}} \end{Bmatrix} \tag{3.96}$$

for $a_{110} > 0, a_{220} > 0$;

$$\begin{Bmatrix} (a_{111}, a_{222}) & (a_{112}, a_{222}) \\ (a_{111}, a_{221}) & (a_{112}, a_{221}) \end{Bmatrix} = \begin{Bmatrix} \underbrace{(\text{SO, SO})}_{\text{source}} & \underbrace{(\text{SI, SO})}_{\text{saddle}} \\ \underbrace{(\text{II : DI, SI})}_{\text{inflection sink}} & \underbrace{(\text{DI : II, SI})}_{\text{inflection sink}} \end{Bmatrix} \tag{3.97}$$

for $a_{110} < 0, a_{220} > 0$,

$$\begin{Bmatrix} (a_{111}, a_{222}) & (a_{112}, a_{222}) \\ (a_{111}, a_{221}) & (a_{112}, a_{221}) \end{Bmatrix} = \begin{Bmatrix} \underbrace{(\text{SI, SI})}_{\text{source}} & \underbrace{(\text{SO, SI})}_{\text{saddle}} \\ \underbrace{(\text{II : DI, SO})}_{\text{inflection source}} & \underbrace{(\text{DI : II, SO})}_{\text{inflection source}} \end{Bmatrix} \tag{3.98}$$

for $a_{110} > 0, a_{220} < 0$;

$$\begin{Bmatrix} (a_{111}, a_{222}) & (a_{112}, a_{222}) \\ (a_{111}, a_{221}) & (a_{112}, a_{221}) \end{Bmatrix} = \begin{Bmatrix} \underbrace{(\text{SO, SI})}_{\text{saddle}} & \underbrace{(\text{SI, SI})}_{\text{source}} \\ \underbrace{(\text{DI : II, SO})}_{\text{inflection source}} & \underbrace{(\text{II : DI, SO})}_{\text{inflection source}} \end{Bmatrix} \tag{3.99}$$

for $a_{110} < 0, a_{220} < 0$,

and the inflection sink and source infinite-equilibrium of $x_2^* = a_{121} = a_{221}$ is

$$\left.\begin{array}{lll}
\underbrace{(\bar{x}_1, a_{221})}_{\text{infinite-equilibrium}} = \underbrace{(\text{DI, SI})}_{\text{decreasing-inflection sink}} & \text{for } \bar{x}_1 \in (-\infty, a_{111}), \\
\underbrace{(\bar{x}_1, a_{221})}_{\text{infinite-equilibrium}} = \underbrace{(\text{II, SI})}_{\text{increasing-inflection sink}} & \text{for } \bar{x}_1 \in (a_{111}, a_{112}), \\
\underbrace{(\bar{x}_1, a_{221})}_{\text{infinite-equilibrium}} = \underbrace{(\text{DI, SI})}_{\text{decreasing-inflection sink}} & \text{for } \bar{x}_1 \in (a_{112}, \infty)
\end{array}\right\} \quad (3.100)$$

for $a_{110} > 0, a_{220} > 0$,

$$\left.\begin{array}{lll}
\underbrace{(\bar{x}_1, a_{221})}_{\text{infinite-equilibrium}} = \underbrace{(\text{II, SI})}_{\text{increasing-inflection sink}} & \text{for } \bar{x}_1 \in (-\infty, a_{111}), \\
\underbrace{(\bar{x}_1, a_{221})}_{\text{infinite-equilibrium}} = \underbrace{(\text{DI, SI})}_{\text{decreasing-inflection sink}} & \text{for } \bar{x}_1 \in (a_{111}, a_{112}), \\
\underbrace{(\bar{x}_1, a_{221})}_{\text{infinite-equilibrium}} = \underbrace{(\text{II, SI})}_{\text{increasing-inflection sink}} & \text{for } \bar{x}_1 \in (a_{112}, \infty)
\end{array}\right\} \quad (3.101)$$

for $a_{110} < 0, a_{220} > 0$,

$$\left.\begin{array}{lll}
\underbrace{(\bar{x}_1, a_{221})}_{\text{infinite-equilibrium}} = \underbrace{(\text{II, SO})}_{\text{increasing-inflection source}} & \text{for } \bar{x}_1 \in (-\infty, a_{111}), \\
\underbrace{(\bar{x}_1, a_{221})}_{\text{infinite-equilibrium}} = \underbrace{(\text{DI, SO})}_{\text{decreasing-inflection source}} & \text{for } \bar{x}_1 \in (a_{111}, a_{112}), \\
\underbrace{(\bar{x}_1, a_{221})}_{\text{infinite-equilibrium}} = \underbrace{(\text{II, SO})}_{\text{increasing-inflection source}} & \text{for } \bar{x}_1 \in (a_{112}, \infty)
\end{array}\right\} \quad (3.102)$$

for $a_{110} > 0, a_{220} < 0$,

$$\left.\begin{array}{lll}
\underbrace{(\bar{x}_1, a_{221})}_{\text{infinite-equilibrium}} = \underbrace{(\text{DI, SO})}_{\text{decreasing-inflection source}} & \text{for } \bar{x}_1 \in (-\infty, a_{111}), \\
\underbrace{(\bar{x}_1, a_{221})}_{\text{infinite-equilibrium}} = \underbrace{(\text{II, SO})}_{\text{increasing-inflection source}} & \text{for } \bar{x}_1 \in (a_{111}, a_{112}), \\
\underbrace{(\bar{x}_1, a_{221})}_{\text{infinite-equilibrium}} = \underbrace{(\text{DI, SO})}_{\text{decreasing-inflection source}} & \text{for } \bar{x}_1 \in (a_{112}, \infty)
\end{array}\right\} \quad (3.103)$$

for $a_{110} < 0, a_{220} < 0$,

as shown in Fig. 3.16. The inflection-source and sink infinite-equilibriums of $x_2^* = a_{122} = a_{221}$ are below the saddle and sink or source equilibriums.

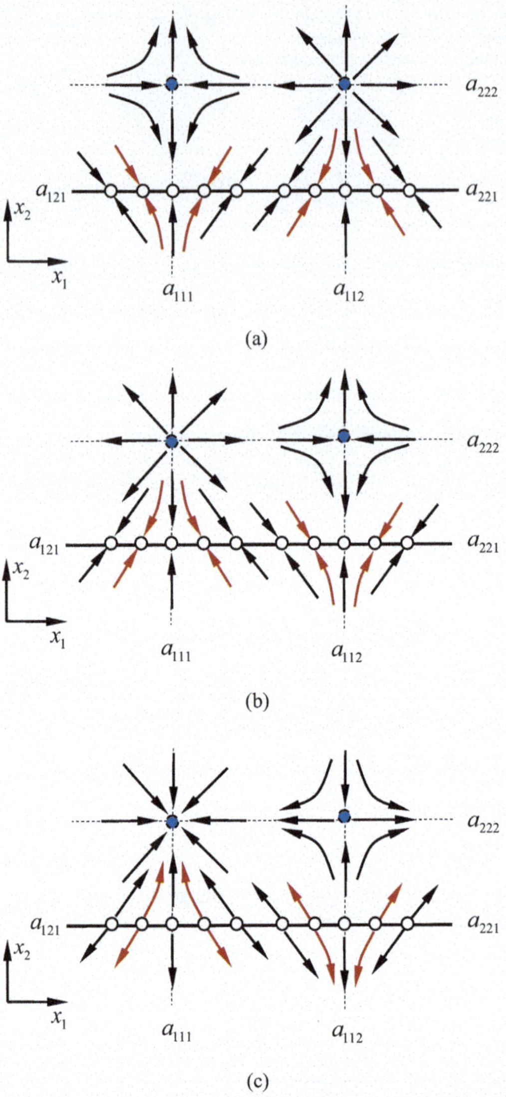

Fig. 3.16 Phase portraits ($a_{121} = a_{221}$) for two-dimensional systems on the x_1-direction with $x_1^* = a_{111}, a_{112}$ and on the x_2-direction with $x_2^* = a_{121} = a_{221}, a_{222}$. The inflection-sink and source infinite-equilibriums with sink, source, and saddles: **(a)** ($a_{110} > 0$, $a_{220} > 0$), **(b)** ($a_{110} < 0$, $a_{220} > 0$), **(c)** ($a_{110} > 0$, $a_{220} < 0$), **(d)** ($a_{110} < 0$, $a_{220} < 0$)

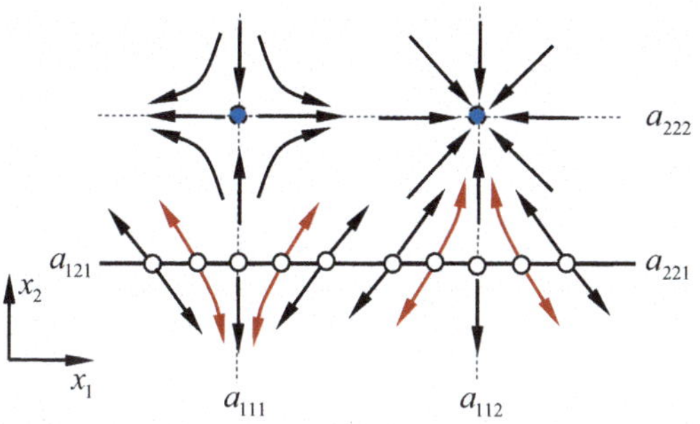

(d)

Fig. 3.16 (continued)

Index

A. C. J. Luo, *Two-dimensional Product-Cubic Systems, Vol. VII*,
https://doi.org/10.1007/978-3-031-48483-4

The manufacturer's authorised representative in the EU is Springer Nature Customer Service Centre GmbH, Europaplatz 3, 69115 Heidelberg, Germany. If you have any concerns regarding our products, please contact ProductSafety@springernature.com

Printed and bound by CPI Group (UK) Ltd, Croydon, CR0 4YY
01/10/2026
02225965-0002